Evolutionary Community Ecology

Evolutionary Community Ecology

MARK A. MCPEEK

PRINCETON UNIVERSITY PRESS

Princeton and Oxford

Published by Princeton University Press,
41 William Street, Princeton, New Jersey 08540

In the United Kingdom: Princeton University Press,
6 Oxford Street, Woodstock, Oxfordshire OX20 1TR

press.princeton.edu

ISBN 978-0-691-08877-8
Library of Congress Control Number: 2017941448

British Library Cataloging-in-Publication Data is available

This book has been composed in Times Roman

Printed on acid-free paper. ∞

Printed in the United States of America

1 3 5 7 9 10 8 6 4 2

This book is dedicated to Granddad, Meema, and Mom,

who never knew what I was doing, and to

Gail, Curtis, and Sarah, who do.

Contents

Acknowledgments

Writing a book like this can be daunting. My goal is to synthesize what are typically very different disciplines and very different literatures. Many ecologists do not read the evolution, speciation, paleontology, or dispersal literature, and have only a passing knowledge of the nuances of ideas and mechanisms in them. The reciprocal is also true of the other disciplines; many evolutionary biologists don't read the ecological literature that would help them understand what it takes for species to coexist, much less to comprehend the centrality of species interactions for determining the fitness of organisms and the consequent dynamics of natural selection. And neither of these groups really read the paleontology and dispersal literatures. My goal is to synthesize all these fundamental ideas and mechanisms that shape local and regional community structure.

Three people showed me how to be a scientist: Earl E. Werner, Robert D. Holt, and Joseph Travis. Earl showed me how to search for the important features of nature that reveal its workings and what to do when I think I've found such a feature. Bob showed me how to extract the essence of a concept from a jumble of ideas and how to expand that essence to make it useful. Joe showed me how embracing the struggle to understand and synthesize seemingly disparate ideas is the greatest fun a scientist can have.

Like most scientists, my intellectual trajectory was set by interactions with the faculty and students I argued with every day in graduate school. I was a graduate student at the Kellogg Biological Station, Michigan State University, in the late 1980s. My most important interactions were with Jonathan M. Brown, Mathew A. Leibold, Robert P. Creed, Thomas E. Miller, Craig W. Osenberg, Susan Kalisz, Stephen J. Tonsor, Gary G. Mittelbach, and Katherine L. Gross. They forcefully challenged me when I didn't make sense, and they always pushed me to go farther when I did make sense. They have remained lifelong friends as they have continued to push me. Much of what is in this book was sparked by discussions with them.

Over a career, one meets a tremendous number of smart people. I have been lucky enough to meet many, and they have all forced me to reconcile how I think about the world with the concepts and data in other disciplines. The friends who had the greatest intellectual impact on shaping the ideas presented here are Sergey Gavrilets, Richard Gomulkiewicz, Richard Harrison, David Jablonski,

Jeremy B. C. Jackson, Jonathan Losos, Winsor Lowe, William J. Resetarits, and John N. Thompson. I thank them all.

One never accomplishes anything alone. I have been blessed to have had great postdocs, graduate students, and undergraduates working with me, particularly Tim Watkins, Robby Stoks, Marjan DeBlock, Julie Turgeon, David Mbora, and Adam Siepielski. I thank them all for their hard work and intellectual stimulation.

My colleagues at Dartmouth—Kevin Peterson, Michael Dietrich, Kathryn Cottingham, John J. Gilbert, and Richard T. Holmes, Hany Farid, Daniel Rockmore—have always been inspirations.

At Princeton University Press, Sam Elworthy was the editor who originally convinced me to undertake this endeavor. Alison Kalett was the editor who got me to finish it. I would also like to thank Lauren Bucca, Leslie Grundfest, and Sheila Ann Dean for getting the book in publishable shape; I greatly appreciate all their diligent work.

A number of friends read various parts of the manuscript, and helped me express these ideas in ways that will hopefully be easier for the reader to understand and appreciate. Joseph Travis, Earl Werner, Robert Holt, Gary Mittelbach, Trevor Price, and Adam Siepielski read the entire book. Winsor Lowe read chapter 6, and Gene Hunt and David Jablonski read chapter 4. I am grateful for all their help.

And finally, this book would not have been possible without the love and support of my family: Curtis Charles Nipp, Isabelle Herbst Nipp, Jean Wilhelmina Nipp McPeek, Gail Suzanne Albert McPeek, Curtis Lincoln McPeek, and Sarah Jean McPeek.

Evolutionary Community Ecology

CHAPTER ONE

Ecological Opportunities, Communities, and Evolution

Spend any time in nature and you quickly notice that even over quite short distances you encounter different creatures in different places. The trees change as you hike from the bottom to the top of a high hill. The understory plants change as you pass from an oak-hickory stand into a pine stand. Different birds are singing in the open field as compared to the forest. The insects in the small creek you jump over are different from those you find in the river miles downstream. The lake at the bottom of the hill has sunfish in it, but the vernal pond at the summit has salamanders instead.

These are the kinds of patterns in nature that fascinate me. Consequently, one question has primarily driven my work: Why do these species live here but not over there, but why do those species live over there and not here? In fact, my very first research project as a budding undergraduate scientist addressed a very simple variant of this question (McPeek et al. 1983). This question continues to beguile me to this day. I am not alone in this quest. Many of those in ecology and evolutionary biology see their task as the explanation of such patterns in the distributions, abundances and diversity of species at one site, at different sites across space, and through time.

To answer this broad and overarching question with appropriate justice, one must consider many different, faceted, and layered questions. The first set of more specific questions necessarily address the ecological bases of these patterns. Are there ecological processes that cause these species assemblages to segregate across the landscape, and if so what are they? In other words, what prevents those particular species from living here? Also, the complementary question is just as essential. What makes species successful (or at least successful enough) to live where you find them? In other words, why do these particular species live here? The answers to these questions establish the ecological processes that maintain these patterns today and into the future.

These answers do not, however, address how these patterns were established in the first place. For this, we need to address a further set of questions. How did these collections of species come to have the properties that influence where they can and cannot live? More generally, where did these species come from in the first place? These are fundamental questions about the evolution of species individually and of the diversification of the clades in which each has been embedded over long time scales. The contemporary ecological context of these taxa form the foundation on which these evolutionary questions must be built, but we must address the longer-term evolutionary dynamics explicitly for a complete answer to the overarching query.

In designing such a research program, I have always been inspired by two grand metaphors. The first is Stephen J. Gould's (1989) metaphor of playing the tape of evolutionary history over. Gould felt that much of what would evolve would be highly contingent, and the course of evolution would be difficult to predict. I am not nearly as pessimistic as he was that we cannot understand or predict what would happen. Certainly, the exact same creatures would not arise, and the stochastic nature of evolution would set Earth's biota down different paths with each replay. However, the governing principles of ecology and evolutionary biology would apply to every replay of the tape; although we may not be able to exactly predict all the particulars, we should be able to understand and interpret what happens. In my own research, I imagine what would happen if I could go back 10 million years into the past to see which types of ponds and lakes the various taxa of eastern North America inhabit, and then watch what happens to them and their descendants until they and I arrive in the present (e.g., McPeek and Brown 2000).

This evolutionary play unfolds in the ecological theater—this is the second grand metaphor, articulated by G. Evelyn Hutchinson (1965), that inspires this book. The evolutionary trajectories of various species are directed by the ecological interactions they experience each day across the landscape. Each species is an actor on this ecological stage, and their interactions define what they experience, and thus how the play proceeds. What new opportunities do the various actors present to one another? How does each attempt to exploit these opportunities and how must they change to exploit these opportunities? Finally, how do those changes affect their distributions, abundances, and diversity across the theater through time?

The structure of this work is inspired by Hutchinson's and Gould's metaphors. Can we organize and link ecological and evolutionary theory to provide a framework in which we can understand and predict the organization of local communities and regional biotas over evolutionary time? For me, the organizing concept in all this is the ecological opportunity. In ecology, we might call this a niche; in evolutionary biology we might call this an adaptive peak—we'll see. Some species can exploit an opportunity, and others cannot. Whether a species or a lineage can do this may have far-reaching consequences for the organization of the local community, the evolutionary trajectory of that taxon, and the structure of the regional biota.

WHAT DO REAL COMMUNITIES LOOK LIKE?

If you stop at any spot along a trail through the woods, you will encounter a dizzying collection of species that are interacting with one another under the abiotic conditions at that site. In fact, an immense amount of work is required to characterize the web of interactions among all the species found in a patch of forest, or in a stream or lake, and the network of connections are typically very dense and complex (e.g., Winemiller 1990, Martinez 1991, Dunne et al. 2002, Bascompte et al. 2003). For example, figure 1.1 reproduces the figure constructed by Winemiller (1990) that depicts the trophic interactions among 63 aggregated taxa found in the Caño Maraca swamp creek in western Venezuela; this work represents one of the best characterized food webs in ecology. Each dot in the figure is either one of these 63 aggregated taxa or 62 common fish species (of the total 121 fish species collected there), and lines connect taxa that are consumer and resource (one species that feeds on another), based on diet information.

Because Winemiller was interested in the fish assemblage, he identified all fish to species. However, nodes representing the aggregate taxa are not species at all, but rather broad taxonomic groups (e.g., diatoms, chironomids, odonates) and a few categories of detritus (fig. 1.1). This figure also omits the bacteria that process the detritus, viruses, pathogens, and parasites that plague all species, as well as the many species such as kingfishers, herons, and raccoons that come to forage but do not live in the swamp. Pictorially depicting this trophic web in a way that illustrates structure clearly shows the complexity of the entire system, even with most of the taxa being pooled into such broad categories. In fact, each dot for an aggregated taxon represents a few (e.g., 35 palaemonid shrimp), to tens (e.g., 49 mosquito larvae), to hundreds (e.g., 47 chironomid larvae) of species (fig. 1.1, and fig. 6 and Appendix A in Winemiller 1990).

While this aggregation is frequently unavoidable, particularly for questions about interaction networks, much of community ecology is directed at explaining features and patterns pertaining to the species themselves. We ask questions about factors influencing the number of species of a particular type that can live in an area, whether some collection of species can coexist with one another, or whether some species can exploit a particular ecological opportunity. Such questions can only be addressed if individuals are identified to species. Moreover, these problems multiply when one realizes that no simple answer exists for the question of what a species is.

Common-Fish Web

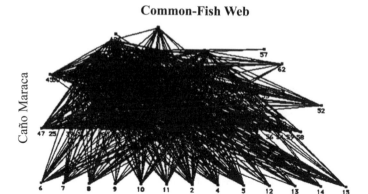

APPENDIX 1

Diet categories used in the fish stomach content analyses.

Numeric code	Web node
2	Fine detritus (organic mud component)
3	Larval anurans (Amphibia)
4	Coarse detritus
5	Vegetative detritus
6	Diatoms
7	Desmids and unicellular green algae
8	Filamentous algae
9	*Chara* sp. (macroscopic filamentous algae)
10	Aquatic macrophytes
11	*Wolffia* spp. and *Lemna* spp.
12	Terrestrial vegetation
13	Fruits (soft tissues) and flowers
14	Seeds and nuts
18	Other Protozoa
19	Difflugiid Protozoa
20	Rotifera
21	Nematoda (non-parasitic forms)
22	Hydracarina (water mites)
23	Nematophora (horsehair worms)
24	Annelida (earthworms)
25	Gastropoda (snails, Mollusca)
26	Bivalvia (clams, Mollusca)
27	Thoracica (shipworms, Mollusca)
28	Copepoda (microcrustacea)
29	Cladocera (microcrustacea)
30	Amphipoda (Crustacea)
31	Eubranchiopoda (Crustacea)
32	Ostracoda (microcrustacea)
33	Isopoda (Crustacea)
34	Unidentified microcrustacea
35	Palaemonid shrimp (*Macrobranchium* spp.)
36	Crabs (*Dilocarcinus* and *Callinectes* spp.)
37	Collembola (springtails)
38	Plecoptera (stonefly nymphs)

Numeric code	Web node
39	Ephemeroptera (mayfly nymphs)
40	Odonata nymphs
41	Other aquatic Hemiptera
42	Corixidae (aquatic Hemiptera)
43	Gerridae (aquatic Hemiptera)
44	Trichoptera larvae
45	Aquatic Coleoptera, larval forms
46	Adult aquatic Coleoptera
47	Chironomid larvae (Diptera)
48	Other aquatic Diptera larvae
49	Mosquito larvae (Diptera)
50	Aquatic Neuroptera larvae
51	Unidentified aquatic insects
52	Unidentified terrestrial insects
53	Hymenoptera
54	Orthoptera
55	Terrestrial Coleoptera
56	Lepidoptera larval forms
57	Hirudinea (leeches)
58	Lepidoptera adults
59	Terrestrial Diptera
60	Terrestrial Hemiptera
61	Isoptera (termites)
62	Arachnida (spiders)
77	Adult anurans (Amphibia) (only eaten at C. Volcan, dry season)
99	Lizard (Gekkonidae) (only eaten at C. Agua Fria, wet season)
122	Turtle flesh (*Trachemys scripta*) (only at C. Maraca, wet season)
123	Bird flesh (only eaten at C. Maraca, wet season)
124	Mammal flesh (only eaten at C. Maraca, wet season)

FIGURE 1.1. Representation of the trophic interaction web for the Caño Maraca swamp creek in western Venezuela. Each node in the web represents an aggregated group of taxa (many of which are identified in the accompanying table) and individual species of fish. Lines identify trophic links between taxa. From figure 6 of Winemiller 1990. *Ecological Monographs* (reprinted with permission from John Wiley and Sons).

Therefore, it would be instructive to actually look at what is inside those aggregated taxa in the trophic web. Because Winemiller (1990) only identified fish to species, we cannot crack open any of his specific aggregated taxa, but species data on some of these taxa in other systems are available. For example, consider the phytoplankton that forms the base of many aquatic communities. In figure 1.1, phytoplankton are represented by three aggregate taxa. Figure 1.2*A* presents data on the number of species per genus for phytoplankton in Peter Lake, northern Wisconsin, United States, in samples taken from 1984 to 1995 (data downloaded 6 April 2016 from https://lter.limnology.wisc.edu/datacatalog/search). In this lake, 257 total species in 123 genera have been identified. Many genera are represented by a single species, but many others have multiple species living together in the lake, including one (*Dinobryon*) with 15 species. Likewise, chironomids (the midge family of dipterans) are typically a species-rich group in aquatic habitats. Figure 1.2*B* presents the same information for chironomids for a 100-meter stretch of a first-order mountain creek in Granada, Spain (Casas and Langton 2008). In this short stretch, 163 species in 61 genera were found in samples taken biweekly over four years. Again, some genera were represented by a single species, but many were represented by multiple species, with one (*Eukiefferiella*) having 11 species living together in the stream reach. Finally, figure 1.2*C* presents the same information from my own work on odonates found in Palmatier Lake, Michigan, United States. I have collected 53 species in 26 genera at this lake over the years (M. A. McPeek, unpubl. data), with almost half of the genera having two or more representative species, and one genus (*Enallagma*) having 12 species living together. Thus, if fully resolved to species, the trophic web depicted in figure 1.1 would have thousands of connected nodes.

A few brave souls try to make sense of these huge webs of interactions, but inquiry at this level is almost exclusively done with taxonomically aggregated nodes (e.g., Winemiller 1990, Martinez 1991, Dunne et al. 2002, Bascompte et al. 2003, Allesina et al. 2008). In contrast to those cited, most community ecologists limit the problems they address to a small subset of interacting species embedded in the broader interaction web—what Holt (1997a) has called the *community module*. As Holt (1997a, p. 333) notes, "the hope is that modules may, at the very least, illuminate general processes and qualitative features of complex communities." It is at this level that questions are addressed regarding coexistence, limits on the numbers and types of species that can live together, and the exploitation of ecological opportunities. Furthermore, if we are interested in understanding how ecological interactions shape the evolution of species, and how the evolution of species in turn shapes the community in which they are embedded, we must limit our inquiry to these smaller pieces. This is the level on which I focus in this book.

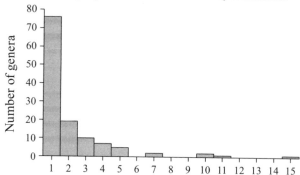

A. Phytoplankton in Peter Lake, Wisconsin

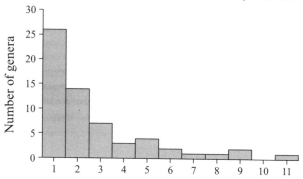

B. Chironomids in 1st order stream, Granada

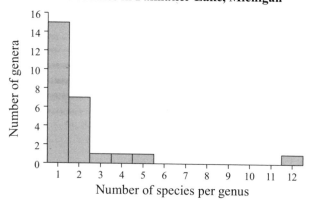

C. Odonates in Palmatier Lake, Michigan

WHAT IS AN ECOLOGICAL OPPORTUNITY?

Each species in the interaction web of a community is potentially exploiting an ecological opportunity. Before the subject of ecological opportunity can be considered, it must be defined. I define an *ecological opportunity* as a functional position within a community. I will come to what I mean by "functional position" in the coming pages. This definition seems simple but implies a number of important ecological and evolutionary issues. From the outset, let me define a *community* as the species found together in a local area that directly or indirectly interact with the local abiotic environment, and with one another, to thereby affect each other's fitness (Fauth et al. 1996). Thus, community in the sense used in this analysis is necessarily a local community. The central issue that this definition raises is species coexistence (MacArthur 1972, Holt 1977, Chesson 2000, Adler et al. 2007, Siepielski and McPeek 2010).

An ecological opportunity may be exploited by a species or may be available but unexploited. The processes by which an ecological opportunity can be exploited by a single or multiple species are of critical importance. These processes include immigration from outside the community under consideration, or the origination of new types of organisms from an existing community member (e.g., genetic differentiation within one species across geographic regions or sympatric speciation to create a new species in one place). How opportunities are exploited over ecological and evolutionary time defines the process of community assembly and shapes community structure on both local and regional geographic scales and through time.

The most proximate concern in evaluating an ecological opportunity is whether a new species can invade a community and establish a viable population to exploit the opening. Some invading species will have a positive population growth rate from the instant they establish a population. Such an invading species, call it species i with abundance N_i, would have a positive population growth rate $dN_i/dt > 0$ at a population size of $N_i \approx 0$. As it increases in abundance, the invading species may simply take its place in the community with only quantitative adjustments in the abundances of other species, or other species may be driven extinct by direct

---

FIGURE 1.2. Numbers of genera with different numbers of species that inhabit communities for representative taxa from three different locations: (*A*) phytoplankton found in Peter Lake, Wisconsin, United States (data from https://lter.limnology.wisc.edu/datacatalog/search); (*B*) chironomid midges (Diptera: Chironomidae) from Río Albuñuelas, a permanent first-order creek in Spain (taken from data in Casas and Langton 2008); and (*C*) the Odonata in Palmatier Lake, Michigan, United States. (Taken from data in McPeek 1990b, 1998; Stoks and McPeek 2003; and M. A. McPeek, unpubl. data.)

interactions with species i, or by the indirect interactions that propagate through the community. By invading, this new community member may also alter the kinds of ecological opportunities for subsequent invaders.

For species that already reside together, the same issue must be considered because these demographic conditions (i.e., $dN_i/dt > 0$ at $N_i \approx 0$) exactly state the criterion of *invasibility*, which is the theoretical benchmark used by community ecologists to evaluate whether an assemblage of species coexist with one another (MacArthur 1972, Holt 1977, Chesson 2000, Adler et al. 2007, Siepielski and McPeek 2010). To an ecologist, *coexistence* has a very precise meaning that differs from the term's more general usage. If the species found living together in a community are coexisting, then they all must be able to pass the invasibility criterion. To experimentally test whether a set of species are coexisting, one would evaluate whether each species in turn can increase in population size when (a) it is removed from the community, (b) the other species come to their new abundances because of its absence, and (c) it is then allowed to reinvade. In practice, invasibility has only been evaluated on the most simple of ecological systems, and so remains mainly a conceptual ideal (Siepielski and McPeek 2010). In the context of ecological opportunities, invasibility will probably also be a conceptual ideal, but it identifies the tangible hurdle that a species faces when invading a new community to exploit some ecological opportunity.

The invasibility criterion of community ecology is equivalent to stating that the average fitness among the individuals in the invading population is greater than replacement; that is, $\ln(\bar{W}_i) > 0$ or $\bar{W}_i > 1$, where $\bar{W}_i$ is the average fitness of the individuals (Lande 1982). When restated in this way, the implications for natural selection and local adaptation become readily apparent. As it increases in size, this invading population may adapt to the natural selection pressures imposed by the local abiotic environment and by the interactions with species in the new ecological regime. In so doing, it may cause further changes in the abundances and phenotypic traits of the other community members.

However, not all invading species may be so fortunate. Other invading species may initially have a negative population growth rate (i.e., $dN_i/dt < 0$, and so $\ln(\bar{W}_i) < 0$), but may nonetheless be able to adapt quickly so as to become a viable population. To exploit an ecological opportunity in this case, species i must evolutionarily adapt to this new environment quickly enough to stop its numbers from declining and begin increasing; in other words, the invader initially could not satisfy the invasibility criterion, but was able to adapt to the local conditions soon enough that it then does satisfy the invasibility criterion. This process necessitates a race between adaptation and population extinction (Gomulkiewicz and Holt 1995). Obviously, the more maladapted the initial colonizing population is (i.e., the more negative $\ln(\bar{W}_i)$ is initially), the less likely that the population will succeed in sufficiently adapting before its demise. Also, greater levels of immigration will impede the necessary local adaptation (Holt and Gomulkiewicz 1997). These considerations additionally imply that we cannot truly speak of the properties of an entire species. Local populations of species may have differentiated to some degree because of adaptation to local ecological conditions in different communities (Thompson 2005). If this differentiation proceeds far enough, speciation may occur (Schluter 2000, Nosil 2012).

These processes over the long term will define evolutionary trajectories for the component species and the phenotypic diversity present in communities and regional biotas. Conceptually, the invasibility criterion is used in community ecology as a retrospective test, in the sense that it is applied to species that are already found living together. For those species found to be coexisting today, we can infer that in the past they each colonized the ecological opportunity that they now exploit. To make this inference, one would need data on contemporary communities to test whether species today are coexisting, as well as evolutionary analyses based on historical, paleontological, or phylogenetic data to test for such an invasion in the past. For example, in my own work, I have used systematics analyses to infer the sequence of habitat shifts that must have occurred in various damselfly lineages to account for the distributions of species along habitat gradients today, and the adaptive evolutionary changes that were associated with each invasion of a new habitat (McPeek and Brown 2000, Stoks and McPeek 2006). Each one of these habitat shifts represented a lineage invading a new ecological opportunity, and by inserting a new functional group into a community each would have changed how the system responded to perturbations. We will consider these matters in more depth in the following pages.

WHO ARE THE MEMBERS OF A COMMUNITY?

The idea of a set of coexisting species will be an important organizing concept throughout these pages. However, this does not mean that each of the thousands of species in the interaction web depicted in figure 1.1 can satisfy its local invasibility criterion. Far from it! By identifying which species are coexisting and what the phenotypic and evolutionary properties of those species must be, one is also identifying the criteria for species that are not coexisting in the strict sense used by ecologists. I use the term *co-occurring* to identify those species in a community that are not coexisting (Leibold and McPeek 2006). All communities are heterogeneous mixtures of many coexisting and co-occurring species.

Moreover, admitting that some (or many) of the species in the community are merely co-occurring immediately forces one to consider that each large and complex local community is embedded in a larger regional system of communities; this system is known as a *metacommunity* (Holyoak et al. 2005), in which dispersal among communities can play a pivotal role in shaping both local and regional community structure. Metacommunities are themselves embedded in an even larger regional biogeographic system. Successful movement of species among metacommunities is quite rare but important when it does happen. I will consider aspects of the consequences of a local community being embedded in larger systems at these two scales.

Four types of species, based on their population dynamical properties, are found in a community: coexisting, neutral, sink, and walking dead species. Conceptualizing these categories is easy, but identifying any particular species to one of these types can be exceedingly difficult in practice. Moreover, some reflection will also identify that these categories are not necessarily mutually exclusive. Also, any particular species will occupy different categories in different communities.

The first and obviously fundamental species type to consider, based on the above discussion, is the *coexisting species*. As stated above, these are the species that can satisfy the invasibility criterion. Note that my definition of ecological opportunity is also based on defining functional groups, rather than species, within a community. I define a single *functional group* as a set of species that are ecologically quite similar to one another and are therefore exploiting the same ecological opportunity. Ecological similarity is defined by the ways individuals respond to and interact with the abiotic conditions and other species in the system. I think of ecological similarity in the context of an Eltonian definition of the niche. Elton (1927) conceptualized the niche of a species as its "occupation"—that is, what individuals of a particular species are capable of doing, and the species' place in the community. I think of these as the abilities of individuals comprising a species to perform in the various interactions with the abiotic environment and with other species, and the resulting demographic consequences of these interactions.

In this context, ecological similarity must be measured in the totality of abiotic and biotic interactions in which a species engages. If two species differ substantially in only one type of interaction (perhaps they eat different resources, or are attacked by different parasites or predators), they are not ecologically similar even if they are identical in all other respects. Obviously, ecological similarity is a measure of degrees, and the degree to which species are similar will be a critical recurring issue throughout this analysis. Species that are too ecologically similar will not coexist in the conceptual sense with which I use the term.

However, strong differentiation in some ecological feature is not a necessary condition of coexistence (e.g., Holt 1977, Tilman 1994, Chesson 2000, McPeek 2012). For example, two species that differ substantially in their abilities to acquire a single essential resource from the environment will not coexist (Tilman 1982).

To coexist, species must differ in particular ways, and these depend on the types of species interactions in which they are engaged. For example, the criteria for these two resource competitors to coexist differ if they have no natural enemies, if they are fed upon by a common predator, or if they each have unique parasites (Tilman 1982; Grover 1994; Holt et al. 1994; McPeek 2012, 2014a). Moreover, how these two species might evolutionarily adapt to acquiring the resource may also differ under these various scenarios. Finally, the number of consumers that can potentially coexist in these scenarios may be quite different. The criteria for coexistence cannot be reduced to platitudes, such as two species must be different to coexist or intraspecific competition must be stronger than interspecific competition. The features favoring coexistence also cannot be generalized across all possible types of species interactions and all community modules. Luckily, as I hope to show, generalities do exist and are based on common features of mechanisms of various species interactions and natural selection.

Species that are ecologically nearly identical to one another comprise the second type of species found in a community—these are *neutral species*. These will follow the neutral dynamics of a random walk in relative frequency, as described by Hubbell (2001). Much of the debate that neutral theory has sparked about the importance of neutral dynamics in communities has constructed the argument as though it were an "either-or" proposition—either all species in a community are coexisting with one another or they are all neutral species. Moreover, almost all analyses addressing these issues construct their tests based on this presumption. If any evidence is found for coexistence of any species in the community, then species in the entire collection are declared to be coexisting, and neutral dynamics are declared completely unimportant.

The conception of community structure in the context of coexisting and neutral species that I want to present is much more nuanced and layered. Specifically, a single functional group may contain multiple neutral species. Again, the definition of ecological opportunity given above does not use the word species—this was intentional. If species in some collection are ecologically identical or nearly so, the rest of the community will experience them as though they represent a single ecological entity. To see this, consider the following simple model of a food web with three trophic levels:

$$\frac{dP}{dt} = P\left(\sum_{j=1}^{q} n_j m_j N_j - x\right)$$

$$\frac{dN_j}{dt} = N_j (b_j a_j R - m_j P - f_j), \qquad (1.1)$$

$$\frac{dR}{dt} = R\left(c - dR - \sum_{j=1}^{q} a_j N_j\right)$$

where R, N_j, and P are, respectively, the abundances of a basal resource, q intermediate-trophic-level consumers ($j = 1, 2, \ldots, q$), and a top predator (see also Leibold 1996, Siepielski et al. 2010). In this model, the basal resource has logistic population growth in the absence of any of the consumers, where c is the intrinsic growth rate when rare, and d is the strength of density dependence in its growth rate. The top predator and all consumers have linear functional responses, where a_j and b_j are the attack rate and conversion efficiency, respectively, of consumer j feeding on the resource; and m_j and n_j are the attack rate and conversion efficiency, respectively, of the predator feeding on consumer j. Additionally, the top predator and consumers have density-independent mortality rates of x and f_j, respectively. If the consumers are all ecologically different from one another, meaning that at least some of the parameters in their population dynamics in equations 1.1 are different, at most two can coexist with the resource and the predator (Holt et al. 1994; Leibold 1996; McPeek 1996b, 2014a). Moreover, they will respond differently to perturbations that affect the equilibrium abundance of the resource or the predator.

However, if all the consumers are ecologically equivalent, meaning that they have exactly the same parameters (e.g., $a_1 = a_2 = \ldots = a_q = a$), notice what happens to this system of equations:

$$\frac{dP}{dt} = P\left(mn \sum_{j=1}^{q} N_j - x\right)$$

$$\frac{dN_j}{dt} = N_i\left(baR - mP - f\right) \quad . \tag{1.2}$$

$$\frac{dR}{dt} = R\left(c - dR - a\sum_{j=1}^{q} N_j\right)$$

In this case, all the ecologically equivalent consumers (i.e., all N_j) in the community act as a single dynamic unit—a single functional group—and the ecological interactions and population dynamics of the community respond to the sum of their abundances (i.e., $\sum_{j=1}^{q} N_j$) (Siepielski et al. 2010). In fact, this system sets the equilibrium abundance of the consumers at $\sum_{j=1}^{q} N_j = x/(mn)$. The relative abundances of the consumers will drift according to a random walk within this constraint on total abundance. Over time, consumer species will be lost randomly from the system based on the rate of drift, which depends on their summed abundances, until only one consumer species remains (Chesson and Huntly 1997, Hubbell 2001). In other words, collections of neutral species that follow the dynamics described by Hubbell (2001) are easily embedded in more complex food webs that follow the strictures of coexistence (Leibold and McPeek 2006).

The presence of neutral species makes the assignment of coexisting and neutral descriptions to each species ambiguous. If the consumers in equations (1.1) are ecologically distinct, and hence only two are present with the predator and resource in this diamond-shaped community, we can easily stipulate that every species in this community is coexisting; that is, each species will satisfy the invasibility criterion (Leibold 1996). However, this is not true if the consumers are ecologically identical. Imagine applying the invasibility criterion to a community described by equation (1.2) with four consumers. Initially, the community would come to a stable equilibrium with $N_1 + N_2 + N_3 + N_4 = x/(mn)$. Then remove consumer 3 from the community and allow the remaining species to re-equilibrate; now $N_1 + N_2 + N_4 = x/(mn)$. When consumer 3 is reintroduced it will have $dN_3/dt = 0$ at $N_3 \approx 0$.

Thus, any single neutral species is not coexisting in this community, which is reassuring given the definition of coexistence. However, the entire collection of neutral species—the functional group that they comprise—*is* coexisting with the rest of the community. To me, functional groups coexist, and so two species that coexist with one another are in two different functional groups exploiting two different ecological opportunities. If all members of a functional group were missing, then the first member that invades would satisfy the invasibility criterion. However, once the first species enters and establishes itself, subsequent invaders that are ecologically identical to the first would have exactly the same demographic performance as the ones already present, and so would not satisfy the invasibility criterion.

My work on the odonates found in lakes across eastern North America sheds some light on this issue. A summary of the taxonomic diversity of the Odonata at any lake across eastern North America from the Gulf of Mexico to northern Canada would look very similar to that for Palmatier Lake, Michigan (fig. 1.2*C*). Palmatier Lake harbors 53 species in 26 genera (McPeek 1990b, 1998; Stoks and McPeek 2003; and McPeek, unpubl. data). Fifteen genera are represented by a single species, whereas the other genera have multiple species present, including one (*Enallagma*) with 12 species. Extensive field experiments on three of these genera show that the 12 *Enallagma* are neutral species occupying a single functional group, whereas genera differ in ways that match the requirements for coexistence (McPeek 1998, Siepielski et al. 2010, Siepielski et al. 2011b, Siepielski and McPeek 2013). In other words, for odonates in lakes, each genus seems to be a different functional group within the community, whereas species within a genus are neutral within the same functional group. Given that the last common ancestor of those 12 *Enallagma* species dates to approximately 10 million years ago, these species must have been neutral members of the local communities for all that time (McPeek and Brown 2000, Turgeon et al. 2005).

I am sure that the taxonomic predictability of coexisting and neutral species I have found in the odonates of lakes is not a general rule. For example, thorough field experiments have shown that two barnacle species in different genera are neutral species in the intertidal zone of the Pacific coast of Chile (Shinen and Navarrete 2014). In contrast, four *Daphnia* species clearly coexist by segregating spatially within the same lakes inhabited by *Enallagma*, and these *Daphnia* species therefore coexist as four separate functional groups in these communities (Leibold and Tessier 1991, Tessier and Leibold 1997). It is probably a loose but general rule that the more phylogenetic and taxonomic distance between two species, the greater the ecological dissimilarity between them. However, as these three examples illustrate, this assumption is by no means inviolate.

The third type of community member are *sink species*, which are present only because of continual immigration from other communities (Shmida and Ellner 1984, Pulliam 1988). If immigration were prevented, sink species would eventually go extinct locally because local average fitness is lower than what is needed to maintain the population (i.e., $\ln(\bar{W}_i) < 0$). A sink species is maintained in the community at a population abundance for which decline due to local average fitness balances immigration rate (Pulliam 1988). Clearly, sink species do not coexist in the local community, since by definition they would have a local negative population growth rate due to local ecological conditions. However, sink species are still interacting with the other species in the community (e.g., utilizing resources, being fed upon by predators, interacting with mutualists) and so can have appreciable effects on the ecological and evolutionary dynamics of other community members.

The presence of sink species highlights the importance of considering how a community is embedded in a larger regional metacommunity (Holyoak et al. 2005). Because a sink species is maintained by continual immigration, those immigrants must come from nearby communities, and these species must have source populations (i.e., $\ln(\bar{W}_i) > 0$ when rare, which means that they are coexisting) in some of those nearby communities. Moreover, the continual gene flow from the source populations would retard or prevent adaptation of the sink population to local ecological conditions under most conditions (Holt and Gomulkiewicz 1997).

Although many extinctions occur via stochastic processes (Raup 1992), many species are driven extinct by ecological interactions with the abiotic environment and other species. This fourth type of species I call the *walking-dead species*, a term first attributed to Daniel Janzen. They have also been called the "extinction debt" of the community (Tilman et al. 1994), and the "living dead" (Hanski 1998). These are species that are slowly being driven extinct by the ecological conditions they experience in the community, and immigration from elsewhere in the metacommunity will not rescue them or prevent their demise. Consequently, they must have a long-term negative population growth rate $\left(\overline{dN_i/dt} < 0 \right)$ and thus a long-term average fitness less than replacement $\left(\overline{\ln(\bar{W}_i)} < 0 \right)$. (Note that in the long term, most neutral species are expected to go extinct, but via a random walk with $\overline{\ln(\bar{W}_i)} = 0$ [Hubbell 2001].) For walking-dead species, extinction has a temporal dynamic specified by their long-term population growth rate, with more negative long-term population growth rates causing species to go extinct more rapidly (Lewontin and Cohen 1969, Turelli 1977, McPeek 2007). Despite their ultimate demise, species that are destined for extinction do not simply give up and throw themselves on their swords. Until the last individual dies, these species will still be present in some number of communities, and they may remain for a very long time if their rate of population decline is quite slow. Spatial and temporal variability in demographic rates may also greatly obscure when species are the walking dead.

A walking-dead species also undergoes another potential outcome; namely, it may adapt to the system sufficiently to experience a positive population growth rate and thereby evolve to be a coexisting or neutral member of the community. Were it to not adapt, the species would eventually be driven extinct. However, adaptation can rescue walking-dead species, and as we will see, this process has been a prevalent mode of ecological speciation.

ARE COMMUNITIES "REAL"?

Ecologists have debated for nearly a century whether there is such a thing as an ecological community. The debate embarked in earnest in the 1910s and 1920s with the views exchanged between Frederic Clements and Henry Gleason regarding plant communities and succession (Clements 1916, Gleason 1926). Clements saw plant communities as integrated units, whereas Gleason argued that plant communities were merely collections of individual species that were all acting largely autonomously. This debate has been a perpetual feature of community ecology and continues to this day (Whittaker 1953, Goodall 1963, Shipley and Keddy 1987, Callaway 1997). In this vein, a critique by Ricklefs (2008) argued for the "disintegration of the community as a central concept." His argument for the abandonment of the concept is that an ecological community is not an integral entity. Instead, the species found at a local site are the result of regional processes that control the "distribution of populations over ecological and geographic gradients." In fact, in chapter 4 I will summarize much paleoecological and phylogeographic work that indicates Gleason and Ricklefs are correct in arguing that communities are not integral entities. Moreover, I wholeheartedly agree with Ricklefs' call for more emphasis on regional, historical, and phylogenetic processes in the study of ecological communities (see also Ricklefs 1987, 1989). Those ideas are at the core of what I am trying to accomplish with this work.

However, I obviously do not agree with the characterization of community as an invalid concept about nature. I hold this view because assemblages of species show regular and predictable patterns of distributions and abundances across the local landscape in all ecological systems. Moreover, myriad experimental studies have shown that local species interactions drive much of this distributional patterning; it is not simply the result of each species' physiological abilities to deal with the abiotic conditions found at some site. In the system I know best, characteristic assemblages of most animal taxa, including snails, leeches, beetles, mayflies, stoneflies, true bugs, damselflies, dragonflies, salamanders, frogs, fish, and so forth, are found segregating along the ecological gradients in ponds and lakes of eastern North America (Wellborn et al. 1996). Taxa do differ in their breadth of distribution along these gradients. For example, different sets of *Enallagma* and *Lestes* damselfly (Odonata: Zygoptera) species are found in ponds and lakes depending on whether fish or dragonflies are top predators and whether the pond has water that continuously or periodically dries up (McPeek 1990b, 1998; Stoks and McPeek 2003, 2006). However, the same *Ischnura* species are found in all these pond types (McPeek 1998). Water bodies with these various ecologies are interspersed across the landscape.

If bluegill sunfish (*Lepomis macrochirus*) are experimentally introduced to a pond where large dragonflies are the top predators, the fish quickly kill all the large dragonflies and the *Enallagma* and *Lestes* species that are specialists living with those large dragonflies (Crowder and Cooper 1982, McPeek 1990b, Werner and McPeek 1994, Stoks and McPeek 2003). Over the next 20 to 30 years the *Enallagma* and *Lestes* species that are specialists in living with fish colonize and increase in abundance (M. A. McPeek, pers. observation). Moreover, removing fish completely from a pond or lake will reverse this process; large dragonflies will colonize, kill the *Lestes* and *Enallagma* species that specialize in living with fish, and the species specialized in living with large dragonflies will colonize (M. A. McPeek, pers. observation). Thus, adding or removing a single species to or from a pond will dramatically and predictably change large suites of other species, and this experimental result explains the natural variation among ponds and lakes. It is true that the same *Ischnura* species will be present in both of these situations, but the source of their mortality and how their population abundances are controlled are substantially changed by switching the top predator (McPeek 1998).

Such patterning of local species assemblages across the landscape is everywhere around us, and innumerable experimental studies over the past 70 years have demonstrated how abiotic factors and species interactions generate these patterns. Zonation patterning of species distributions in the marine intertidal are classic examples of the maintenance of this strong patterning, and these are over mere centimeters of tidal height (Connell 1961, Payne 1966, Dayton 1971, Menge

1976, Sousa 1979). Plant assemblages in local areas strongly depend on whether grazers have access to forage in these areas (Harper 1969). Similar patterning to what I described for damselflies in the littoral zones of ponds and lakes also occurs for the zooplankton found in the pelagic zones of these same lakes (Brooks and Dodson 1965, Dodson 1970, Sprules 1972, Vanni 1988).

I take the fact that we can make regular and predictable changes in the composition and abundances of many species across disparate taxa at a very local scale by experimentally manipulating the presence/absence of one or a few species as strong evidence that local species interactions are a primary determinant of the distributions and abundances of species across the landscape. Note that I say *a* primary determinant, and not *the* primary determinant. It is this importance of local species interactions that causes most ecologists to think the concept of a community has significant utility. The fact that species composition does not change wholesale or that the local assemblage does not form a completely integrated unit seems immaterial to whether local species interactions play a substantial role in determining which species are there and what their abundances are. How could it be expected to change wholesale when composed of not only coexisting, but also neutral, walking-dead, and sink species? Also, like *Enallagma* and *Ischnura* damselflies, different species will be members of different collections of local communities. The definition of community needs no statement about integration or boundaries.

Moreover, that which shapes the population abundance of a species at each locality within the geographic region must control the "distribution of populations over ecological and geographic gradients." These controls are precisely the set of abiotic conditions and other species that are found at a local position in space and time and with which the species in question directly and indirectly interacts. A species is found in a local area if it can support a population because of favorable local demographic conditions (i.e., coexisting or neutral species), by continual immigration from other local areas (i.e., sink species), or if it cannot support a population there but has not yet gone extinct (i.e., walking-dead species). Sorting through these possibilities requires one to consider ecological issues at both local and regional scales.

Rather than dismissing local interactions and the community concept, I think a better approach is to consider how processes operating at the local and regional scales interact. As with issues of coexistence and neutrality, this is not an "either-or" proposition. Both are important in fundamental ways, and I want to shift the argument to say that you cannot understand one without the other. I find it hard to imagine a scenario in which a species will be found in a region unless it has at least one local source population within that region—that is, it is a coexisting or neutral member of a functional group in at least one community (but see Roy et al. 2005). Moreover, explaining why a species is present in some local areas may necessitate considering other areas (i.e., sink species). I hope in this work I can make a more concrete connection between local and regional processes that shape patterns of biodiversity across the landscape and through time.

WHAT ARE THE SOURCES OF NEW SPECIES?

The biodiversity present today in a community comes from two sources. Species may have immigrated in from outside, or the species may have arisen within. Thus, biogeographic-scale immigration and speciation are the two critical sources of species to fill ecological opportunities, or at least to try and fill them.

Today, the field of conservation biology is quite concerned with "invasive" species, and the surrounding ecological issues (Elton 1958, Davis 2009, Simberloff 2012, Valéry et al. 2013). In the present work, I will consider the ecological and evolutionary dynamics caused by immigrating species. This would be equivalent to a "foreign" invading species, but this is not a statement about the debate that continues in conservation biology regarding what constitutes an "invasive species" (e.g., Davis 2009, Simberloff 2012, Valéry et al. 2013). Fundamentally, the species that are considered by invasion ecology today are undergoing the same ecological and evolutionary processes that taxa filling ecological opportunities in the past also experienced (Richardson and Pysek 2006, Strauss et al. 2006). Therefore, while not the focus of this work, I hope that the models and results discussed here can contribute to studies of "invasive species," however they are defined.

More broadly, I will take a more thorough view of deep time to consider how local ecological interactions spread across a landscape shape local community development and regional patterns of biodiversity; this view involves invasions and internal lineage diversification over long expanses of evolutionary time. The source of a new species will influence how similar it will be on average to those already present, and thus will affect the likelihood of initial invasion success. For example, an invader from outside the system will have phenotypic properties and ecological performance capabilities that are essentially random with respect to what is needed to fill some ecological opportunity. Thus, an approach of adding species with random phenotypes to a system would suffice for this immigration scenario.

Species do not have to be from some other biogeographic region to essentially be from outside the system. Consider, for example, the *Enallagma* damselflies I described above. Ponds and lakes with fish as the top predator harbor one set of *Enallagma* species, whereas ponds and lakes with large dragonflies as the top predators (and therefore lacking fish) harbor a different set of *Enallagma* species (Johnson and Crowley 1980; McPeek 1990b, 1998). For many (but not all) species, these two habitats operate as separate metacommunities, despite the fact that they are interspersed across the landscape. Shifts between them are exceedingly rare over evolutionary time (e.g., three are apparent over the *Enallagma* clade's ~10 million year history, and two over the *Lestes* clade's ~6 million year history), despite the fact that these habitats may often be separated by only a few meters, and founder populations are probably being established each year across these boundaries in both directions (McPeek and Brown 2000, Turgeon et al. 2005, Stoks and McPeek 2006).

These invasions from outside the community may result in a new species being formed if the invading populations must significantly adapt to the new ecological. milieu and in so doing become reproductively isolated from their ancestral populations. This is essentially the process of ecological speciation (Schluter 2000, Nosil 2012).

However, ecological differentiation is not the only mode of speciation, and not all new species come from outside the system. In fact, speciation is caused by myriad processes that generate reproductive isolation, and these processes generate varying levels of ecological differentiation, from producing new coexisting species to those that are ecologically identical but reproductively isolated (Dobzhansky 1937a, Mayr 1942, Coyne and Orr 2004). As you might imagine, the degree of ecological differentiation at the time of speciation can have a tremendous effect on the resulting structure of communities and regional assemblages as well as the macroevolutionary structure of diversifying clades (McPeek 2007, 2008).

An analogous argument to the "local versus regional" perspective of community ecology also can be made about considerations of the short-term versus long-term evolutionary dynamics of populations, species, and clades—"micro-" and "macro-" evolutionary dynamics, respectively. The distinction that is usually made to distinguish these two is that microevolution encompasses the evolutionary processes that occur within a species (i.e., mutation, genetic drift, gene flow, and natural selection), and macroevolution involves the processes driving speciation, extinction, and thus long-term clade diversification. Paleontologists have debated for years whether macroevolutionary dynamics are simply the natural extension of microevolution or whether there are processes unique to macroevolution (Erwin 2000). I believe there are unique events and processes that are only seen on a macroevolutionary scale, but here again, I think the important issue is not the primacy of one over the other, but rather how they interact to produce a biota. Ecological interactions, adaptation, dispersal, speciation, and extinction all go hand in hand. This perspective will be the premise of this work.

NATURAL SELECTION, COEVOLUTION,
AND COMMUNITY STRUCTURE

Darwin and Wallace (1858) recognized the power of ecological interactions to change the traits and abilities of species over time. Since this original description of evolution by natural selection, evolution has traditionally been assumed to occur very gradually in the wild—almost imperceptibly over the course of a human lifetime—and so population dynamics were assumed to occur on a fast ecological time scale, meaning that species could be assumed to be fixed in their traits and abilities. Important evolutionary change was assumed to occur on a much, much slower evolutionary time scale (Darwin 1859, Mayr 1942, Simpson 1944).

However, we have known for a very long time that this distinction in time scales is not always true; important evolutionary change can happen very rapidly. Darwin's original descriptions highlighted that domesticated animals and plants could be rapidly altered in several to dozens of generations by artificial selection (Darwin 1868). Selection in the wild can be observed to change species' traits significantly in only a few generations (e.g., Reznick 1982, Reznick and Endler 1982, Hendry and Kinnison 1999, Grant and Grant 2002, Ozgul et al. 2009), and significant phenotypic selection can be measured in almost any species one cares to observe (Endler 1986, Kingsolver et al. 2001). Moreover, a dominant pattern of character change in the fossil record is the one termed punctuated equilibrium in which the trait of a new species appears to change almost "instantaneously," at least given the fossil record's temporal resolution (Eldredge and Gould 1972). Finally, some mechanisms of speciation require only a single generation to create a new species (Rieseberg 1997, Otto and Whitton 2000, Mallet 2007).

The explicit recognition that substantial ecological and evolutionary dynamics can occur simultaneously has renewed interest in the exploration of their joint effects on population dynamics (e.g., Fussman et al. 2000, Yoshida et al. 2003), community structure (e.g., Travis et al. 2014), and ecosystem functioning (e.g., Harmon et al. 2009, Bassar et al. 2012). This has also spawned the coining of many new terms to describe the joint actions of ecological and evolutionary dynamics, including, for example, eco-evolutionary dynamics, eco-evo feedbacks, or ecogenetic links (e.g., Fussmann et al. 2007, Kokko and López-Sepulcre 2007, Schoener 2011). Given that these linkages have always been understood, I see no reason to generate a new jargon and so will refrain from using these terms.

However, this book is about fundamental aspects of the joint dynamics of ecology and evolution in communities. Because these dynamics can be exceedingly complicated, I focus specifically on the joint dynamics of (1) species interactions that regulate their abundances in a community context, and (2) natural selection that shapes the trajectories of their trait evolution. I simplify the analysis by assuming the standard basis of quantitative inheritance (Lande 1979, Falconer and Mackay 1996). In the following chapters, I explore the theoretical connections among community ecology, evolutionary adaptation, dispersal, and speciation and extinction to understand how local and regional patterns of community structure develop across space and through time. This is obviously a large and expansive undertaking, and I have tried to organize the presentation in a way that makes the interconnections obvious. Also, I cannot cover all types of species interactions. In particular, I do not consider the action of pathogens and disease-causing organisms in a community context (e.g., Ostfeld et al. 2008).

In chapter 2, I develop the purely ecological dynamics of interacting species in many types of community modules. The goal of the chapter is to explore the criteria required for different types of species to coexist in each kind of community module, and also to see how these criteria change as more species are added to some functional position and as new functional groups are added to change the interaction network to a new type of module. By defining what is required for coexistence, we are also defining what types of species would also be walking-dead and sink species.

In chapter 3, I explore how species in simple community modules evolve to adapt to one another. To understand adaptation, I first characterize the three general types of traits that underlie the trophic interactions between predators and their prey. As we will see, the type of traits involved in species interactions have substantial effects on the dynamics of adaptation, the likelihood that species will ecologically differentiate from one another (e.g., character displacement of competitors), and the structure of the resulting communities. I then explore the underlying dynamics of adaptation when all the species in these simple community modules can coevolve, and the influences of various system features on the outcome of this coevolution.

In chapter 4, I evaluate the main processes that operate at the regional and biogeographic scales to ultimately shape local community structure—namely, speciation and biogeographic mixing of taxa. I first consider the definition of a "species." Next I review the various mechanisms that create new species. In this analysis, I focus on the degree to which the speciation process directly results in differences in ecologically important traits between the progenitor and daughter species, and the geographic configuration of progenitors and daughters relative to one another. The phenotypic differences generated at the time of speciation determine the type of community member the new species begins as, and the geographic configuration of speciation determines whether it is introducing new species into any local communities. Finally, I review paleontological and phylogeographic data to show that past climate change has had an enormous effect on current local and regional community structure by periodic forcing of mass movements of species across Earth and sparking spikes in speciation and extinction rates. Current community structure cannot be understood without incorporating the massive perturbations that have routinely and regularly occurred over the past few million years.

In chapter 5, I develop explicitly the evolutionary dynamics that follow mixing different types of species together in a community either by invasion or by perturbation, as well as community mixing due to climate change. Specifically, this chapter focuses on the features that promote or retard ecological differentiation of species. Specifically, when will initially neutral species differentiate from one another to convert them into a set of coexisting species, and when will initially differentiated species converge to become ecologically more similar? Here again, we will see that the types of traits that underlie species interactions influence the likelihood and outcome of differentiation. Moreover, trait types also fundamentally shape the community structure that results when species do differentiate from one another.

In chapter 6, I consider the ecological and evolutionary dynamics of species across a metacommunity, and how these dynamics shape regional community structure. I first consider the evolution of dispersal—specifically, when movement between local communities is and is not favored by natural selection, and what these various movement patterns do to local community structure. Necessarily, this is the main area where the existence and evolution of sink species is considered. I then combine consideration of the evolution of dispersal rates among communities along with local adaptation within each to explore a few simple scenarios of how these two trait sets will evolve in a correlated fashion.

In chapter 7 I reflect on what I see as the important insights that emerge from this synthetic analysis of evolution in communities. The dynamics of species' abundances and traits are jointly driven by the same processes, and these dynamics can only be understood in the context of a community of interacting species.

Finally, I have created a website to accompany this book (http://press.princeton .edu/titles/11175.html). There you will find animations and games for a visual experience of the abundance and trait dynamics of interacting species that I discuss throughout the book. I hope you will interact with these web games and animations while you read. I also plan to add many features over time to the website, so please visit it periodically for new features.

CHAPTER TWO

The Community of Ecological Opportunities

Reflect on your walks through nature for a moment. Stare into the water of the lake at the bottom of the hill, and you will see a bounty of life. Innumerable species of algae form the base of the food web, harvesting sunlight to make their own biomass. Many species of cladocerans, copepods, snails, caddisflies, chironomids, frog larvae, and other herbivores eat the algae. Other species of caddisflies, flies, annelids, crayfish, additional arthropods, and frog larvae feed primarily on heterotrophic bacteria growing on allochthonous inputs such as leaves and woody debris and decaying material from within the lake. Above them in the food web, a large collection of predators feed on the herbivores, the scavengers, and one another. These predators are a diverse taxonomic lot, including dragonflies, beetles, hemipterans, salamanders, reptiles, birds, mammals, and fish.

The taxonomic diversity that is evident within these broad trophic categories hints at the huge functional diversity that is represented. Some algae species are fast growing and poorly defended against particular types of herbivores, while others have strong defenses against certain kinds of herbivore feeding strategies. Likewise, within the herbivore and predator groups, a bewildering array of predation defenses are effective against some enemies and not against others; all the defenses may limit other activities such as foraging for their own food. These differences in the types and strengths of interactions among the species found in the lake are what define the structure of the community.

Walk to the pond closer to the top of the hill that lacks some of the top predators (e.g., no fish), and stare into it. Here you will also see a bounty of life, and most of the same taxonomic groups found in the lake below are present. However, if you watch carefully, you will notice that many of the species in each group are different, and they look and act differently than their relatives in the big lake. Many of the species here are much bigger than their fish-lake relatives, and many are more active. Careful inspection will also reveal a diversity of feeding strategies and antipredator defenses, but many taxa are doing different things to find food and thwart predators here, as compared to the lake below.

The same is true of the communities of organisms that live in the field versus the understory of the forest, or any comparable habitats you have traversed on your walk. Each community offers a diverse array of ecological opportunities to species, and those opportunities are created by the abiotic conditions found there and the other species that are present. Differences in species composition among these communities you came across on your walk are primarily caused by the differences in the ecological opportunities available in the different settings.

An ecological opportunity is the fundamental basis for a species to persist somewhere in nature. When a species originates, it must be able to persist and potentially thrive somewhere. That somewhere is where an ecological opportunity exists for it to exploit. This chapter explores what ecological opportunities are available to species in different positions within a biological community. Most ecologists think of ecological opportunities only as available resources to exploit, but I hope to convince the reader that a much more expansive conception is warranted, including the tolerance of various abiotic conditions and natural enemies (i.e., herbivores, predators, omnivores, pathogens, parasites, parasitoids) and the benefit of mutualists. Community structure results from the diversity of ecological opportunities that are available. Here, I am not interested in exploring this overall community structure, but rather considering what is needed for an ecological opportunity to exist. Specifically, I will focus on what performance capabilities a species must have to fill various types of ecological opportunities.

Because I define an ecological opportunity as a coexisting functional position within a community, this presentation will focus on what is required for invasibility of species into different functional positions in a community. Invasibility, as discussed earlier, is the criterion for determining whether a species has some long-term advantage in a community that permits it to persist. Specifically, invasibility asks whether a species can increase when it is rare and all other species in the community are at their long-term demographic steady states (either point equilibrium, limit cycle, or chaos) when the species in question is absent (MacArthur 1972, Holt 1977, Chesson 2000, Siepielski and McPeek 2010). In other words, invasibility asks whether a species can invade or reinvade the community. Such ecological analyses are typically preoccupied with the resulting dynamics of the system—stable equilibrium point, limit cycles, chaos—after invasibility criteria have been satisfied. In this presentation, I touch on the types of dynamics that result, but my main focus is not whether a stable point equilibrium, stable limit cycle, or chaos results. For any of these, the species will still be present, and that is my main concern here. The reader will be directed to the original papers for the specifics of the locations of equilibria and attractors, the conditions under which different dynamics result, and the mathematical proofs of these results.

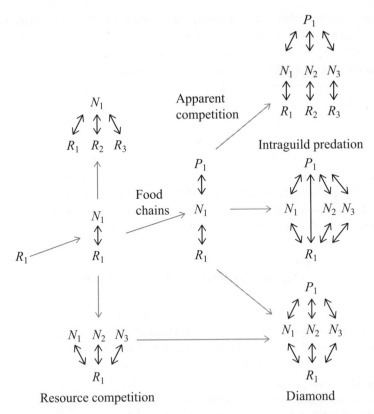

FIGURE 2.1. Connections among species that define various community modules, and the transitions from one module to another generated by species additions. Begin with a basal resource R_1 that can maintain a population on its own. Add a single consumer N_1 to this community, and a food chain with two trophic levels results. Add a predator P_1 that feeds on the consumer, and a three-trophic-level food chain results. Resource competition is a community module where multiple consumers feed on a single resource (or multiple predators feed on a single consumer). Apparent competition is the opposite module in which a single consumer or predator feeds on multiple prey. A diamond community module has multiple intermediate-trophic-level consumers feeding on a single resource and being fed upon by a single predator. Intraguild predation modules involve predators engaging in resource competition with their prey. Any given community can contain multiple modules simultaneously; for example, the diamond-shaped module involves both resource and apparent competition modules simultaneously.

I organize the exploration of ecological opportunities using Holt's (1997a) concept of a community module. As discussed earlier, a community module is a small set of interacting species that has a defined network configuration (fig. 2.1). One can think of a community module as a small interaction network that is embedded in a larger food web. I start with the simplest community module, explore what is needed to have these species coexist, and then investigate what is required to add more species (both while maintaining the same module); the next step is to add species that would move the system to a new module (fig. 2.1). I see this as the natural framework for understanding community assembly.

This chapter also focuses on the analysis of one local community and what is required for a particular type of species to invade and coexist in the system. Spatial dynamics and multi-patch results will be deferred to later. Also, by defining the criteria for invasibility, we establish the criteria that also define what types of species cannot invade and coexist. If these co-occurring species are present, they must be either neutral, sink, or walking-dead species. Moreover, remember that coexisting species here in reality define possible functional groups that may be filled by a single species or a set of neutral species. The interactions among the evolutionary actors determine which can be on any particular ecological stage.

MODELING SPECIES INTERACTIONS

One of the major problems with ecological theory is that population dynamics and species interactions have no universal mechanistic framework underlying them. As a result, many different modeling frameworks have been utilized to explore the dynamics of communities and the development of community structure caused by interactions among species.

The basic framework of modeling of interactions among species that was introduced independently by Vito Volterra and Alfred Lotka has been a venerable approach (Volterra 1926; Lotka 1932a, 1932b), and many modifications and derivative approaches based on this basic model have been used over the years. This framework describes a phenomenological model in which the interactions among species are abstracted to the "interspecific" effect of the abundance of one species on the per capita growth rate of another, and the "intraspecific" effect of the abundance of the species on its own per capita growth rate (May 1973). These per capita effects are assumed to be linear; that is, the per capita effect of species 1 on species 2 changes linearly as the abundance of species 1 changes. These abstracted effects may include competitive interactions, predator-prey interactions, or mutualisms among species. Simply changing the sign of the term representing the per capita effect of one species on another can shift among these types of interactions (May 1973).

These models have provided many insights and predictions about community structure. One major prediction to emerge is that coexistence of two competitors requires that intraspecific effects via negative density dependence must be greater than negative interspecific effects (May 1973, Chesson 2000, Adler et al. 2007).

However, this prediction is specific to competitive interactions among species. The basic model can be expanded to include many species, as in the community matrix (May 1973, Pimm 1982). From the matrix, predictions emerged that stable communities are a small subset of all possible configurations, with a number of general relationships about the strength of species interactions and the connectedness of species within communities (May 1973, Paine 1980, Pimm 1982, Paine 1992). Chesson (2000) has also used this framework to develop a generalized relationship between the degree of overall average fitness similarity among species (i.e., equalizing effects) relative to the density-dependent fitness differences (i.e., stabilizing effects) that foster coexistence among competitors.

The focus on resource competition as the main species interaction structuring communities also led MacArthur and Levins (1967) to think of the problem as one of limiting similarity (e.g., Roughgarden 1974; Abrams 1975, 1983). That is, for a collection of resource competitors feeding on a distribution of resources, how similar can their resource utilizations be while they still coexist with one another? For example, imagine multiple species of Darwin's finches feeding on seeds that cover a large range of sizes (Grant and Grant 1982, Schluter and Grant 1984). How similar can their beak morphologies be while they still coexist with one another? Again, these issues really only pertain to the coexistence of resource competitors. Moreover, this approach abstracts the dynamics of the multiple species that constitute the distribution of resources; these are in fact typically a large collection of species, each of which has their own population and evolutionary dynamics.

Thus, many of these alternative approaches are centered specifically on competition among species and abstract the real mechanisms of how species interact with one another. While resource competition is one important type of species interaction, it is actually one specific example of a more general set—namely, trophic interactions between consumers and resources. All species must consume resources from their environment, be they abiotic (e.g., water, phosphates, nitrogenous compounds, silica) or other species. Most species are also a resource for other consumers, be they herbivores, carnivores, omnivores, parasites, parasitoids, or pathogens. In other words, most species are simultaneously predator and prey.

The structure of a community is typically defined by how species are linked in the web of trophic interactions. For example, resource competition at its essence is simply multiple consumer species simultaneously feeding on the same set of resource species. Omnivory and intraguild predation are also simply variations on webs of interactions (Holt and Polis 1997, Diehl and Feißel 2000). In addition, most mutualistic interactions between species are at their core trophic interactions; for example, pollinators consume pollen as food, and the plants thus serve as a resource for the pollinators while simultaneously deriving some benefit (Holland et al. 2002, Holland et al. 2005, Holland and DeAngelis 2010, Jones et al. 2012). The basic ways that species can be linked by trophic interactions define the various community modules (Holt 1997a) that I will explore (fig. 2.1).

Thus, throughout this book I will take trophic interactions as the basis of interactions between species. I will also use more mechanistic models than the Lotka-Volterra or community matrix approaches to explore the ecological and evolutionary dynamics of a community. I will mainly use a modification of the Rosenzweig and MacArthur (1963) model of consumer-resource interactions as a base. I use this framework for three reasons: many different mechanistic features of species interactions can be incorporated into this basic model, multiple species can be explicitly modeled in relatively complex webs of interactions, and it describes the dynamics of average fitness for each species in the interaction web. All the results presented here can generally be derived using other modeling frameworks as well, such as discrete difference equations (Hassell 1978, Murdoch et al. 2003), but I choose continuous differential equations for their ease of analysis, and because they provide a standard framework for most of the topics I will address in this book. (The only deviation will be in chapter 6 when I will use difference equations to model dispersal among communities.)

In this model, predators consume their prey according to a saturating functional response, and all species may have some form of intraspecific density dependence. For a three-trophic-level food chain with multiple species at each level, the model is

$$
\begin{aligned}
\frac{dP_k}{dt} &= P_k \left(\frac{\sum\limits_{i=1}^{p} w_{ik} v_{ik} R_i + \sum\limits_{j=1}^{q} n_{jk} m_{jk} N_j}{1 + \sum\limits_{i=1}^{p} v_{ik} u_{ik} R_i + \sum\limits_{j=1}^{q} m_{jk} l_{jk} N_j} - x_k - y_k P_k \right) \\
\frac{dN_j}{dt} &= N_j \left(\sum\limits_{i=1}^{p} \frac{b_{ij} a_{ij} R_i}{1 + \sum\limits_{i=1}^{p} a_{ij} h_{ij} R_i} - \sum\limits_{k=1}^{s} \frac{m_{jk} P_k}{1 + \sum\limits_{i=1}^{p} v_{ik} u_{ik} R_i + \sum\limits_{j=1}^{q} m_{jk} l_{jk} N_j} - f_j - g_j N_j \right) \cdot (2.1) \\
\frac{dR_i}{dt} &= R_i \left(c_i - d_i R_i - \sum\limits_{j=1}^{q} \frac{a_{ij} N_j}{1 + \sum\limits_{j=1}^{q} a_{ij} h_{ij} R_i} - \sum\limits_{k=1}^{s} \frac{v_{ik} P_k}{1 + \sum\limits_{i=1}^{p} v_{ik} u_{ik} R_i + \sum\limits_{j=1}^{q} m_{jk} l_{jk} N_j} \right)
\end{aligned}
$$

Table 2.1 summarizes the state variables and parameters for the model. The basal resource species, $R_i(i = 1, 2, \ldots, p)$, have logistic population growth in the absence of other species, with c_i as the intrinsic growth rate when rare, and d_i as the strength of density dependence in growth rate. (I prefer this formulation of logistic growth over the more conventional "r-K" formulation since it expresses the form of intraspecific density dependence in a straightforward fashion as a per

TABLE 2.1. State variables and parameters used in the population
dynamics model (equations 2.1 and 2.11) of species interactions

State Variable	Description
i	Subscript indexing resource species
p	Total number of resource species available to colonize community
R_i	Population abundance of resource i
j	Subscript indexing consumer species
q	Total number of consumer species available to colonize community
N_j	Population abundance of resource i
k	Subscript indexing predator species
s	Total number of predator species available to colonize community
P_k	Population abundance of resource i

Parameter	Description
a_{ij}	Attack coefficient of consumer j on resource i
b_{ij}	Conversion efficiency for consumer j eating resource i
h_{ij}	Handling time for consumer j eating resource i
v_{ik}	Attack coefficient of predator k on resource i
w_{ik}	Conversion efficiency for predator k eating resource i
u_{ik}	Handling time for predator k eating resource i
m_{jk}	Attack coefficient of predator k on consumer j
n_{jk}	Conversion efficiency for predator k eating consumer j
l_{jk}	Handling time for predator k eating consumer j
c_i	Intrinsic population growth rate of resource i when rare
d_i	Strength of intraspecific density dependence for resource i
f_j	Death rate of consumer j when rare
g_j	Strength of intraspecific density dependence on death rate for consumer j
x_k	Death rate of predator k when rare
y_k	Strength of intraspecific density dependence on death rate for predator k
z_j	Intraspecific interference coefficient for Beddington-DeAngelis functional response for consumer or mutualist j

capita demographic rate that is consistent with intraspecific density dependence in other types of species.) The intermediate-trophic-level consumer species, $N_j(j=1,2,\ldots,q)$, feed on the resources according to Holling's (1959) disc equation, which models a saturating functional response, where a_{ij} is the attack rate, b_{ij} is the conversion efficiency, and h_{ij} is the handling time of consumer j feeding on resource i. The top predator species, $P_k(k=1,2,\ldots,s)$, also feed on the consumers according to the disc equation, where m_{jk} is the attack coefficient, n_{jk} is the conversion efficiency, and l_{jk} is the handling time; they can similarly feed on the basal resources, but with attack rate v_{ik}, conversion efficiency w_{ik}, and handling time u_{ik}.

The consumers and predators also have intraspecific density-dependent mortality rates. For consumer j, f_j is the per capita death rate when it is rare, and g_j specifies the rate at which the per capita death rate increases with its abundance (i.e., the strength of intraspecific density dependence); for predator k, the corresponding parameters are x_k and y_k, respectively (Gilpin 1975; Gatto 1991; Caswell and Neubert 1998; Neubert et al. 2004; Amarasekare 2008; McPeek 2012, 2014a). Many different types of interactions among conspecifics can generate such negative intraspecific density dependence (Tanner 1966), including cannibalism (Fox 1975, Polis 1981), physiological stress responses that reduce fecundity and survival (Marra et al. 1995, Lochmiller 1996, McPeek et al. 2001a), territoriality and despotic habitat filling (Pulliam and Danielson 1991, McPeek et al. 2001b), mate access, mate finding, and mate harassment (Bauer et al. 2005, M'Gonigle et al. 2012).

In this chapter, I assume that the phenotypic traits of species do not change, so that we can focus on the demographic effects of species interactions and what is required for invasibility and coexistence of species. In the following chapters, I expand the model to allow the species to evolve so that I can explore the joint demographic and evolutionary dynamics of community modules.

Obviously, many other functional forms could be utilized in this modeling framework, and additional population regulation effects and types of species interactions are poorly characterized by the approaches I am using. For example, many other types of dynamics may be important for renewal of the basal resource besides logistic growth (e.g., Michalis-Menten, Monod, chemostat dynamics). Also, I will not consider Allee effects, which can be important for population dynamics at low abundance (Allee and Bowen 1932, Kramer et al. 2009), model formulations that are more appropriate for infectious disease dynamics (e.g., Ostfeld et al. 2008), or features associated with detritivores and the "brown" portions of food webs (e.g., Wolkovich et al. 2014). One cannot consider everything. However, the models examined here do cover a wide diversity of community module configurations encompassing many different classes of species interactions (fig. 2.1), as well as many different mechanistic features of these interactions for consumers, resources, and mutualists.

Throughout this work, I analyze various forms of this model using the standard methods of solving for the isoclines of the system, identifying whether stable point equilibria and limit cycles result; box 2.1 presents the basics of such analyses for a simple one resource-one consumer model of equation (2.1). However, my focus is not on what dynamics result, but rather when particular types of species will be present and what the general pattern of abundance is, regardless of the specific dynamical features. (I leave it to the reader to consult the original literature for evaluations of the stability of these equilibria and limit cycles; for a general treatment of these methods, see Hirsch et al. 2012.) Animations and

BOX 2.1.

ANALYSIS OF A CONSUMER-RESOURCE MODEL

I illustrate the analyses being performed on models in this chapter here for a single consumer and a single resource in which the consumer has a linear functional response and both species experience intraspecific density dependence. The model is then

$$\frac{dR_1}{dt} = R_1 \left(c_1 - d_1 R_1 - a_{11} N_1 \right)$$

$$\frac{dN_1}{dt} = N_1 \left(a_{11} b_{11} R_1 - f_1 - g_1 N_1 \right)$$

The isoclines for the two species are found by setting each of these equations equal to zero and simplifying. Doing so gives

$$\frac{dR_1}{dt} = 0: \quad N_1 = -\frac{d_1}{a_{11}} R_1 + \frac{c_1}{a_{11}}$$

&

$$\frac{dN_1}{dt} = 0: \quad N_1 = \frac{a_{11} b_{11}}{g_1} R_1 - \frac{f_1}{g_1}$$

These two functions are both lines. The resource isocline $\left(dR_1 / dt = 0 \right)$ has a positive intercept on the N_1 axis at c_1 / a_{11} and a negative slope, and so it intersects the R_1 axis at c_1 / d_1. The consumer isocline $\left(dN_1 / dt = 0 \right)$ has a negative intercept on the N_1 axis at $-f_1 / g_1$ and a positive slope, so that it intersects the R_1 axis at $f_1 / (a_{11} b_{11})$. These isoclines intersect each other in the first quadrant of the phase plane (i.e., where both species have positive abundances) if $f_1 / (a_{11} b_{11}) < c_1 / d_1$.

If the two isoclines intersect in the first quadrant, three equilibrium abundances for the two species exist. An equilibrium is any point where the system can be started, and it will remain there indefinitely. In this case, they are the points where both species are absent [0, 0], where the resource is at its equilibrium abundance but the consumer is absent $\left[c_1 / d_1, 0 \right]$, and the point where the two isoclines intersect. The point where the two isoclines intersect is found by setting the isoclines equal to one another and solving for the two species abundances. In this case,

$$R_1^* = \frac{c_1 g_1 + a_{11} f_1}{a_{11}^2 b_{11} + d_1 g_1} \quad \& \quad N_1^* = \frac{c_1 a_{11} b_{11} - d_1 f_1}{a_{11}^2 b_{11} + d_1 g_1}.$$

The stability of the equilibria is found by evaluating the Jacobian matrix for this system at each of the equilibria separately (Hirsch et al. 2012). The Jacobian matrix

(Box 2.1 continued)

is the matrix of partial derivatives of each species' equation taken with respect to the abundance of each, which in this case is:

$$
\begin{bmatrix}
\dfrac{\partial\left(dR_1/dt\right)}{\partial R_1}\bigg|_{R_1=R_1^*,N_1=N_1^*} & \dfrac{\partial\left(dR_1/dt\right)}{\partial N_1}\bigg|_{R_1=R_1^*,N_1=N_1^*} \\[3mm]
\dfrac{\partial\left(dN_1/dt\right)}{\partial R_1}\bigg|_{R_1=R_1^*,N_1=N_1^*} & \dfrac{\partial\left(dN_1/dt\right)}{\partial N_1}\bigg|_{R_1=R_1^*,N_1=N_1^*}
\end{bmatrix}
$$

$$
=\begin{bmatrix}
c_1-2d_1R_1^*-a_{11}N_1^* & -a_1R_1^* \\[2mm]
a_{11}b_{11}N_1^* & a_{11}b_{11}R_1^*-f_1-2g_1N_1^*
\end{bmatrix}
$$

If the real parts of the eigenvalues of this matrix when evaluated at a specific equilibrium are all less than or equal to zero and not all are zero, then the equilibrium is stable (Hirsch et al. 2012), meaning that if the system is started at a point near the equilibrium, it will eventually move to the equilibrium. If at least one eigenvalue is positive, the equilibrium is unstable, meaning that the system will move away from this equilibrium when it is started near it. Finally, if all eigenvalues are zero, the equilibrium is neutral, meaning that the system has no tendency to move toward or away from this equilibrium.

First, evaluate the equilibrium at the origin (i.e., $[R_1^*,N_1^*]=[0,0]$). The eigenvalues of the Jacobian matrix for this equilibrium are $\lambda=c_1$ and $\lambda=-f_1$. This equilibrium is unstable if the resource can increase when rare (i.e., invade) in the absence of the consumer (i.e. $c_1>0$).

The equilibrium at which the resource is present but the consumer is absent (i.e., $[R_1^*,N_1^*]=[c_1/d_1,0]$) is similarly unstable for the biologically reasonable situation. The eigenvalues of the Jacobian matrix for this equilibrium are $\lambda=-c_1$ and $\lambda=(c_1a_{11}b_{11}/d_1)-f_1$. This equilibrium is stable or unstable based on the second eigenvalue. Note that this equilibrium is positive if $c_1/d_1>f_1/(a_{11}b_{11})$, which is the criterion for the consumer being able to invade when the resource is at this equilibrium. Therefore, if the consumer can invade the system, the equilibrium where only the resource is present is unstable.

The calculations for the stability of the equilibrium where both species are present (i.e., $[R_1^*,N_1^*]=[(c_1g_1+a_{11}f_1)/(a_{11}^2b_{11}+d_1g_1),(c_1a_{11}b_{11}-d_1f_1)/(a_{11}^2b_{11}+d_1g_1)]$) are much more tedious. The eigenvalues are given by the roots of the characteristic equation, which can be simplified to give

$$
\lambda^2+(d_1R_1^*+g_1N_1^*)\lambda+a_{11}^2b_{11}R_1^*N_1^*=0.
$$

Substituting the equilibrium values and then finding the roots of this quadratic gives very complicated functions, with both real and complex parts. However, the real parts of the two roots are negative when

$$-d_1(c_1 g_1 + a_{11} f_1) < g_1 (c_1 a_{11} b_{11} - d_1 f_1).$$

If all the parameters are greater than or equal to zero, this inequality is always true. Also, note that the two quantities in parentheses are the numerators of the equilibrium resource and consumer abundances. Thus, this equilibrium is always stable if it exists. Because the eigenvalues have complex parts, the system will spiral on its approach to this equilibrium.

games illustrating how model parameters change the shapes of the species' isoclines and the resulting dynamics of the community can be found at http://press.princeton.edu/titles/11175.html. Because most of the material presented in this chapter is presented in detail in the ecological literature, I will keep the mathematics here to a minimum and use phase portraits of the models to illustrate the key issues for these analyses. I direct the reader to the pertinent literature for the mathematical details.

TWO TROPHIC LEVELS

To develop intuition about the requirements for species coexistence, and therefore invasibility, consider first the simplest possible community: a single consumer N_1 eating a single resource R_1. The simplification of equation (2.1) for this situation is

$$\frac{dN_1}{dt} = N_1 \left(\frac{b_{11} a_{11} R_1}{1 + a_{11} h_{11} R_1} - f_1 - g_1 N_1 \right)$$

$$\frac{dR_1}{dt} = R_1 \left(c_1 - d_1 R_1 - \frac{a_{11} N_1}{1 + a_{11} h_{11} R_1} \right) \tag{2.2}$$

This is a form of the familiar Rosenzweig-MacArthur consumer-resource model (Rosenzweig and MacArthur 1963). To further build intuition about how the various features of this interaction influence community dynamics, I start with the simplest version of this interaction. Thus, set $d_1 = h_1 = g_1 = 0$. These assumptions give the resource exponential growth in the absence of the consumer, and the consumer feeds on the resource with a linear functional response—this is the basic Lotka-Volterra predator-prey model (Volterra 1926, Lotka 1932a). The resource isocline intersects the N_1 axis c_1/a_{11} at and is parallel to the R_1 axis, and the consumer isocline intersects the N_1 axis at $f_1/(a_{11}b_{11})$ and is parallel to the N_1 axis (fig. 2.2A). The points where the isoclines intersect the axes of the other

A $d = 0, g = 0, h = 0$

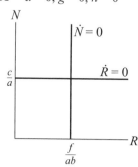

B $d > 0, g = 0, h = 0$

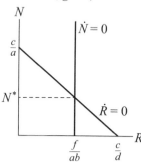

C $d = 0, g > 0, h = 0$

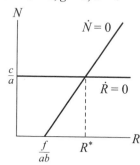

D $d > 0, g > 0, h = 0$

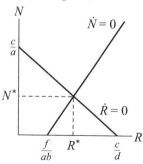

E $d = 0, g = 0, h > 0$

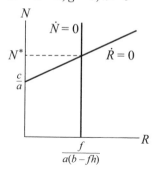

F $d > 0, g = 0, h > 0$

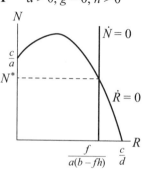

G $d > 0, g > 0, h > 0$

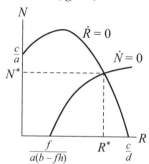

species are critical for understanding invasibility because these points define the abundances of the other species where this species' birth and death rates exactly balance when it is rare.

This system also has two equilibria. The origin (i.e., both species absent) is an unstable equilibrium, and if both species are present, their abundances cycle in a neutral orbit around the other equilibrium at $[R_1^*, N_1^*] = [f_1/(a_{11}b_{11}), c_1/a_{11}]$ where the isoclines intersect (Case 1999). The resource can invade the system if $c_1/a_{11} > 0$, and the consumer can invade the system if the resource is present and if $f_1/(a_{11}b_{11}) > 0$. In other words, because these two species can invade when rare and the other species is at its demographic equilibrium in its absence, these two species coexist.

Coexistence also implies that the isoclines intersect one another at positive abundances for both species, giving $R_1^* > 0$ and $N_1^* > 0$, but it does not imply that this equilibrium point is stable. Species will coexist either if a stable equilibrium point exists that the system will approach, or if the system will approach a stable or chaotic limit cycle that orbits an equilibrium (i.e., an "attractor"). Limit cycles may increase the probability of extinction of one or both species due to stochastic events that occur when the species are rare, but for our purposes we will consider limit cycles to imply coexistence.

Now systematically make d_1, h_1, and g_1 nonzero to explore what each of these features of the interaction does. First, introduce intraspecific density dependence to the resource by allowing $d_1 > 0$. This causes the resource isocline to pivot downward so that it now intersects the R_1 axis at c_1/d_1 but still intersects the N_1 axis at c_1/a_{11}; this introduces a new equilibrium at this point of intersection (fig. 2.2B). Density dependence in the resource limits its population abundance from increasing to infinity in the absence of the consumer. The resource abundance now follows logistic growth in the absence of the consumer with an equilibrium at c_1/a_{11}, and the resource can invade and exist in the system if $c_1/d_1 > 0$.

FIGURE 2.2. Isocline portrait of a simple one consumer-one resource community as specified by equations (2.2) with various zero and nonzero parameter combinations to illustrate the effects of different mechanistic features on the shapes of the isoclines for each species. The resource isocline is identified with $\dot{R} = 0$ and the consumer isocline with $\dot{N} = 0$. The panels illustrate different combinations of linear ($h = 0$) or saturating ($h > 0$) functional responses with intraspecific density-dependent population growth rates in the resource ($d > 0$) and consumer ($g > 0$). A linear functional response is combined with intraspecific density dependence in (A) neither species, (B) only the resource, (C) only the consumer, and (D) both the resource and consumer. A saturating functional response is combined with intraspecific density dependence in (E) neither species, (F) only the resource, and (G) both species. (Panel G is redrawn from figure 3 of McPeek 2012, with permission.)

This limitation on resource abundance also causes two fundamental changes to the dynamics of the system. First, the point where the two isoclines intersect is now a stable equilibrium—if it exists. The existence of this equilibrium is the first major place we see the emergence of an ecological opportunity because of species interactions. If $f_1/(a_{11}b_{11}) > c_1/d_1$, the consumer cannot invade the system with the resource present at its equilibrium, and so it cannot exploit this potential ecological opportunity. Only consumers with $f_1/(a_{11}b_{11}) < c_1/d_1$ can exploit this opportunity. In words, the consumer can invade if adequate resource abundance is present to support its population. Second, the change in the position of this equilibrium because of changes in resource attributes (c_1 or d_1) results in the same resource abundance but a different consumer abundance, and the equilibrium consumer abundance decreases with a lower resource growth rate when rare (i.e., decreasing c_1) or stronger resource density dependence (i.e., increasing d_1) (fig. 2.2B).

Next, introduce intraspecific density dependence to the consumer by allowing $g_1 > 0$, but keep $d_1 = 0$ (fig. 2.2C).

Increasing g_1 causes the consumer isocline to pivot away from the consumer axis (fig. 2.2C). This limitation of the consumer's abundance causes the two-species equilibrium to shift position such that resource abundance increases with increasing values of g_1. This limitation also makes the two-species equilibrium stable. Here again, density dependence in one species changes the abundance of the other (fig. 2.2C). When both species experience intraspecific density dependence, the resource abundance increases, and the consumer abundance decreases at the two-species equilibrium (fig. 2.2D). In addition, intraspecific density dependence in both species contributes to the stability of the two-species equilibrium. However, note that the criteria for invasibility of the two species have not changed.

Thus, *invasibility and stability are not synonymous*. Invasibility merely asks whether a species could invade or reinvade the community, independent of the type of long-term dynamics that the community will display if it can invade. Stability, in turn, questions whether any equilibrium point for the community is a stable or unstable equilibrium. Even if any particular equilibrium is unstable, species may still not go extinct if the system displays limit cycles or chaotic dynamics.

Shifting the functional response from linear to saturating by allowing $h_1 > 0$ alters the position of the consumer isocline and the shape of the resource isocline (fig. 2.2E). The consumer isocline shifts along the R_1 axis to greater values with increasing h_1 but remains parallel to the N_1 axis. The resource isocline pivots away from the R_1 axis with increasing h_1 but remains anchored at c_1/a_{11} on the N_1 axis; this now represents positive density dependence in the death rate of the resource. As a consequence, the two-species equilibrium shifts position to increase both the resource and consumer abundances with increasing h_1. This equilibrium is now unstable, with the system spiraling away from it in ever-increasing cycles until one or both species become extinct. This is the one case where the consumer cannot coexist with the resource, even though it has $f_1/(a_{11}(b_{11}-h_{11}f_1))>0$; the unstable equilibrium causes ever-increasing spirals away from it, and the consumer or both species eventually go extinct.

Combining intraspecific density dependence in the resource ($d_1 > 0$) with a saturating functional response ($h_1 > 0$) causes the resource isocline to become non-linear, with a negative second derivative (i.e., bowed up); but it still intersects the R_1 axis at c_1/d_1 and the N_1 axis at c_1/a_{11} (fig. 2.2F). If h_1 is large enough for the resource isocline to have a maximum value of N_1 at a value of $R_1 > 0$, the stability of the two-species equilibrium becomes conditional. If the isoclines intersect at a higher R_1 value than this peak (i.e., to the right of the peak) in the resource isocline, the two-species equilibrium is stable; but if they intersect at or to the left of the peak, a stable limit cycle exists around this equilibrium (Rosenzweig and MacArthur 1963, Rosenzweig 1969).

Finally, combining all these effects gives the isocline portrait in figure 2.2G. In particular, the combination of a saturating functional response and consumer intraspecific density dependence causes the consumer isocline to become an asymptote at a maximum N_1 abundance. The two-species equilibrium is stable, and changes in position with changes in parameter values as described above for the individual parameters (Gatto 1991, McPeek 2012).

As an aside, other forms of intraspecific density dependence in the consumer lead to comparable results. For example, interference among the consumers leading to intraspecific density dependence can alternatively be included in the model by utilizing a Beddington-DeAngelis functional response (Beddington 1975, DeAngelis et al. 1975). Here, an extra term that includes the abundance of the consumer is added to the denominator of the functional response; for example, the denominators of the functional responses in equation (2.2) might be $(1 + a_{11}h_{11}R_1 + z_1N_1)$. Including consumer intraspecific density dependence in this way has the same effect on the consumer isocline as the linearly increasing term for per capita death rate; both cause the consumer isocline to rotate away from the consumer axis. However, a Beddington-DeAngelis functional response also acts against the handling time effect on the resource isocline—that is, increasing interference among the consumers decreases the upward bend in the resource isocline caused by the handling time (Beddington 1975, DeAngelis et al. 1975, Arditi and Ginzburg 2012). However, the qualitative forms of the isoclines with a standard saturating functional response and density dependence in the consumer death rate and with a Beddington-DeAngelis functional response are identical (cf. fig. 2.2G here with fig. 8 in DeAngelis et al. 1975).

In almost all of these situations, the resource and consumer are coexisting. Only in the situation with a saturating functional response and no intraspecific density dependence in either species will one or both species almost invariably go extinct (fig. 2.2E). Intraspecific density dependence in the resource causes its abundance to equilibrate at an abundance of c_1/d_1 in the absence of the consumer. The consumer can then invade if its isocline intersects the R_1 axis below this point. In mathematical terms, this means that the predator will have a positive population growth rate when it is rare and the resource is at its equilibrium abundance if

$$\frac{c_1}{d_1} > \frac{f_1}{a_{11}(b_{11} - f_1 h_{11})}. \tag{2.3}$$

In words, this means that an adequate resource abundance must be available for the consumer to have a positive population growth rate when it invades. The parameters of this interaction represent the abilities of the consumer to eat the resource. If this condition is met, the consumer and resource will come to either a stable point equilibrium or a stable limit cycle.

The ecological opportunities for these two species are quite easy to understand in this very simple community. The ecological opportunity for the resource is to have a place to support a population, and to do so, it must simply be able to increase when it is rare (i.e., $c_1 > 0$). Note that if these are the only two possible species, the resource will only have to invade the system when the consumer is absent since the consumer will equilibrate at $N_1^* = 0$ in that case. If the resource is present, the consumer has an ecological opportunity to utilize this unexploited resource if it is abundant enough to give the invading consumer a positive population growth rate, given the other local impingements on its demography that are encapsulated in its death rate terms. Although these criteria are very simple, they represent two of the essential features of ecological opportunities in more complex community modules we will encounter below.

Resource Competition

Given the very simple consumer-resource community module above (fig. 2.2), we can now query whether any opportunities exist for other consumer or resource species in this system. First, consider whether any more consumers can be added to this system with only one resource, and if so, under what circumstances such ecological opportunities are available. This is essentially a question of whether any other consumers (i.e., resource competitors) can be added to utilize the single resource that is present. Because all the conceptual issues are apparent when the consumers have linear functional responses, $h_{ij} = 0$ for all consumers throughout this section for the sake of simplicity (McPeek 2012).

Whether ecological opportunities exist here for other consumers depends on whether the consumer that has a positive population growth rate when it is rare at the lowest resource abundance also experiences any degree of intraspecific density dependence (McPeek 2012). Note that the resource abundance at which the consumer has a positive population growth rate when it is rare is just to the right of the point where its isocline intersects the R axis. First, consider the shape of the consumer isoclines when two consumers are competing for the resource (fig. 2.3). Because no direct interactions between the consumers occur, the isocline of each is parallel to the axis of the other, and no matter how many consumers are present, the isocline of each is parallel to the axes of all other consumers. The general principle is that an isocline is parallel to the axis of species that do not have any direct influence on its abundance.

If the consumers experience no intraspecific density dependence themselves (i.e., $g_j = 0$ for all j species), the isocline of each is parallel to its own axis as well since it does not influence its own abundance. In this case, the consumer isoclines are all parallel planes that each intersect the R axis at different points given by $f_j/(a_{1j}b_{1j})$ (fig. 2.3A). Each consumer's population increases when the resource abundance is higher than this intersection point (i.e., to the right of its isocline) and decreases when the resource abundance is lower than the intersection point (i.e., to the left). Thus, the species with the lowest value of $f_j/(a_{1j}b_{1j})$ will equilibrate the resource abundance at a level at which no other consumer can maintain a population. This is another formulation of the R^* rule; that is, the consumer with the lowest $R^* = f_j/(a_{1j}b_{1j})$ is the superior resource competitor and in this case will indirectly drive all others extinct (Tilman 1982). This is because consumer 1 depresses the resource abundance to a level at which no other consumer can support a population.

Intraspecific density dependence in the consumers ($g_j > 0$) causes a change in the shapes of the consumer isoclines that permit more than one to coexist on one resource (McPeek 2012). With greater levels of intraspecific density dependence, remember that the consumer's isocline pivots away from the consumer axis (fig. 2.2D), but remains parallel to the axes of all other consumers (fig. 2.3B). As a convention, identify consumers by numbering them in order of increasing values of $f_j/(a_{1j}b_{1j})$, so that

$$\frac{f_1}{a_{11}b_{11}} < \frac{f_2}{a_{12}b_{12}} < \ldots < \frac{f_j}{a_{1j}b_{1j}} < \ldots \frac{f_q}{a_{1q}b_{1q}}. \tag{2.4}$$

Imagine that the resource is initially present with no consumers at $R_1^* = c_1/d_1$, and then allow consumer 1 to invade (fig. 2.3B). Once these two species come to their stable equilibrium, call the resource abundance at this equilibrium $R_{1(1)}^*$ to signify that this is the equilibrium abundance of resource 1 when consumer 1 is present

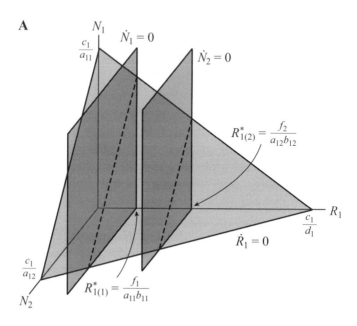

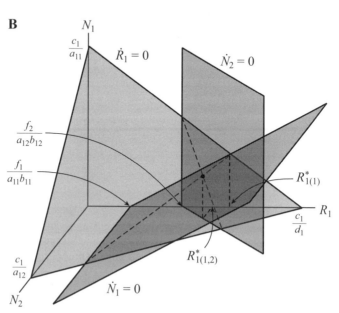

(fig. 2.3B). Stronger consumer intraspecific density dependence (i.e., larger values of g_1) will increase $R^*_{1(1)}$, because the consumer isocline will pivot farther away from its own axis. (Note that figure 2.2.D is a two-dimensional slice through figure 2.3B at $N_2 = 0$, the $R_1 - N_1$ face.)

Any consumer, j, with

$$\frac{f_1}{a_{11}b_{11}} < \frac{f_j}{a_{1j}b_{1j}} < R^*_{1(1)} \tag{2.5}$$

can invade this community and stably coexist with consumer 1 and the resource (McPeek 2012). Imagine that consumer 2, the consumer with the next highest $f_j/(a_{1j}b_{1j})$, is the next to invade. Once these three species reach their new equilibrium (i.e., the point where their three isoclines intersect), the resource will be depressed to a new lower level, $R^*_{1(1,2)}$ (fig. 2.3B). Any of the remaining consumers that have $f_j/(a_{1j}b_{1j}) < R^*_{1(1,2)}$ can invade this community. With each successive successful invader, the resource abundance will be depressed further, which will further limit which other consumers can subsequently invade. In fact, the order of consumer invasion does not matter (McPeek 2012). What matters is the ordering of consumers with respect to the minimum resource level at which each has a positive population growth rate, as expressed in equation (2.4). With each subsequent consumer invading the community, resource abundance will be depressed to a new level. Consumer invasion can continue until no remaining consumer has a positive population growth rate at the current resource abundance (fig. 2.4). The final consumer guild whose members can coexist are the set of consumers with the lowest $f_j/(a_{1j}b_{1j})$ values that are lower than the equilibrium resource abundance when they all are present (fig. 2.4).

The ecological opportunity available to an invading consumer in this case involves whether the resource abundance is at a sufficient level to support a population of the invaders. The ability of each consumer to exploit this opportunity is determined by how its phenotype establishes $f_j/(a_{1j}b_{1j})$. If the system can support a population of the consumer, it will have a positive population growth rate when rare; that is, the species satisfies its invasibility criterion (fig. 2.4). The ecological

---

FIGURE 2.3. Isocline system for a resource competition module in which two consumers (N_1 and N_2) feed on a single resource (R_1). The isocline of the resource is shaded *green*, and those of the two consumers are shaded *blue*. (A) If neither consumer experiences any intraspecific density dependence ($d_1 = d_2 = 0$), their isoclines are parallel, and the consumer with the isocline that intersects the R_1 axis at the lower value will drive the other consumer extinct by depressing the resource to that level. (B) If the consumers experience intraspecific density dependence, they can coexist on the single resource if they can meet the criteria described in the text. (This figure is redrawn from figure 1 of McPeek 2012, with permission.)

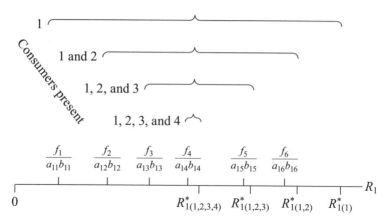

FIGURE 2.4. Isolating the effects of multiple consumers competing for a single resource on the R_1 axis in a resource competition module. The various points identified by $f_j/(a_{ij}b_{ij})$ are where the corresponding consumer isoclines intersect the R_1 axis, and the points identified by $R^*_{1(1...j)}$ are the equilibrium resource abundance with consumers $1-j$ present. The brackets above the axis show the range of resource abundance in which the isocline of any additional consumer must intersect the R_1 axis if it is to coexist, assuming the consumers are identified according to equation (2.4). (This figure is redrawn from figure 2 of McPeek 2012, with permission.)

opportunities available in this case depend critically on the degree of intraspecific density dependence that the consumers with lower $f_j/(a_{1j}b_{1j})$ also experience and the collection of other consumers that can potentially invade. These set the scope of what resource levels are available for each potentially invading consumer (fig. 2.4); remember that greater consumer intraspecific density dependence will cause the resource to equilibrate at a higher abundance (figs. 2.2 and 2.3). Moreover, once a consumer has invaded and established itself in the community, it limits the ecological opportunities available for subsequent invaders by reducing resource abundances. This will be a recurring theme in this chapter.

Apparent Competition

Now consider again the very simple predator-prey community module represented in various forms in figure 2.2, and ask whether any ecological opportunities are available for additional resource species if only one consumer is present. This is a question of what limits the number of "apparent" competitors and was first addressed by Holt (1977). To do this, I will assume that each new resource added to the community exploits its own set of resources, and so no direct interactions occur among the resources. All interactions among the resources are indirect and mediated through the consumer. Thus, as in the case of resource competition above, the isocline of each resource is parallel to the axes of all other resources that can be added to the system.

Initially, assume that the consumer has a linear functional response for feeding on all resources ($h_{i1} = 0$ for all i) and experiences no intraspecific density dependence of its own ($g_1 = 0$); I will relax these assumptions shortly. Analogous to resource competition, the critical defining relationship in the case of apparent competition is the ordering of the points at which the isoclines of the various invading resources intersect the N axis—c_i/a_{i1} (Holt 1977). Following Holt (1977), number resources in decreasing values of this quantity, such that

$$\frac{c_1}{a_{11}} > \frac{c_2}{a_{21}} > \ldots > \frac{c_i}{a_{i1}} > \ldots > \frac{c_p}{a_{p1}}. \tag{2.6}$$

Again, in this case the isoclines of all the resources are parallel planes. If the consumer's abundance is above the isocline of a resource, the death rate of the resource will be higher than its birth rate, and the resource's abundance will then decline. Consequently, if resource 1 experiences no intraspecific density dependence ($d_1 = 0$), it will be the only resource that can survive with the consumer; here, the isocline of no other resource will intersect that of resource 1, and so all other resources will go extinct because resource 1 can support a population at a higher consumer abundance than any other resource; see figure 2.5A (Křivan and Eisner 2006). Intraspecific density dependence is just as critical for multiple resources to coexist with one consumer, as it was for multiple consumers to coexist with one resource. In fact, as we will see, the two processes are mirror images of one another.

Assuming all $i = 1, 2, \ldots, p$, resources experience some degree of intraspecific density dependence (i.e., $d_i > 0$), now consider a sequence of invasions by these resource species. Begin with resource 1 and consumer 1 coexisting at their stable point equilibrium with $N^*_{1(1)}$ to signify that resource 1 is present at this equilibrium (again identified inside the parentheses in the subscript); see figure 2.5B, and note that figure 2.2B is the $R_1 - N_1$ face at $R_2 = 0$ of figure 2.5B. The equilibrium consumer abundance will be lower if the resource experiences stronger intraspecific density dependence. (Note that the consumer abundance will also be lower if the consumer experiences greater levels of intraspecific density dependence itself [cf. figs. 2.2B and D].) Any resource that has a positive population growth rate when rare at this consumer abundance can now invade. These would be any of the resources that have

$$\frac{c_1}{a_{11}} > \frac{c_i}{a_{i1}} > N^*_{1(1)}. \tag{2.7}$$

A

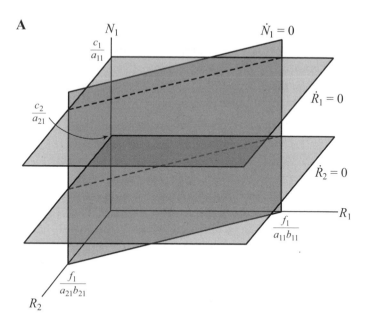

B

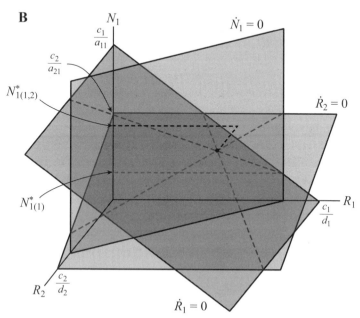

For the sake of simplicity, assume that the next one to invade is resource 2 and that this resource satisfies inequality (2.7). The consumer will increase in abundance to $N^*_{1(1,2)}$ because it has two resources on which to feed (fig. 2.5B). As a consequence, the next resource to invade must meet a more stringent condition on consumer abundance to have a positive population growth rate (Holt 1977) (fig. 2.6). The consumer abundance will increase with every successive invasion by a new resource until no further resource species have c_i/a_{i1} greater than the consumer's abundance (fig. 2.6). Here too, the final community composition does not depend on the order of resource invasions (Holt 1977). With linear functional responses, this system always comes to a stable point equilibrium (Holt 1977). Moreover, notice that as each successive resource invades, the abundance of the other resources decline, because of the increase in the consumer's abundance with more to feed on—this negative relationship between resource abundances mediated through the consumer is why Holt (1977) originally dubbed this module *apparent competition*. Thus, with apparent competition, coexistence of resources requires that they not inflate the consumer's abundance above a level at which any of them can support a population.

If the consumers' functional responses are saturating ($h_{i1} > 0$ for all i) and handling times are sufficiently large to cause the isoclines of at least some of the resources to be hump-shaped (fig. 2.2F), the system will come to a stable limit cycle (Abrams and Matsuda 1996, Abrams et al. 1998, Křivan and Eisner 2006) (fig. 2.7A). If the consumer has similar handling times for all resources so that the maximum heights of the resource isoclines have the same rank order as c_i/a_{i1}, the isocline of resource 1 still sets a lower limit defining what other consumers can invade. Increasing the handling time of the consumer on all resources decreases this limit and thus lets more resources persist in the system (fig. 2.7B). Notice that for the parameter values used to generate figure 2.7, the average resource abundances increase with handling time and as successive resources invade the system. This indirect effect of increasing the average abundance of one resource when another invades has been called "apparent mutualism" in this case, because the resource already present apparently benefits (i.e., its abundance increases)

FIGURE 2.5. Isocline system for an apparent competition module in which a consumer (N_1) feeds on two resources (R_1 and R_2). The isoclines are identified as in figures 2.1 and 2.3. (A) If neither resource experiences any intraspecific density dependence, their isoclines are parallel, and the resource with the isocline intersecting the N_1 axis at the higher value (i.e., R_1) will drive the other resource extinct by inflating the consumer's abundance above a level at which the other resource can survive. (B) If the resources experience intraspecific density dependence, they can both coexist with the consumer—if they can meet the criteria described in the text.

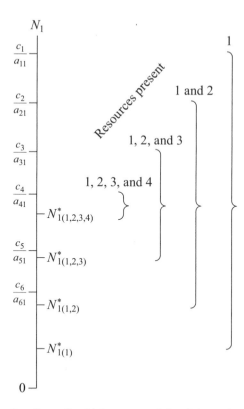

FIGURE 2.6. Isolating the effects of multiple resources being fed upon by a single consumer on the N_1 axis in an apparent competition module. The various points identified by c_i/a_{ij} are where the corresponding resource isoclines intersect the N_1 axis, and the points identified by $N^*_{1(1...i)}$ are the equilibrium consumer abundance with resources $1...i$ present. The brackets above the axis show the range of consumer abundance in which the isocline of any additional resource must intersect the N_1 axis if it is to coexist, assuming the resources are identified according to equation (2.6).

from the other invading the community (Holt 1977, Abrams and Matsuda 1996, Abrams et al. 1998).

If the consumer constrains its own abundance through intraspecific density dependence ($g_1 > 0$), the dynamics of the system again return to a point equilibrium instead of a limit cycle. In addition, this self-constraint on the consumer's abundance reduces $N^*_{1(...)}$, the lower limits on the criteria that new invaders must meet; see figures 2.2D and G (Holt 1977). This reduces the growth rate needed for a new resource to invade; the $N^*_{1(...)}$ values in figure 2.6 all decrease as intraspecific density dependence in the consumer increases. If intraspecific density dependence

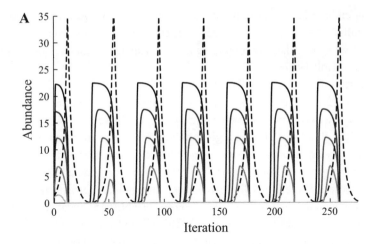

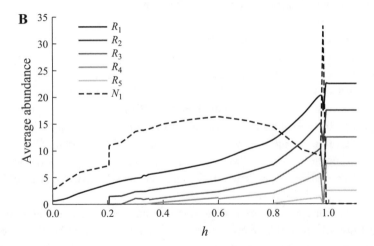

FIGURE 2.7. Simulation results illustrating an "apparent mutualism" in an apparent competition module. (A) One simulation in which one consumer feeds on five resources is illustrated to show the resulting six-dimensional stable limit cycle. For this simulation, all resources have the parameters of $d = 0.2$, $a = 1.5$, $b = 0.3$, $h = 0.81$, and the consumer has $f = 0.3$. The various resources differ in their intrinsic birth rates, with $c_1 = 4.5$, $c_2 = 3.5$, $c_3 = 2.5$, $c_4 = 1.5$, and $c_5 = 0.5$. (B) The average abundances of the six species are shown over a gradient of handling times (the consumer has the same handling time for all resources in a simulation), with all other parameters as above. Species abundances are averaged over one complete cycle. Consumers with larger handling times will increase the scope for more resources to coexist. This system cycles from $h = 0.01$ to $h = 0.97$. At $h = 0.98$, the system comes to a stable equilibrium point with a very high consumer abundance. At $h \geq 0.99$, the consumer cannot coexist with the resources, and so each resource is at its equilibrium abundance (i.e., c_i/d_i) in the absence of the consumer.

is strong enough in the consumer, its abundance will change very little and thus the mortality rate it imposes on the resources will be a constant; consequently, any resource with a population growth rate c_i greater than this constant mortality rate can invade (Holt 1977).

Interestingly, the combination of consumer intraspecific density dependence and a saturating functional response causes the consumer isocline to become an asymptote at a maximum consumer abundance; see figure 2.2G (Gatto 1991, Holt 1997a, McPeek 2012). Thus, a high handling time and a moderate level of consumer intraspecific density dependence make the consumer abundance essentially unresponsive to the addition of new resources once more than a few resources are present; therefore, the lower limits on resource population growth rate needed to invade stop increasing as new invaders become established. Because any resource with c_i/a_{i1} greater than this asymptotic value could invade, this would further increase the number of resources that would have ecological opportunities in this community. Saturating functional responses can also generate chaotic dynamics in some areas of parameter space (Křivan and Eisner 2006).

In the case of apparent competition, the ecological opportunity that is potentially available to an invader is the chance to utilize an unexploited resource (i.e., the density dependence in each resource implies that each has a unique resource to exploit itself), but it must be able to do so while experiencing predation from another species higher in the trophic hierarchy. To invade, the benefit to population growth from exploiting this unique resource must be greater than the mortality it experiences due to predation when it is rare. This mortality constraint is greatest when the consumer has a linear functional response and experiences no intraspecific density dependence itself, and becomes less stringent as either or both of these are increased.

The combined lessons learned from these analyses of resource and apparent competition illustrate decisively how resource availability from below and predation pressure from above can interact to limit the types of species that can exploit specific ecological opportunities. Imagine an ecological situation where 20 different resources are available to be exploited by 20 different specialist consumers. However, if a predator is also present that can potentially feed on these 20 consumers, many of the resources may go unexploited because the predator is too abundant for them to invade. (Obviously, a consumer that was completely immune to the predator could invade if it could exploit one of the available resources.) The fact that some resources go unexploited in a community does not mean that they could not be exploited; whatever predators are present in the community may be too abundant for potential exploiters to support a population. These unexploited resources would only be ecological opportunities if some species could invade and maintain a population in the face of predation pressures as well.

THREE TROPHIC LEVELS

The above analyses of communities with only two trophic levels illustrate how indirect interactions that alter the abundances of key species will define whether an ecological opportunity is available for a potential invader. Furthermore, these analyses isolate the effect of predator and resource abundance in shaping any particular ecological opportunity. In effect, one can consider the results for two trophic levels at the top of a food web, since the same results are obtained if all the resources feed on their own exclusive resource below them. In this section, these analyses will be extended to three trophic levels so that species at intermediate levels might have ecological opportunities that are simultaneously determined by both resource and apparent competitive effects.

As with two trophic levels, begin by considering a simple three-trophic-level food chain with the simplest assumptions possible for equations (2.1)—namely, no species has any intraspecific density dependence ($d_1 = g_1 = y_1 = 0$), all functional responses are linear ($h_{11} = l_{11} = w_{11} = 0$), and the predator does not eat the basal resource ($v_{11} = 0$). The isoclines for a community of these species are given in figure 2.8A. (Note that for this and all other situations discussed below, the $R_1 - N_1$ faces in the various panels of figure 2.8 are identical to the corresponding panels in figure 2.2 except that the predator isoclines are missing in the latter.) The resource isocline intersects the N_1 axis at c_1/a_{11}, and is parallel to both the R_1 and P_1 axes (fig. 2.8A). The consumer isocline intersects the R_1 axis at $f_1/(a_{11}b_{11})$, is parallel to the N_1 axis, and tilts away from the predator axis (fig. 2.8A); this shape of the consumer isocline is the geometric statement of the transduction of resource biomass to the predator through the consumer. The predator isocline intersects the N_1 axis at $x_1/(m_{11}n_{11})$ and is also parallel to both the R_1 and P_1 axes (fig. 2.8A). As a result, no equilibrium is possible because the three isoclines can never intersect simultaneously at a single point or line, and the predator cannot invade a system of the resource and consumer in a neutral limit cycle (McPeek 2014a). Here again, we see that *density dependence is required in at least one species for multiple species to coexist.*

Adding intraspecific density dependence for the resource ($d_1 > 0$) pivots the resource isocline to intersect the R_1 axis at c_1/d_1 (fig. 2.8B). Remember that this creates a stable point equilibrium, $N^*_{1(1)}$, in the absence of the predator (fig. 2.2B). The predator can invade this community if its isocline intersects the N_1 axis below this point, as in

$$N^*_{1(1)} > \frac{x_1}{m_{11}n_{11}}. \tag{2.8}$$

This is the mathematical statement that the consumer is present at a sufficient abundance to support the predator population. When the predator invades, the

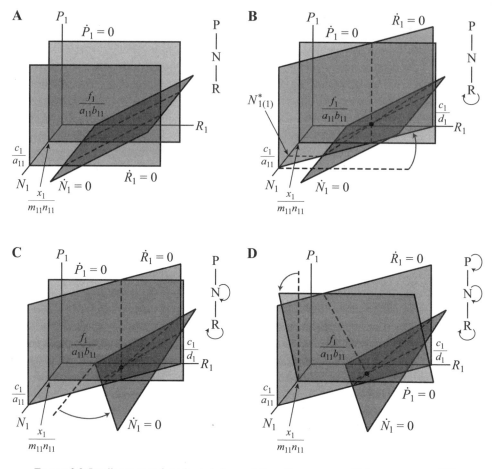

FIGURE 2.8. Isocline system for a food chain consisting of one resource (R_1), one consumer (N_1), and one predator (P_1). The resource isocline is shaded *green*, the consumer isocline *blue*, and the predator isocline *red*. The consumer and predator have linear functional responses in all panels, and the panels show intraspecific density dependence in (A) none of the species, (B) only the resource, (C) the resource and the consumer, and (D) all three species. (Panels A–C of this figure are redrawn from figure 1 of McPeek 2014a, with permission.)

system comes to a new equilibrium with lower consumer and higher resource abundances, because the intersection between their isoclines forces this inverse relationship between their equilibrium abundances (fig. 2.8B).

Intraspecific density dependence in the consumer ($g_1 > 0$) causes its isocline to pivot away from its own axis (cf. figs. 2.2C and 2.8C), but the intersection of the consumer isocline with the $R_1 - P_1$ face does not change (fig. 2.8C). As g_1 is

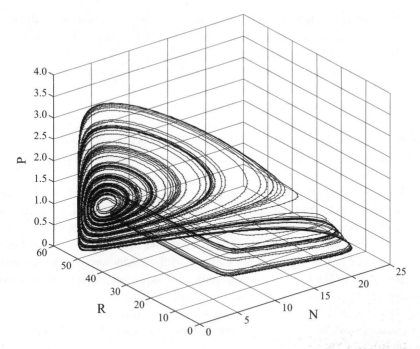

FIGURE 2.9. Three-dimensional chaotic trajectory of a three-trophic-level food chain with one species at each trophic level. The parameters for this simulation are as follows: $c = 5.0$, $d = 0.1$, $a = 0.5$, $b = 0.1$, $h = 0.2$, $f = 0.2$, $g = 0.0$, $m = 0.2$, $n = 0.1$, $l = 0.2$, $x = 0.05$, and $y = 0.0$.

increased, the equilibrium abundance of both the predator and consumer decrease and the resource increases. Moreover, higher values of g_1 would limit the predators that could invade, because $N^*_{1(1)}$ would be decreased, and available food for any invading predator would consequently be lower. Intraspecific density dependence at lower trophic levels seems to generally reduce the scope of ecological opportunities by decreasing available food for invading predators.

Intraspecific density dependence in the invading predator ($y_1 > 0$) has no effect on whether the predator can invade the system, because it does not influence the point where the predator isocline intersects the N_1 axis (fig. 2.8D). Greater values of y_1 will, however, shift the equilibrium point to decrease the abundance of the predator, increase the consumer, and decrease the resource.

Permitting saturating functional responses in the predator-prey interactions ($h > 0, l > 0$) introduces the possibilities of chaos in the dynamics in particular areas of parameter space (fig. 2.9; Hastings and Powell 1991, McCann and Yodzis 1994, Duarte et al. 2008, Visser et al. 2012), but largely does not affect the likelihood of predator invasion. Chaos and stable limit cycles are still possible when intraspecific density dependence in either the consumer or predator is added, but the range of parameter space in which they occur is reduced (Caswell and Neubert 1998).

Multiple Species at Each Trophic Level

With this knowledge of the requirements for invasibility and the resulting dynamics of having one species per trophic level, now consider what is required to have additional species invade the system at each trophic level.

Adding resource species at the bottom of the food chain is a case of apparent competition as described above, but the criteria for invading resources are relatively relaxed. A predator feeding on the consumer will limit the its abundance just like intraspecific density dependence, and so the consumer abundance is constrained in the amount it can increase with the addition of more resource species (Holt 1977). If the predator permits little or no increase in the consumer abundance, the consumer essentially becomes a constant mortality source that is unresponsive to the number or abundances of the resources below. Consequently, any resource species with c_i/a_i greater than the consumer's fixed abundance can invade (Holt 1977).

Adding predator species at the top of the food chain follows the general strictures described in the section on resource competition with two trophic levels above. Grover (1994) showed how multiple consumers that each has its own specialist predator can coexist while competing for a single resource. Specialist predators and intraspecific density dependence generate essentially the same mechanisms permitting multiple consumers to coexist on one resource (cf. Grover 1994, McPeek 2012).

Adding consumer species to a food chain at the intermediate trophic level adds new problems, because both resource and apparent competition affect the invasibility criteria of each new consumer. Multiple intermediate-trophic-level consumers would result in the diamond community module (fig. 2.1). First consider the situation where all consumers and the predator have linear functional responses and no intraspecific density dependence ($h_{j1}=l_{j1}=g_j=y_1=0$ for all j consumers). In this case, two consumers at most can coexist at a stable equilibrium with a single predator and a single resource (Levin 1970, Holt et al. 1994, Leibold 1996, McPeek 1996b). Begin with the phase portrait shown in figure 2.8B, and remember that with no intraspecific density dependence, the consumer's isocline will be parallel to its own and all other consumer axes. As a consequence, for the two consumer isoclines to intersect at positive abundances, the isocline of the consumer with the lower $f_j/(a_{1j}b_{1j})$ must have a steeper slope in the $R_1 - P_1$ face (fig. 2.10). (Maintain the identity relationships among species defined in inequality (2.4).) Thus, for the two consumer isoclines to intersect, the following criterion must be satisfied:

$$\frac{m_{11}}{m_{21}} > \frac{a_{11}b_{11}}{a_{12}b_{12}} > \frac{f_1}{f_2} \tag{2.9}$$

(Leibold 1996, McPeek 2014a). Because the two consumer isoclines are parallel to all consumer axes, they intersect at the same values of R_1 and P_1 for all combinations of consumer abundances. Thus, the equilibrium resource and predator abundances are fixed if the two consumers are present. Also, this criterion specifies a trade-off in the abilities of the two consumers; the consumer that is the better resource competitor (i.e., $f_1/(a_{11}b_{11}) < f_2/(a_{12}b_{12})$) must suffer the mortality imposed by the predator disproportionately to its own ability to acquire and utilize the resource (i.e., $m_{11}/m_{21} > a_{11}b_{11}/(a_{12}b_{12})$).

One other criterion must be satisfied for the four-species equilibrium to exist:

$$\frac{m_{11}n_{11}}{a_{11}} > \frac{x_1}{c_1 - d_1 R^*_{1(1,2)}} > \frac{m_{21}n_{21}}{a_{12}} \tag{2.10}$$

(Leibold 1996, McPeek 2014a). To understand how criteria (2.9) and (2.10) define coexistence, consider the positions of the four isoclines in the $R_1 - N_1 - N_2$ subspace (fig. 2.10). The predator and resource isoclines are independent of the P_1 axis and so their positions in this subspace do not change with different values of P_1. However, the two consumer isoclines intersect the R_1 axis at larger values for larger values of P_1 (fig. 2.10). When $P_1 = 0$, the consumer isoclines are at their respective $f_j/(a_{1j}b_{1j})$ values. As P_1 is increased, the consumer isoclines slide to higher values of R_1 (fig. 2.10).

Criterion (2.9) ensures that consumer 1's isocline moves to higher values of R_1 faster than that of consumer 2, so that the two consumer isoclines are coincident at a single value of $R_1 = R^*_{1(1,2)}$, and criterion (2.10) ensures that this coincident point occurs within the range on the R_1 axis where the resource and predator isoclines also intersect (fig. 2.10; McPeek 2014a). The lower limit of this range is $R^*_{1(1)}$ and the upper is $R^*_{1(2)}$, both with P_1 also present at its equilibrium abundance; these are the equilibrium resource levels in the respective three-species food chains with only one consumer present (Holt et al. 1994, Leibold 1996, McPeek 2014a). Consumer 1 can invade because it is better at acquiring and utilizing the resource despite suffering greater predator-inflicted mortality, and consumer 2 can invade because it suffers proportionately less predator-inflicted mortality despite being poorer at competing for the resource.

If the consumers experience intraspecific density dependence ($g_j > 0$ for all j consumers), more than two consumers can coexist with one resource and one

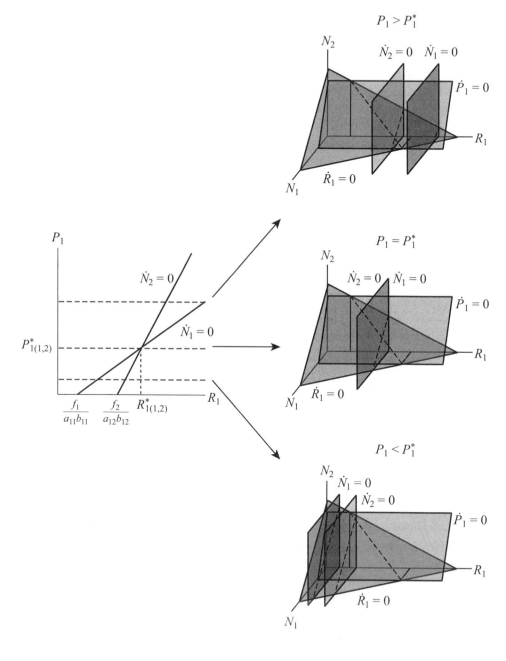

predator (McPeek 2014a). Because intraspecific density dependence causes each consumer's isocline to pivot away from its own axis but remain parallel to all other consumer axes (fig. 2.3B), the two consumer isoclines will now intersect over a range of R_1, and therefore also over P_1 values. With two consumers present, the system will come to a four-species equilibrium where all four of their isoclines intersect at a single point (fig. 2.11). As with only resource competition and consumer density dependence, an invading consumer will decrease the equilibrium resource abundance, but some amount of resource will remain that other consumers may exploit (fig. 2.4); and as with apparent competition, each new consumer will increase the predator abundance, but in this case not to the same degree (McPeek 2014a). Any other consumer that has a positive population growth rate at the combination of $R_{1(\ldots)}^{*}$ and $P_{1(\ldots)}^{*}$ can then invade the system.

No set of criteria can be specified to say which consumers can possibly invade or which set of consumers will be present because both apparent and resource competition simultaneously determine the set of coexisting species (McPeek 2014a). Consumer 1, defined by having the lowest $f_1/(a_{11}b_{11})$, may not be present if it experiences too much predation, given the other set of consumers present in the community. Likewise, the consumer with the highest ratio of its own resource consumption ability relative to predator mortality (i.e., the "best" apparent competitor with the highest $a_{1j}b_{1j}/m_{j1}$, which is the ratio by which consumers are ranked according to their apparent competitive ability in this model) may still not be present if the resource is not adequately abundant relative to the predator mortality that it would experience. Numerical simulations also indicate that allowing saturating functional responses ($h > 0, l > 0$) and density dependence in the predator ($y_1 > 0$) also provides greater scope for more consumers to coexist at stable point equilibria; this is due to the same effects these mechanisms have on apparent and resource competition alone (M. A. McPeek, unpubl. results).

The ecological opportunities available to consumer species here are much more limited because these species face both resource and apparent competition.

FIGURE 2.10. Isocline system for a diamond module of two consumers feeding on one resource and being fed upon by one predator, in which only the resource experiences intraspecific density dependence. The isoclines are as identified in previous figures in this chapter. The *left* panel shows the relationship between the two consumer isoclines in the two dimensional $R_1 - P_1$ space; because these two isoclines are parallel to both consumer axes, this relationship holds for all values of N_1 and N_2. The three panels in the *right column* show the positions of the four isoclines at various values along the P_1 axis (as identified in the *left* panel) in the 3D $R_1 - N_1 - N_2$ subspace. For higher values on the P_1 axis, the consumer isoclines intersect the R_1 axis at higher values (note the blue isoclines at higher values on the R_1 axis for higher values of P_1). (This figure is redrawn from figure 2 of McPeek 2014a, with permission.)

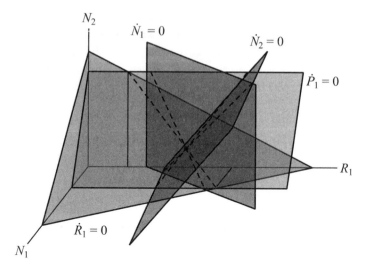

FIGURE 2.11. Isocline system for a diamond module of two consumers feeding on one resource and being fed upon by one predator, where the resource and both consumers experience intraspecific density dependence. The isoclines are as identified in previous figures in this chapter. This figure shows the isoclines of the four species in the reduced 3D $R_1 - N_1 - N_2$ subspace. (This figure is redrawn with from figure 4 of McPeek 2014a, with permission.)

Successful invaders will be those species that can adequately compete for resources while suffering mortality from the predator. However, "adequate" in this case is defined by the collection of consumers already present in the community. Moreover, being exceptional in one of these interactions does not guarantee a place in the community. The balance of a species' abilities to engage in resource and apparent competition is what that determines whether an ecological opportunity exists for it in a community. Thus, a species will be able to invade and coexist if another does not deflate resource abundance or inflate predator abundances to levels at which the invader cannot maintain a population.

INTRAGUILD PREDATION/OMNIVORY

Intraguild predation, also known as omnivory, is the situation where a predator is also a resource competitor with its prey (fig. 2.1) (Holt and Polis 1997; Diehl and Feißel 2000; Křivan and Diehl 2005; Abrams and Fung 2010a, 2010b; McPeek 2014a). This represents the final complication introduced by the structure of community modules. Here, intermediate-trophic-level consumers face resource and apparent competition with other consumers as well as resource competition with their predator. How are the ecological opportunities further limited for these consumers and further enhanced for the predator?

As implied by the simple network diagram of species interactions (fig. 2.1), the geometry of the phase portraits for intraguild predation are very similar to those of the diamond module. Again, begin with only the resource experiencing intraspecific density dependence and with linear functional responses for all predator-prey interactions. The two differences in geometry between the diamond and intraguild modules are that with intraguild predation the resource isocline intersects the predator axis at c_1/v_{11}, and the predator isocline intersects the resource axis at $x_1/(v_{11}w_{11})$ (fig 2.12A). Because the predator can coexist with the resource without eating any of it (i.e., fig. 2.8B), the predator's isocline need not intersect the R_1 axis at a point less than c_1/d_1. However, if the predator's isocline intersects the R_1 axis at a point less than $R^*_{1(1)}$, the predator will depress the resource abundance to a level at which the consumer cannot support a population; in other words, the predator must be a poorer resource competitor than the consumer for them to coexist (fig. 2.12A) (Diehl and Feißel 2000, Křivan and Diehl 2005, Amarasekare 2008, Abrams and Rueffler 2009).

Feeding on the resource does, however, relax another constraint on the predator to some degree. Remember that for the predator to invade the community when it could not eat the resource, its isocline had to intersect the N_1 axis at a point below $N^*_{1(1)}$ (fig. 2.8B). However, when the predator can feed on the resource, its isocline can intersect the consumer's axis above $N^*_{1(1)}$. There is a maximum on the N_1 axis for this intersection point that is set by where the predator's isocline intersects the N_1 and N_1 axes (McPeek 2014a). The constraint is that the predator isocline must intersect the line of intersection between the R_1 and N_1 isoclines at a positive value for P_1. Thus, the predator isocline must lie in the $R_1 - N_1$ face such that it is closer to the origin than the point where the R_1 and N_1 isoclines intersect in the $R_1 - N_1$ face (fig. 2.12B). Consequently, the predator can invade even though it can support a population on neither the resource nor consumer alone (i.e., the predator can coexist with the resource and consumer if its isocline in the $R_1 - N_1$ face is closer to the origin than the point $[R_1, N_1, P_1] = [f_1/(a_1b_1), N^*_{1(1)}, 0]$, even with $x_1/(v_{11}w_{11}) > c_1/d_1$ on the R_1 axis and $x_1/(m_{11}n_{11}) > N^*_{1(1)}$ on the N_1 axis).

By the predator now being able to also exploit the resource, the three-species equilibrium point shifts so that the consumer abundance decreases and the resource abundance actually increases, all other parameters remaining constant (cf. figs. 2.8B and 2.12B). Intraspecific density dependence in the consumer increases the resource and predator abundance while decreasing the consumer abundance (fig. 2.12C), and intraspecific density dependence in the predator increases the consumer and decreases the resource abundances (fig. 2.12D). Interestingly, these simple

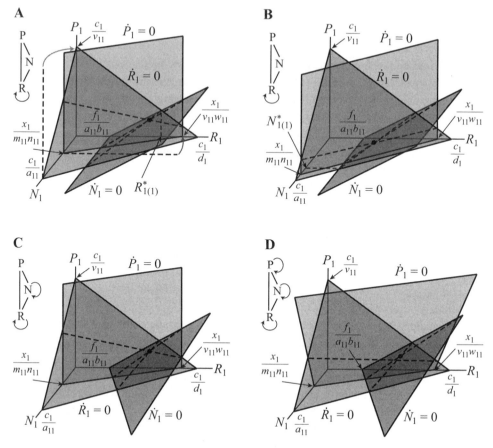

FIGURE 2.12. Isocline system for a intraguild predation module of one resource (R_1), one consumer (N_1), and one predator (P_1). The isoclines are as identified in previous figures in this chapter. Panel A illustrates how the various isoclines are altered by the introduction of intraguild predation, as compared to the isocline system for a linear food chain, without intraguild predation and with only the resource experiencing intraspecific density dependence (i.e., fig. 2.8B). Panel B illustrates how intraguild predation can support the predator even though the predator could not support a population from feeding on the consumer alone. The other two panels illustrate the isocline system when the (C) consumer and (D) both the consumer and predator experience intraspecific density dependence as well as the resource. (Panels A and C of this figure are redrawn from figure 1 of McPeek 2014a, with permission.)

three-species systems with planar isoclines can express limit cycles if the equilibrium abundance of the resource when alone is high enough (Holt and Polis 1997, Diehl and Feißel 2000, Revilla 2002, Křivan and Diehl 2005, Tanabe and Namba 2005). Permitting saturating functional responses also increases the complexity of the system dynamics (Křivan and Diehl 2005; Abrams and Fung 2010a, 2010b).

The conditions for additional consumers to invade mirror those of the diamond module in areas of parameter space where stable point equilibria result. In the absence of intraspecific density dependence in the consumers, two consumers at most can coexist for exactly the same reasons as with the diamond module; since the consumer isoclines are unchanged with the predator eating the resource, only two consumer isoclines can intersect at some combination of resource and consumer abundances (McPeek 2014a). We can therefore apply the same type of analysis as in figure 2.10, but with intraguild predation, the position of the resource isocline changes as the predator abundance changes (fig. 2.13). Consequently, the upper and lower bounds for which the resource and predator isoclines intersect in the $R_1 - N_1 - N_2$ subspace increase as the predator abundance increases (fig. 2.13). Thus, the conditions for the four-species equilibrium to exist are more stringent than for the diamond module (McPeek 2014a). Permitting intraspecific density dependence in the consumers ($g_j > 0$) also allows multiple consumers to coexist for the same reasons as in the diamond module, but again with more restrictive conditions because of the narrower and moving boundaries for the intersecting resource and predator isoclines (McPeek 2014a). In contrast, in areas of parameter space where stable limit cycles result, more than two consumers can invade and coexist even in the absence of intraspecific density dependence (McPeek 2014a).

With intraguild predation, the ecological opportunities for predators and consumers in the system move in opposite directions. For a top predator, the ability to exploit species at multiple trophic levels increases its ecological opportunity to invade because it can possibly support a population from multiple food sources, even when this is not possible by exploiting only one. In contrast, for consumers, resource competition with the predator narrows the scope of available ecological opportunities by limiting the ranges of resource and predator abundances in which consumer isoclines must intersect (cf. figs. 2.10 and 2.13).

MUTUALISM

Up until now I have considered only predator-prey interactions as direct effects of one species on another and the indirect consequences of consumer-resource interactions such as resource and apparent competition. Mutualisms in which both species benefit from interacting with one another is another prevalent form of species interaction (Bronstein 1994, Callaway 1995, Jones et al. 2012). Early model formulations of mutualistic interactions searched for conditions in which species abundances would not both increase to infinity—what May (1981) called the "orgy of mutual benefaction." A number of models, most developed in a Lotka-Volterra competition framework, showed that mutualisms were not inherently unstable

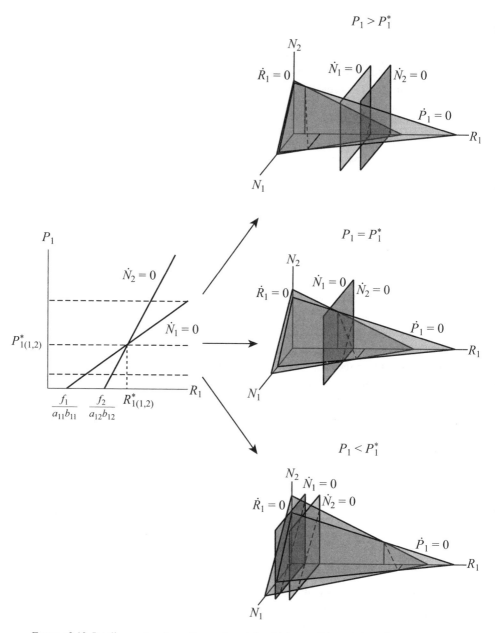

FIGURE 2.13. Isocline system for a diamond module with intraguild predation of two consumers feeding on one resource, and being fed upon by one predator that also eats the resource. The panels of this figure are comparable to those of figure 2.10, except for the inclusion of intraguild predation. With intraguild predation, the position of the resource isocline changes as the predator abundance changes in the $R_1 - N_1 - N_2$ subspace, along with the isoclines of the two consumers. (This figure is redrawn from figure 5 of McPeek 2014a, with permission.)

interactions (Vandermeer and Boucher 1978, Goh 1979, Travis and Post 1979, Dean 1983, Wolin and Lawlor 1984, Boucher 1985, Wright 1989). Even May (1981) illustrated how mutualists can stably coexist and stated that model assumptions in which mutualist abundances increased without limits were "silly solutions."

Within the last decade, a number of scientists have recognized that many mutualisms can be understood as consumer-resource interactions in their own right (e.g., Holland et al. 2005, Jones et al. 2012). For example, consider the typical interaction between a plant and a bee that pollinates the plant (this description could be translated for many types of mutualisms in nature). The bee's actions of pollinating the plant are only an incidental consequence of foraging for pollen. Plants have a fantastic diversity of flower morphologies and arrangements of life history events to ensure that pollen gets onto pollinators as they forage. To the bee, flowers are simply a source of food—a resource. The plant benefits from the actions of the bee pollinating ovules, but this comes at the cost of pollen loss to the bees. Because of the abundance of pollen made relative to what is needed to fertilize ovules, this cost is typically very small in many systems, particularly when considered against the benefits that accrue. In addition, because the number of ovules produced by plants are limited by other factors, the benefits of bee pollination saturate at high bee abundances. As a consequence, the equilibrium abundance of plants will be higher in the presence of the bees than in their absence, because more ovules are fertilized; but plant abundances will not increase without bounds in the "orgy of mutual benefaction."

The study of the dynamics of mutualist interactions is rapidly increasing, and many different possible configurations of community modules involving both mutualistic and antagonistic interactions are being explored (Jones et al. 2012). For present purposes, I will only consider a simple mutualism community module to give the reader a flavor of how mutualisms are related to consumer-resource interactions.

In a mutualism, the benefits to one partner are also typically formulated using a saturating functional response term but one that is positive in sign instead of negative; the costs are also formulated as saturating functional response terms that depend on the abundance of the interactor in the denominator (e.g., Holland and DeAngelis 2010). In some systems, such as yuccas and yucca moths, and figs and fig wasps, the pollinator is also a seed predator. Therefore, the costs to the plant of interacting with the pollinator become quite steep at high pollinator abundances because of the direct loss of ovules (Holland et al. 2002, Holland and DeAngelis 2010, Wang et al. 2011). In these systems, the form of the isoclines can become quite complicated but typically result in coexistence at stable equilibria or stable limit cycles (Holland et al. 2002, Holland and DeAngelis 2010, Wang et al. 2012). Both the plant and the bee benefit from their interaction—a mutualism—and their abundances are higher than in the absence of the other. These interactions are also frequently characterized as obligate or facultative, depending on whether the presence of the partner is required for the species to occur, or if the presence of the partner only enhances the density of the species, respectively.

To provide some idea of the relationship between consumer-resource and mutualism interactions, consider a model of a plant-pollinator interaction where a single plant species (R_1) can support a population in the absence of pollinators, but its abundance is enhanced if pollinators (N_j) of various species are present (following Holland et al. 2002, Fishman and Hadany 2010, Holland and DeAngelis 2010, Wang et al. 2012). The abundance of each pollinator species (N_j for $j = 1$, $2, \ldots, q$ species) depends on how many resources (e.g., nectar, pollen) it garners from the plant population (R_1), and the interaction between the plant and each pollinator is defined by a saturating functional response (Holling 1959). In addition, the pollinators interact with conspecifics while foraging, which reduces their foraging rate. This interference with conspecifics introduces negative intraspecific density dependence into the population birth rate of each pollinator species. As a result, the overall functional response is characterized by the Beddington-DeAngelis functional form (Beddington 1975, DeAngelis et al. 1975, Fishman and Hadany 2010). Each pollinator species also potentially experiences some form of negative intraspecific density dependence not involved in this interaction with the plant (e.g., competition for nest sites).

These assumptions imply the following model for the community module of this guild of species competing for the benefits supplied by a mutualist partner:

$$
\begin{aligned}
\frac{dN_j}{dt} &= N_j \left(\frac{b_{1j} a_{1j} R_1}{1 + a_{1j} h_{1j} R_1 + z_j N_j} - f_j - g_j N_j \right) \\
\frac{dR_1}{dt} &= R_1 \left(c_1 - d_1 R_1 + \sum_{j=1}^{q} \frac{a_{1j} N_j}{1 + a_{1j} h_{1j} R_1 + z_j N_j} \right)
\end{aligned}
\tag{2.11}
$$

All the terms in equations (2.11) have the same meaning as in equations (2.1) except for the interpretations of the attack coefficient and the interference term in the denominator of the functional response. In this formulation, the "attack coefficients," a_{1j}, in the functional responses describe the basic rate at which the net benefits (e.g., pollination) accrue to the plant via the foraging activities of the pollinators. Thus, the functional response term in the equation describing the plant population dynamics is positive (cf. equations (2.1) and (2.11)). The population birth rate of each pollinator is then this functional response times a conversion efficiency coefficient b_{1j}, which translates that foraging into new pollinator offspring. The z_j coefficient in the denominator of the functional response scales the rate of foraging saturation by the pollinators via interference competition among them while foraging.

ONE POLLINATOR ON ONE PLANT

Begin by considering the mechanisms fostering coexistence between one plant species and one pollinator species described by equations (2.11). Negative intraspecific density dependence in both species is critical for preventing each from growing without limits. With no negative intraspecific density dependence in the plant (i.e., $d_1 = 0$), the population increases exponentially with $c_1 > 0$; and if the pollinator is present, this rate accelerates with increasing pollinator abundance (i.e., the orgy of mutual benefaction). Also, with no negative intraspecific density dependence in the pollinator (i.e., $z_j = 0$ and $g_j = 0$), it will increase without bounds if the plant is at an abundance of $R_1 > f_1 / (a_{11}(b_{11} - f_1 h_{11}))$.

Coexistence at a stable equilibrium with both species present is only possible when they experience some degree of intraspecific density dependence that reduces their per capita growth rates as their respective abundance increases. This implies that $d_1 > 0$ for the plant species, and for the pollinator species this means that either $g_1 > 0$ or $z_1 > 0$. First consider the case of pollinator density dependence outside the interaction (i.e., $g_1 > 0$ and $z_1 = 0$); for simplicity also assume initially a linear functional response with $h_{11} = 0$. The isoclines for this set of parameters are both linear (fig. 2.14). A stable equilibrium occurs at the point of their intersection if the following criterion holds:

$$\frac{g_1}{a_{11}} > \frac{a_{11}b_{11}}{d_1} > \frac{f_1}{c_1}. \tag{2.12}$$

This criterion ensures that the pollinator can invade if the plant is at its equilibrium abundance in the absence of the pollinator (i.e., $c_1/d_1 > f_1/(a_{11}b_{11})$), and the slope of the plant's isocline is greater than that of the pollinators (i.e., $d_1/a_{11} > a_{11}b_{11}/g_1$) so that the isoclines intersect where both species have positive abundances (fig. 2.14A). Inequality (2.12) assures that this is a stable equilibrium. At this stable equilibrium, the plant abundance is now greater than its equilibrium abundance in the absence of the pollinator (i.e., $R^*_{1(1)} > c_1/d_1$). The orgy of mutual benefaction is still possible if the slope of the pollinator's isocline is greater than that of the plant (i.e., $b_{11}a_{11}/g_1 > d_1/a_{11}$); that is, the pollinator can invade, but no equilibrium exists. Also, increasing the strength of intraspecific density dependence for either species decreases the equilibrium abundances of both. Most important, the addition of the mutualist pollinator to the community module increases the abundance of the plant species above what it is in the pollinator's absence.

Adding pollinator satiation while foraging (i.e., $h_{11} > 0$) to pollinator density dependence outside the interaction ensures that the two species will coexist if the pollinator can invade. Pollinator satiation makes its invasibility criterion more stringent

A $d_1 > 0, h_{11} = 0, z_1 = 0, g_1 > 0$

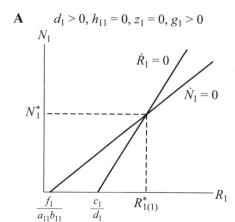

B $d_1 > 0, h_{11} > 0, z_1 = 0, g_1 > 0$

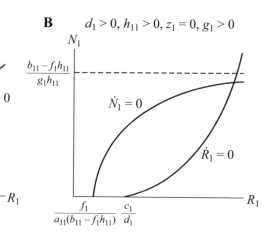

C $d_1 > 0, h_{11} > 0, z_1 = 0, g_1 > 0$

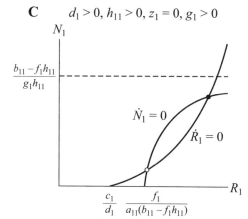

D $d_1 > 0, h_{11} > 0, z_1 > 0, g_1 = 0$

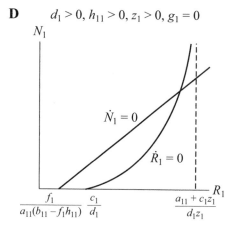

E $d_1 > 0, h_{11} > 0, z_1 > 0, g_1 > 0$

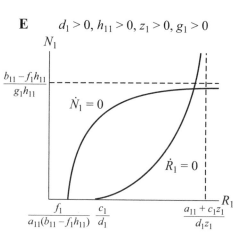

$$\frac{f_1}{a_{11}(b_{11} - f_1 h_{11})} < \frac{c_1}{d_1}, \tag{2.13}$$

and causes its isocline to become an asymptote at a maximum N_1 value (fig. 2.14B). The plant's isocline also now bends upward as a quadratic with positive first and second derivatives, which ensures that the two isoclines will intersect at a stable equilibrium if the pollinator can invade when rare (fig. 2.14B). A stable equilibrium is still possible if the pollinator cannot invade when rare (e.g., because its intrinsic death rate, $[f_1]$, is higher). The isoclines may still intersect, but now two equilibria exist—an unstable equilibrium with both at lower abundances and a stable equilibrium at higher abundances (fig. 2.14C; see also Wang et al. 2012). However, this stable equilibrium of coexistence would be attainable only if the pollinator could reach an adequately large abundance to enter this equilibrium's domain of attraction (e.g., via immigration from some other population).

If the pollinator's negative intraspecific density dependence is generated only while foraging on the plant (i.e., $z_1 > 0$ and $g_1 = 0$), the its isocline remains linear, but the plant's isocline now becomes an asymptote at a maximum R value; this also precludes the possibility of an orgy of mutual benefaction (fig. 2.14D). Pollinator satiation (i.e., h_{11}) does not change the fundamental shapes of the isoclines in this case but does change their positions. The invasibility criterion for the pollinator remains unchanged—see criterion (2.13)—but the asymptotic nature of the plant's isocline ensures that the two isoclines will intersect at a stable equilibrium if the pollinator can invade. Again, even if the pollinator's invasibility criterion is not satisfied, a stable equilibrium may still exist if the isoclines do cross (as in fig. 2.14C; see also Wang et al. 2012).

If the pollinator experiences negative intraspecific density dependence both as part of, and outside of, the interaction with the benefactor (the plant), coexistence is assured if the benefiter (the pollinator) can invade (fig. 2.14E). With no benefiter satiation, the pollinator's isocline is a quadratic with positive first and negative second derivatives (not shown); and with pollinator satiation, its isocline becomes an asymptote at a maximum N value (fig. 3B).

FIGURE 2.14. Isocline portraits for the mutualistic interaction between one plant species (R_1) and one pollinator species (N_1). Each panel shows the configuration of the isoclines that result in a stable and feasible equilibrium of these two species for different parameter combinations in equations (2.11). Parameters are specified above each panel. If the isocline of a species becomes an asymptote at a particular value, the asymptotic value is identified with a dashed line. Panel C also illustrates an isocline configuration that results in one stable (*filled circle*) and one unstable (*open circle*) equilibrium. All isoclines are identified as in figure 2.2.

A few general results are highlighted by this simple analysis of a mutualistic interaction between two species. Their stable coexistence requires that both species experience some degree of negative intraspecific density-dependent population regulation. If interactions among the species while foraging for benefits from the other species generates their intraspecific density dependence (i.e., $z_1 > 0$), the abundance of the species supplying the resources has a maximum. Likewise, if the foraging species becomes satiated at a high abundance of the other species (i.e., $h_1 > 0$) and experiences intraspecific density dependence outside the interaction (i.e., $g_1 > 0$), the foraging species also has a maximum abundance set by the interaction. Either of these latter two conditions ensures that the species can coexist if both can invade when rare.

MULTIPLE POLLINATORS ON ONE PLANT

Now consider what is required for additional pollinator species to invade and coexist while foraging for benefits from the same plant species. This analysis contrasts with that of multiple resource competitors feeding on a single resource presented above. Without loss of generality, again identify the pool of possible pollinator species based on the order that their isoclines intersect the R-axis, from smallest to largest:

$$\frac{f_1}{a_{11}(b_{11} - f_1 h_{11})} < \frac{f_2}{a_{12}(b_{12} - f_2 h_{12})} < \cdots < \frac{f_j}{a_{1j}(b_{1j} - f_j h_{1j})} \qquad (2.14)$$
$$< \cdots < \frac{f_q}{a_{1q}(b_{1q} - f_q h_{1q})}.$$

(Note that this is identical to inequality (2.4), but here I am considering the case with $h_{1j} > 0$.) Only some of these pollinators will satisfy their equivalent of inequality (2.13), meaning that only some of the pollinators can coexist with this plant species by themselves. However, their prospects become better if other pollinator species are already coexisting with the plant.

Because the pollinator species only interact indirectly with one another through their shared interactions with the plant, the isocline of each pollinator species is again parallel to the axes of all other pollinators, regardless of the parameter set being considered. The simplest situation to visualize is the case when pollinators experience negative intraspecific density dependence outside the interaction with the plant and do not satiate when foraging (i.e., all $g_j > 0$, and all linear functional responses with all $h_{1j} = 0$ and $z_j = 0$). Figure 2.15 illustrates two different situations for two pollinator species interacting with the plant.

With these parameters, the isocline of each species is a plane. The R_1 isocline intersects the R_1 axis at c_1/d_1, has a slope of d_1/a_{11} in the N_1-R_1 face, and a slope of d_1/a_{12} in the N_2-R_1 face. The N_1 isocline intersects the R_1 axis at $f_1/(a_{11}b_{11})$ in this case (i.e., $h_1=0$) and increases linearly along the R_1 axis with slope $a_{11}b_{11}/g_1$ (as in fig. 2.14A), but it does not intersect the N_2 axis because it runs parallel to this axis. Again, note that figure 2.14A is the two-dimensional representation of these two isoclines in the N_1-R_1 face of this three-dimensional depiction. Likewise, N_2's isocline intersects the R_1 axis at $f_2/(a_{12}b_{12})$, with a slope of $a_{12}b_{12}/g_2$ along this axis, and runs parallel to the N_1 axis.

First, assume that N_1 and R_1 come to equilibrium in the absence of N_2. The plant's abundance $R^*_{1(1)}$ again determines whether any of the other pollinators can invade, which is possible if

$$\frac{f_j}{a_j(b_j-f_jh_j)} < R^*_{1(1)} \tag{2.15}$$

(note that this criterion is identical to that for resource competition in criterion (2.5) if $h_{1j}>0$). Coexistence of these three species at a stable equilibrium requires that an additional criterion be satisfied in this case. The line of intersection of the N_1 and R_1 isoclines has a positive slope along all axes. For the three species to coexist at a stable equilibrium, the N_2 isocline must intersect the line of intersection of the N_1 and R_1 isoclines (fig. 2.15). This requires that

$$\frac{f_j}{a_j(b_j-f_jh_j)} < R^*_{1(1)}. \tag{2.16}$$

The term on the left of inequality (2.16) is the slope of the intersection line between the N_1 and R_1 isoclines projected on the N_2-R_1 face, and the term to the right of the inequality sign is the slope of the N_2 isocline in the N_2-R_1 face (fig. 2.15). If both inequalities (2.15) and (2.16) are satisfied, the three isoclines intersect at a single point that is a stable equilibrium with all three species at positive abundances. Because the lines of intersection of all three isoclines have positive slopes when projected on all faces, the addition of N_2 to the community causes the abundances of both N_1 and R_1 to increase. This is a general result: the addition of a new benefiter species to the community causes the abundances of all species in the community module to increase. If inequality (2.15) is satisfied but inequality (2.16) is not, N_2 will invade and all three species will increase without bounds (i.e., a three-species orgy of mutual benefaction).

Comparison of figures. 2.15A and 2.15B illustrates how *facilitation* arises in these mutualisms. Facilitation is the presence of one species in a community permitting another to invade and support a population. In figure 2.15A, both N_1 and N_2 can coexist alone with R, because both of their isoclines intersect the R_1 axis

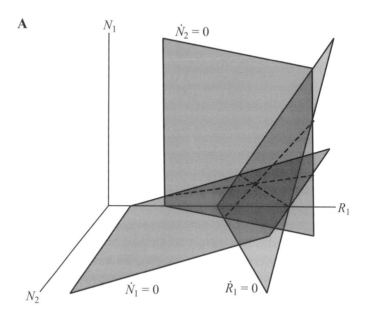

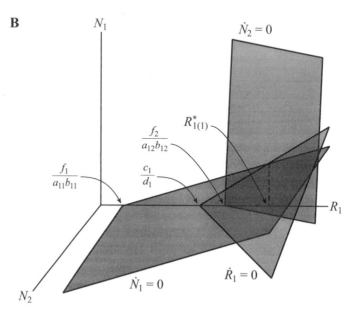

below the intersection of the R_1 isocline (i.e., both satisfy their respective inequality (2.13) criteria). Consequently, the order of invasion of these two benefiters does not matter to the assembly of the three-species community module. In contrast, figure 2.15B illustrates a case where N_2 cannot coexist with R_1 alone because its isocline intersects the R axis above the intersection of the R isocline. Thus, without N_1 present, R_1 is not abundant enough to supply adequate resources to support a population of N_2. Therefore, N_2 can only invade after N_1 has inflated the abundance of R_1 from $R_1^* = c_1/d_1$ to $R_{1(1)}^*$. In other words, N_1 must be present in the community to facilitate the invasion of N_2 by inflating the abundance of R_1 (fig. 2.15B).

The addition of either intraspecific density dependence as part of the interaction (i.e., all $z_j > 0$) or pollinator satiation (i.e., all $h_{1j} > 0$) does little to alter whether new pollinators can invade, and in fact both make coexistence easier. Remember that $z_j > 0$ means that the R_1 isocline becomes an asymptote at a maximum R value in the $N_j - R_1$ face, which ensures that the N_j and R_1 isoclines will intersect if N_j can invade (fig. 2.14). Likewise, satiation (i.e., $h_{1j} > 0$) combined with intraspecific density dependence outside the interaction (i.e., $g_j > 0$) causes the N_j isocline to become an asymptote at a maximum N_j value, which also ensures that the N_j and R_1 isoclines will intersect if N_j can invade (fig. 2.14). Thus, with $z_j > 0$, or with $h_{1j} > 0$ and $g_j > 0$, visualizing the multidimensional isoclines is difficult. However, the results are simple; if a pollinator species can invade a community module, it will always coexist at a stable equilibrium with all the species already present, since their isoclines are assured of intersecting. Moreover, the abundances of all species will increase with each new pollinator that invades. Nevertheless, the asymptotes will cause diminishing levels of increase of all species with the addition of each new pollinator.

In other words, with these greater complexities to the species interactions, only criteria comparable to criterion (2.15) determine whether a new pollinator species can invade. To determine which pollinator species can coexist, all that

FIGURE 2.15. Isocline portraits for the mutualistic interactions among one plant species (R_1) and two pollinator species (N_1 and N_2), when the pollinators experience intraspecific density dependence only outside the interaction with the plant ($g_j > 0$ and $z_j = 0$), and the pollinators do not satiate while foraging ($h_{1j} = 0$). All features are identified as in figure 2.3. (A) The configuration of the isoclines in which both pollinators can coexist with the plant in the absence of the other pollinator. The three dashed lines identify the lines of intersection between the three isoclines, and the single point of intersection of these three lines (black circle) is the stable point equilibrium for these three species. (B) The isocline configuration in which only N_1 can coexist with the plant by itself. The dashed line in this panel identifies the equilibrium abundance of the plant ($R_{1(1)}^*$) when it is coexisting with only N_1. The lines of intersections between the isoclines are not shown in B.

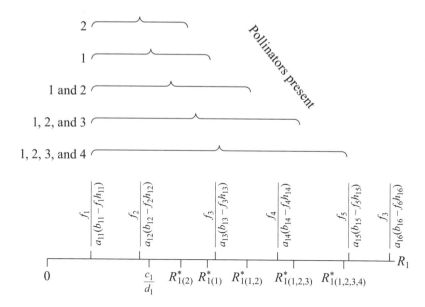

FIGURE 2.16. The configuration of isocline intersection points and equilibrium plant abundances along the R_1 axis for a single plant potentially interacting with a pool of six pollinator species $(N_1 - N_6)$. The *hashes* above the axis identify the points where the isoclines for each pollinator species intersect the R_1 axis. The *hashes* below the axis identify the points where the benefactor isocline intersects the R_1 axis (c_1/d_1) and the equilibrium abundances of the plant when coexisting with various combinations of pollinators as in figure 2.4.

is required is knowledge about where their individual isoclines intersect the R_1 axis, and the equilibrium abundance of R_1 with various combinations of coexisting pollinators. Figure 2.16 illustrates these important ecological parameters for a pool of six potential pollinator species interacting in a community module with R_1—the reader should note the similarity of this figure to figures 2.4 and 2.6. Of the six pollinators, only N_1 and N_2 can coexist alone with R_1 because their isoclines intersect the R_1 axis below c_1/d_1 (fig. 2.16). The pollinator N_3 cannot invade and coexist if only N_1 and N_2 are present, because its isocline intersects the R_1 axis above both $R^*_{1(1)}$ and $R^*_{1(2)}$. However, N_3 can invade and coexist if N_1 and N_2 are both already present, because its isocline intersects the R-axis below $R^*_{1(1,2)}$. Continuing this exercise, we see that of this pool of potential invaders, the final community will consist of N_1, N_2, N_3, and N_4 all interacting with their mutualistic benefactor R_1. Moreover, the presence of N_1 and N_2 facilitate the presence of N_3, and N_3 facilitates the presence of N_4.

From this, it should be clear that the same basic issues arise for exploiting ecological opportunities whether interactions are antagonistic or mutualistic.

For example, for a second plant species to invade an established plant-pollinator mutualism, the new plant may have to successfully compete for the benefits of the pollinator, and criteria for coexistence require that adequate pollination services be available to acquire these benefits (Johnson and Amarasekare 2013). Likewise, it is easy to imagine competition for pollen among multiple pollinators limiting the number of species that could coexist on one plant as described above. An invader still needs to acquire enough resources relative to the potential mortality sources it faces to have a positive population growth rate when rare. The major difference between antagonistic and mutualistic interactions is that they can have opposite effects on the abundances of other members of the community and the likelihood of creating ecological opportunities for other species through facilitation. For example, the actions of pollinators make a plant more abundant, and thereby increase the total amount of resource available to the herbivore species that feed on that plant; whereas the action of an herbivore make the plant less abundant, thereby reducing the ecological opportunities for pollinators. Clearly, these are intuitive descriptions of what can happen. Much more formal theoretical work needs to be done to explore how the presence of mutualisms embedded in community modules and food webs may influence the coexistence of multiple species in various types of interactions and modules scattered throughout the food web (e.g., Bronstein et al. 2003, Morris et al. 2003, Wilson et al. 2003, Jones et al. 2012, Wang et al. 2012, Johnson and Amarasekare 2013).

GENERAL THOUGHTS ON INVASIBILITY AND COEXISTENCE

Given the great diversity of species interactions and the potential for species with myriad different properties to invade a local community, making general statements about the structure of communities and the types and diversity of organisms that will inhabit it is daunting. Some patterns are apparent when one focuses on particular types of interactions, or on very general relationships between connectedness of species and interaction strengths (e.g., May 1973, Paine 1980). Moreover, recent studies of food webs claim to identify very gross patterns in how food webs are arrayed (e.g., Allesina et al. 2008).

However, much of the focus of community ecology is on finer scales, particularly on whether, and how, functionally similar species can live together in a community. Going back to Hutchinson (1958), the overwhelming focus has been on the coexistence of resource competitors, but many other types of species interactions and community configurations abound. I believe Holt's (1997a) focus on community modules systematizes how we can think about the critical features of species interactions in a much richer conceptual context. It organizes features of basic predator-prey coexistence (Rosenzweig and MacArthur 1963, Rosenzweig 1969, Murdoch and Oaten 1975, Murdoch and Stewart-Oaten 1989) with both resource and apparent competition (Holt 1977, Holt et al. 1994, Leibold 1996, McPeek 1996b, Abrams 1999, Křivan 2014, McPeek 2014a), as well as intraguild predation (Holt and Polis 1997, Křivan 2000, Křivan and Diehl 2005, Rudolf 2007, Amarasekare 2008). In addition, many mutualisms in which a partner provides a demographic benefit can be characterized conceptually as simply another resource (Holland and DeAngelis 2010, Jones et al. 2012). Thus, mutualistic species interactions also form the basis of many types of resource competition; each partner receives a demographic benefit from the mutualism partner, but each partner may also compete for access to mutualism partners with other potential partners (Bronstein et al. 2003, Morris et al. 2003, Wilson et al. 2003, Holland and DeAngelis 2010, Jones et al. 2012, Johnson and Amarasekare 2013).

Myriad mechanisms can promote the coexistence of species in a local community and exclude others from this same community (e.g., Tilman and Pacala 1993, Chase and Leibold 2003). The set of mechanisms considered in this chapter only provides a glimpse of the diversity of these mechanisms. However, this synthesis does suggest the general features of most mechanisms promoting local species coexistence: *Species will coexist with one another when no one species in the set can deplete resource abundances/mutualist benefits or inflate enemy abundances to levels at which the other species cannot invade and support populations.* This is simply a general statement of the ecological conditions required for invasibility, the fundamental criterion for coexistence (MacArthur 1972, Chesson 2000). In other words, these are the general ecological conditions that permit each coexisting species to have a positive population growth rate when it is rare and the other species are at their demographic steady states given its absence. Those species for which this is not true will either be excluded from the local community altogether, or exist in the community as sink or walking-dead species.

It sounds mildly tautological to say that a species will be able to maintain a population anywhere that resources (e.g., abiotic elements, forage, prey, mutualism partners) are sufficiently plentiful, and that enemies (e.g., herbivores, carnivores, omnivores, parasites, diseases, various mutualism partners that also inflict costs) are sufficiently scarce. However, the devil is in the details, and those details are the panoply of ecological mechanisms that community ecologists study. Are two competitors able to coexist because they segregate to the use of different resources, because intraspecific density-dependent processes also limit their abundances, because their functional responses cause their abundances to cycle, or because of some other limitation on their ability to deplete resources? Are two prey able to coexist because they have alternative antipredator defenses to thwart their shared predator directly, because their antipredator defenses generate trade-offs that limit their abilities to exploit various resources, or because of some other process?

One important demographic interaction that is frequently overlooked as a condition for coexistence is species self-limitation—namely, negative intraspecific density dependence. Phenomenological models of species interactions (e.g., Lotka-Volterra competition models such as Volterra 1926; MacArthur 1970, 1972); community matrices (May 1973, Yodzis 1981, Pimm 1982); and more recent analyses (Chesson 2000, Adler et al. 2007), in which explicit mechanisms of species interactions are not modeled per se but rather abstracted as interspecific effects, all routinely include intraspecific effects of species on their own demographics. The general conclusion of these analyses are that coexistence results from mechanisms that cause "species to limit themselves more than they limit others" (Adler et al. 2007, p. 96). Typically, analyses of mechanistic models of species interactions, as considered in this chapter, omit such intraspecific density-dependent limitations from species at all trophic levels except the basal resources.

The analyses presented here also highlight the importance of intraspecific density dependence in shaping community structure, but not in the ways suggested by these more phenomenological models. First, some species in the community must experience some degree of intraspecific density dependence if more than a single resource and consumer are to coexist. In addition, intraspecific density-dependent limitation in some species enhances the likelihood that other species at the same and other trophic levels can invade. However, strong density-dependent self-limitation does not provide a blanket guarantee of coexistence, as the analyses of phenomenological models would suggest. To exploit a particular ecological opportunity in a community, a species must meet stringent criteria for that position in the interaction network. Intraspecific density-dependent limitation creates a broader range of opportunities for successful species, but the qualitative criteria for coexistence remain unchanged, and these criteria vary for different functional positions. For example, in the apparent competition module, each new resource species can only invade if its population growth rate when rare is greater than the consumer's abundance (fig. 2.6). Stronger intraspecific density dependence in the resources that are already present reduces the consumer's abundance so that a greater range of resources with lower population growth rates when rare would be able to invade. In contrast, in the resource competition module, each new consumer species can only invade if its population growth rate when rare is less than the resource's abundance (fig. 2.4). Stronger self-limitation in the consumers already present reduces the amount by which each will depress the resource's abundance; this permits a greater range of consumers to coexist.

The presentation here has also been explicitly focused on conceptual and theoretical issues. However, a few words about empirically testing these issues is warranted. An empirical research program that involves experimental tests to verify the operation of relevant processes and to measure changes in underlying demographic (i.e., survival, fecundity, and individual growth) rates over gradients of pertinent resource, enemy, and mutualist abundances will generate strong evidence for or against any hypothesized mechanisms of coexistence discussed here. Testing for coexistence fundamentally involves testing whether a species can increase when rare (i.e., invasibility). Direct experimental tests of invasibility are exceedingly rare and for most systems logistically impossible to perform in the first place (Siepielski and McPeek 2010). However, invasibility can be tested indirectly by performing experiments that generate data to estimate population growth rates over gradients of absolute and relative abundances of community members; this is experimental demography (e.g., Caswell 1989, 1996; McPeek and Peckarsky 1998; Caswell 2001; Goldberg et al. 2001; Levine and HilleRis-Lambers 2009; Caswell 2010).

Life table response experiments that incorporate gradients of species abundances or comparisons across different localities (e.g., Angert 2006) and that compare projected population growth rates when rare would be ideal (Caswell 2001, 2010). However, capturing responses in multiple demographic rates simultaneously to estimate overall population growth rate may also be logistically challenging in many systems, but capturing responses in key demographic traits may provide strong insights. For example, Levine and HilleRisLambers (2009) constructed replicated experimental plots where each of ten serpentine plant species were planted over a gradient of relative abundances within the community. In each plot, they estimated the number of seeds produced per plant and seed bank carryover, and found that seeds entering the population each year were significantly higher when a species was rare for six of the ten species; this is consistent with the interpretation that the species can invade the community because their population growth rate is higher when they are rare (Levine and HilleRisLambers 2009).

Knowing that a species will increase when rare is critical to a full test of coexistence, but it is rather vacuous without knowing the mechanism or mechanisms of species interactions that influence its demographic performance in the community. Tests of invasibility should be coupled with experiments that quantify the importance of key species interactions to these demographic responses, and that identify the phenotypic properties of species that influence their performances in these interactions. In other words, experimental evidence is also needed to explain why adequate resources are available and why enemies have sufficiently low impacts on a species when it is rare.

As we will see in the next few chapters, understanding these mechanisms are also critical to understanding the evolution of species in their community context. Species evolve in response to one another, and the dynamics of that evolution differs for species at different positions in community modules. Moreover, the types of traits involved in determining the performance of species in those interactions with other species also critically shape whether and how species will evolutionarily respond to one another. Understanding mechanisms at this level is key to our understanding of both the ecological and evolutionary dynamics of interacting species in communities.

CHAPTER THREE

Evolving in the Community

On your hikes through nature, if you stop and ponder the features of any species you encounter, you will quickly realize that many of those features are critical for how it interacts with its environment. Many traits permit the organism to cope with the physical features of the environment it encounters. Waxy cuticles of leaves and insects reduce evaporative water loss. Lamellar gills permit aquatic insects to extract oxygen from the water.

Other traits shape how the individual will interact with individuals of its own and other species. The long proboscis of a butterfly and the extended bill of a hummingbird allow them to extract nectar from flowers with long corollas. Snails have shells that protect them from many predators, but the crushing pharyngeal jaws of a pumpkinseed sunfish permit it to feast on those snails. Tadpoles of many frog species have noxious chemicals that make them distasteful to many predators, but these toxins do not deter the few predators that lack taste receptors for the chemicals. Some prey can move rapidly to evade attacking predators, but others remain motionless and cryptic in order to not be seen.

The abilities of species to engage in interactions with the environment and other species are defined by the phenotypes they possess. Presumably, many of these taxa acquired their collection of traits through evolution in response to the pressures of natural selection generated by interactions with their physical environment and other species in this community. In other words, these species evolved traits to exploit the ecological opportunities available to them.

These ecologically important traits are what determine the parameters of the models we considered in chapter 2. The butterfly with a longer proboscis will be able to extract more nectar from flowers with longer corollas, and will therefore have a higher attack coefficient for harvesting this resource. Likewise, a damselfly larva that moves very little to remain cryptic will have a lower attack coefficient from foraging fish than a damselfly larva that moves more and thus is seen more easily by the fish. However, the damselfly larva that moves less will also have a lower attack coefficient on its own prey because it encounters them at a lower rate than the one that moves more.

Thus, a reciprocity exists between the ecological structure in which a species is embedded, and the evolutionary dynamics of it and all the other species with which it interacts. The phenotype of a species defines how successful, in demographic terms, a species will be in the various interactions in which it must engage. These performance abilities in turn are what determine all the parameters of the models we considered in chapter 2 that delineate whether the species can coexist in a particular community. In addition, variation in demographic success among individuals within that species also defines the nature of natural selection acting on that species, and so defines the evolutionary trajectory of that species. Because all species are evolving in response to one another, we must think of this as a coevolutionary dynamic (Thompson 1994, 2005). As one species evolves, its demographic impact on other species in the community change, and in turn alters their evolutionary trajectories. While coevolutionary dynamics may not be ongoing—the system may have reached stable evolutionary equilibria for the phenotypes of all species—the interactions among species are what determine where these evolutionary equilibria are located. Thus, the reciprocal coevolutionary responses among species and the ultimate evolutionary outcomes are defined by the ecological structure of the community. The changing nature and abilities of the actors in the ecological theater animate the evolutionary play.

THE ECOLOGICAL BASIS OF NATURAL SELECTION

Natural selection, the struggle for existence of different types within a species, as first outlined by Charles Darwin and Alfred Russel Wallace (Darwin and Wallace 1858), is essentially the demography of phenotypes and genotypes within and among populations. The process of natural selection has two components (Endler 1986). The first is phenotypic selection, in which the phenotypic distribution in the parental generation is changed because of differential survival or reproduction based on the phenotypic properties of individuals (or a collection of individuals if higher-level selection is being considered). The second is the genetic response to this phenotypic selection.

In most theoretical considerations of natural selection, the focus is placed squarely on the mechanisms involved in the genetic response to selection, and many simplifying assumptions are made about phenotypic selection and its underlying cause. This is typically done so that complex genetic interactions can be explored. In general, how the fitnesses of various members of the population are determined is ignored completely by merely assigning constant fitness values to various genotypes or phenotypes in the population.

However, the fitness of an individual is determined by the ecological conditions in which the individual is embedded and will therefore change as those ecological conditions change. Consequently, the relationship between fitness and phenotype has dynamics that are governed by the ecology of the system. "The ecology of the system" is not some external set of conditions imposed on these species but rather is defined by how the phenotypes of the interacting species determine the parameters of their interaction. For example, the ability of a dragonfly to catch a damselfly depends on the phenotypic traits of these two individuals, and the population level parameters determining the attack coefficient of the dragonfly predator on its damselfly prey depend on the distribution of phenotypes in the populations of both these interacting species.

Moreover, as the phenotypic distribution of one species evolves, the fitness consequences of interacting with that species will also change. If over one generation the damselfly population evolves to swim faster because of selection pressures imposed by dragonfly predation, in the next generation, the attack coefficient between this predator and prey will be smaller. As a result, the fitnesses of dragonflies with a particular value of a phenotypic trait for feeding on these damselflies will decrease. If the dragonfly population then evolves to be faster at pursuing the damselfly in the next generation, the attack coefficient between them will increase again.

These considerations lead inexorably to the conclusion that the ecological dynamics of the evolutionary process are fundamental to understanding the outcomes of that process. Therefore, in the analyses presented here, I turn the tables by allowing the ecological mechanisms defining natural selection to have full reign, and having the simplifying assumptions made about the genetics of the system. This is in essence the basis of quantitative genetic analyses (Falconer and Mackay 1996, Lynch and Walsh 1998). Thus, I focus on how the ecological dynamics of natural selection influence what will coevolve in a community of interacting species.

For natural selection to occur, three criteria must be met. First, some trait or traits expressed by individuals in a population must influence their survival or reproduction (i.e., their fitness). Second, individuals in the population must vary in these traits that cause fitness differences among them. Third, these trait differences must have a heritable genetic basis. The first and second criteria identify the conditions needed for phenotypic selection to occur, and the third criterion establishes that the population will genetically change in response to phenotypic selection.

The general term "fitness" is used in many, many different ways, and the debate about what is the "correct" fitness measure to consider often obscures the issues more than clarifies. When the ecological dynamics of natural selection are explicitly explored, the *absolute fitness* of an individual is the foundational metric underlying the dynamics of selection. This is because absolute fitness is the central mediating parameter between the evolutionary dynamics caused by natural selection and the demography of a population. Absolute fitness is defined as the number of offspring contributed to the next generation by an individual.

An individual's absolute fitness is the demographic consequence of the interaction of the individual's phenotype with its ecological environment. The environment includes the abiotic conditions experienced by the individual as well as all the interactions with conspecifics and other species. Fundamentally, *absolute fitness is an ecological property*. If all individuals in a population have identical phenotypes, the fitness of each individual would precisely describe the dynamics of the population. In other words, the overall dynamics of the population would be simply the fitness of each individual times the number of individuals in the population, or

$$\frac{dN_i}{dt} = N_i \ln(W_i) = N_i \frac{dN_i}{N_i \, dt}, \tag{3.1}$$

where N_i is the number of individuals in the population of species i, and $\ln(W_i)$ is the logarithm of the absolute fitness of each individual; this product is equivalent to the per capita population growth rate $dN_i/N_i \, dt$ (see chapter 2). (Because the modeling framework utilized here is a continuous time frame using differential equations, absolute fitness is measured on a log scale, where individuals are just replacing themselves at $W_i = 1$ so that $\ln(W_i) = 0$.) One interpretation of these equations implies that the population dynamic models used in chapter 2 assume that each species is composed exclusively (or at least predominantly) of only one phenotype. (See table 3.1 for a complete list of state variables and parameters used in models in this chapter.)

However, a little algebra shows that even if the population is composed of individuals that vary in their demographic rates because their phenotypes vary, the overall population growth rate has a rational interpretation. First, define the number of species i individuals having phenotypes in the infinitesimal range $z_i + dz_i$ to be $n_i(z_i)$, total population size to be $N_i = \int n_i(z_i) dz_i$, and the absolute fitnesses of individuals in this infinitesimal phenotypic range are $W_i(z_i)$ (as in Lande 2007). The total population growth rate is then given by

$$\frac{dN_i}{dt} = \int n_i(z_i) \ln(W_i(z_i)) dz_i. \tag{3.2}$$

Defining the frequency of individuals in each narrow phenotypic range to be $p_i(z_i) = n_i(z_i)/N_i$, we can arrange this equation to be

$$\frac{dN_i}{dt} = \frac{N_i}{N_i} \int n_i(z_i) \ln(W_i(z_i)) dz_i = N_i \int p_i(z_i) \ln(W_i(z_i)) dz_i. \tag{3.3}$$

The integral in this equation is the average fitness in the population: $\ln(\bar{W}_i) = \int p_i(z_i) \ln(W_i(z_i)) dz_i$. For completeness, this means that

$$\frac{dN_i}{dt} = N_i \int p_i(z_i) \ln(W_i(z_i)) dz_i = N_i \ln(\bar{W}_i) = N_i \frac{dN_i}{N_i \, dt}. \tag{3.4}$$

TABLE 3.1. Additional state variables and parameters in the evolutionary models of species interactions presented. Variables and parameters that are common to multiple species types are shown for only the resource species. All other variables and parameters are as listed in table 2.1.

State Variable	Description
z_R, z_N, z_P	Traits of the resource, consumer, and predator species, respectively
$\bar{z}_R, \bar{z}_N, \bar{z}_P$	Mean trait values of species
$R(z_R), N(z_N), P(z_P)$	Population abundances of species with the associated trait values
$W_R(z_R), W_N(z_N), W_P(z_P)$	Absolute fitnesses of species with the associated trait values
$\Delta = z_N - z_R$	Difference between the consumer and resource trait values
$\Omega = z_P - z_N$	Difference between the predator and consumer trait values
$\Sigma = z_P - z_R$	Difference between the predator and resource trait values

Parameter	Description
$V_{z_R}, V_{z_N}, V_{z_P}$	Additive genetic variation for traits in the three species
c_0	Maximum value for the resource species' intrinsic birth rate
$\tilde{z}_R^c$	Optimal trait value for the intrinsic birth rate of the resource species
γ	Scaling parameter for the underlying selection strength on the resource's intrinsic birth rate
d	Density-dependent rate of decrease in the resource's birth rate
f_0, x_0	Minimum value for the intrinsic death rates of the consumer and predator, respectively
θ, δ	Scaling parameters for the underlying selection strengths on the intrinsic death rate of the consumer and predator, respectively
g, y	Density-dependent rates of increase in the intraspecific death rates of the consumer and predator, respectively
$\tilde{z}_N^f, \tilde{z}_P^x$	Optimal trait value for the intrinsic death rates of the consumer and predator, respectively
a_0	Maximum value of the attack coefficient of the consumer feeding on the resource
$\varepsilon_i, \alpha, \beta$	Scaling parameters for the rate of change in the attack coefficient of the consumer feeding on the resource for the unidirectional-independent, unidirectional-dependent, and bidirectional-dependent trait interactions, respectively
m_0	Maximum value of the attack coefficient of the predator feeding on the consumer

(continued)

TABLE 3.1. (*continued*)

State Variable	Description
η_i, ρ, ϕ	Scaling parameters for the rate of change in the attack coefficient of the predator feeding on the consumer for the unidirectional-independent, unidirectional-dependent, and bidirectional-dependent trait interactions, respectively
v_0	Maximum value of the attack coefficient of the predator feeding on the resource
κ_i, τ, ψ	Scaling parameters for the rate of change in the attack coefficient of the predator feeding on the resource for the unidirectional-independent, unidirectional-dependent, and bidirectional-dependent trait interactions, respectively

In words, the per capita population growth rate of the population is equivalent to the average fitness of individuals in the population: $dN_i/N_i \, dt = \ln(\bar{W}_i)$ (Lande 2007).

In chapter 2, we saw how the population growth rate of a species was influenced by interactions with other species in the local community, but none of the species could evolve in response to one another. Equation (3.4) implies that these influences are mediated through their effects on the absolute fitnesses of individuals that constitute the local population of that species. Moreover, because the fitnesses of those individuals are determined by how their phenotypes demographically translate these interactions into absolute fitness, overall population growth rate depends on the phenotypic composition of the population, $p_i(z_i)$. *The per capita population growth rate is the average absolute fitness of individuals in the population.* Given these relationships, it should be apparent that we can use the ecological machinery describing the population dynamics of interacting species from chapter 2 as a descriptor of how species interactions influence the absolute fitness of each species within a local community, and thus the ecological dynamics of natural selection for each. This provides the fundamental link between ecological and evolutionary dynamics, since the basis of both are defined by how ecological interactions shape absolute fitness.

If we expand our expression for absolute fitness to represent all the influential species interactions, the complexity of fitness dynamics becomes apparent. For example, in this framework we can represent the per capita effects of various species interactions on the absolute fitnesses of individuals with phenotype z_i within species i as

$$\ln(W_i(z_i)) = \sum_j N_j \int p_j(z_j) f_{ij}(z_i, z_j) dz_j. \tag{3.5}$$

Here $f_{ij}(z_i, z_j)$ are functions describing the per capita effect of various phenotypic classes within species j on the absolute fitness of an individual of species i, and $p_j(z_j)$ are the frequencies of individuals in the various phenotypic classes. These per capita effects may depend on the phenotypes of both species. Note that equation (3.5) includes both the effects of species i on itself (i.e., intraspecific effect when $j = i$) and the effects of other species on species i (i.e., interspecific effects when $j \neq i$). The effect of species j on the fitness of individuals of species i represents a *fitness component* that contributes to the *overall absolute fitness* of species i. Fitness components are typically considered to be associated with life stages, but this interpretation separates fitness into components due to the action of different selective agents on various demographic rates (e.g., survival of resource i due to predation by consumer j).

Equation (3.5) also describes the *fitness surface* defined by the ecological environment in which the population of species i is embedded at any given instant. This equation immediately identifies that the shape of the fitness surface of each species will vary with both the phenotypes and abundances of all species in the community, meaning that analyses of natural selection that assume fixed fitnesses associated with various types (genotypes or phenotypes) in the population ignore the rich ecological dynamics that govern the process. This equation also highlights the fact that the shape of the overall fitness surface depends on the contributions of the various underlying fitness components (Arnold and Wade 1984b, Travis 1989, Wade and Kalisz 1990, McPeek 1996a). The relative importance of each species interaction to determining the shape of the overall fitness surface will depend on both the magnitudes of the per capita effects and the abundances of the various species.

Many different modeling approaches can be taken to explore the dynamics of trait change in interacting populations. In principle, one could take a population genetic approach, but this becomes exceedingly opaque and cumbersome when fitnesses are density and frequency dependent (Nagylaki 1992), which is why I will make simplifying assumptions about the genetics of the system. An alternative that has many appealing features is the adaptive dynamics approach (Dieckmann and Law 1996, Doebeli 2011). With adaptive dynamics, a population is assumed to contain one genetic type of individual. At each step in time, individuals with slightly different phenotypes (and genotypes) are assumed to invade the population at low frequency (e.g., as mutations from the dominant type), and the population changes if these invading individuals have fitnesses higher than the dominant type. Adaptive dynamics approaches have been applied to questions of behavioral choice among individuals in a population (e.g., Eshel 1981a, 1981b), adaptive evolution in interacting species (e.g., Dieckmann and Law 1996), and sympatric speciation (e.g., Doebeli and Dieckmann 2000).

I will utilize the approach developed by Lande (1982) and elaborated by Iwasa et al. (1991) and Abrams et al. (1993) (see also Lande 2007, Barfield et al. 2011), which models the evolution of quantitative traits in the same framework used by empirical biologists to study natural selection in the wild (Lande 1979; Lande and Arnold 1983; Arnold and Wade 1984b, 1984a). Thus, results of these analyses should provide testable predictions about species interactions and natural selection that can be directly tested in the field. Under the standard assumptions of the genetic basis of quantitative traits (i.e., many loci, each of small effect) and of traits and breeding values being normally distributed, the dynamics of natural selection are closely approximated by the dynamics of the average phenotypic *trait value* in the population ($\bar{z}_i$); since population dynamics are defined by the average fitness (equation (3.5)), this assumption further associates the average fitness of the population with the average phenotype in the population (Lande 1982, 2007). This framework can be further simplified by assuming that the effects of interactions with species j are primarily defined by the effects of individuals with the average phenotype ($\bar{z}_j$). Thus, equation (3.5) becomes

$$\ln(W_i(z_i)) = \sum_j N_j f_{ij}(z_i, \bar{z}_j). \tag{3.6}$$

In this framework, the evolutionary dynamics of the mean trait in the population is given by

$$\frac{d\bar{z}_i}{dt} = V_{z_i} \left. \frac{\partial \ln(W_i(z_i))}{\partial z_i} \right|_{z_i = \bar{z}_i} = V_{z_i} \left(\sum_j N_j \left. \frac{\partial f_{ij}(z_i, \bar{z}_j)}{\partial z_i} \right|_{z_i = \bar{z}_i} \right), \tag{3.7}$$

where the partial derivatives with respect to z_i are evaluated at the mean trait value $\bar{z}_i$, and V_{z_i} is the additive genetic variance among individuals in the population for the trait (Lande 1982, Iwasa et al. 1991, Abrams et al. 1993). The appendix in Iwasa et al. (1991) provides a clear and lucid presentation of the assumptions and derivation of this approach. The entire summation in parentheses of equation (3.7) is the overall *selection gradient* on the phenotype—this is the dynamical descriptor of phenotypic selection. This quantity defines the overall strength and direction of natural selection on the average phenotype. Each term in this summation is the selection gradient associated with each fitness component of the species, which defines the strength and direction of phenotypic selection impinging on each.

Equation (3.7) can be used to describe changes in the trait caused by either adaptive evolution across generations of a population or the adaptive plasticity of individuals (i.e., individuals modify their phenotype in response to environmental conditions) within a generation. For adaptive evolution, V_{z_i} represents the additive genetic variation in z_i and defines the rate of the genetic response to phenotypic selection, and its value is set to a small value (Lande 1982, Abrams et al. 1993). For adaptive plasticity, V_{z_i} is set to a large value, so that trait changes occur very quickly, or assumes a more complex functional form (Abrams et al. 1993). Thus, this framework can be used to explore both trait-mediated indirect effects of species interactions via adaptive plasticity (Werner and Peacor 2003, Křivan and Schmitz 2004, Ohgushi et al. 2013) and adaptive evolution (Lande 1982, Iwasa et al. 1991). The results are typically much the same.

Interpreted as a model of adaptive evolution, as I do here, equation (3.7) is simply a continuous time version of the standard breeders' equation describing the change in a quantitative trait due to natural or artificial selection (Lande 1979, Lande and Arnold 1983, Arnold and Wade 1984b, Lande 2007). Obviously, this derivation ignores within-population variation in phenotypes (e.g., Slatkin 1980, Taper and Case 1985, Price and Kirkpatrick 2009, Schreiber et al. 2011), and assumes that the dynamics of the mean trait value is a good approximation for the dynamics of evolution by natural selection (Lande 1982, 2007). Equation (3.7) can be extended to multivariate phenotypes (Lande 1982) and complex life cycles (Barfield et al. 2011), but for my purposes here, the main points can be made by considering only one trait per species with a simple life cycle. I will leave it to the reader and to future analyses to explore more complicated phenotypes and life histories in the contexts I explore here.

Equation (3.7) also identifies another key feature of the dynamics of natural selection that is little appreciated in the general literature—namely, the dynamical equilibria of natural selection (the peaks and nadirs of the fitness surface) also depend on the relative strengths of selection gradients operating separately on the various fitness components for each species. Each term in the summation in equation (3.7) describes how the contribution of that fitness component changes with a modification in the traits and abundances of all the species in the community; the magnitude of the selection gradient associated with a fitness component is the measure of the *strength of selection* on that same fitness component. This implies that the phenotypic trait value that is favored by selection overall will be more influenced by fitness components that experience stronger selection (McPeek 1996a). At an evolutionary equilibrium, whether it is stable or unstable, the various selection gradients must balance, and hence the selection strengths on the various fitness components weighted by the abundances of the interacting species must sum to zero (i.e., the terms in parentheses of equation (3.7) must sum to zero).

This framework also highlights that the members of a community evolve in a coevolutionary context that depends on both the abundance dynamics and trait dynamics of the interacting species. In a theoretical context, the entire system may reach a point equilibrium where all abundances and traits approach a single point in multidimensional abundance–trait space. At this point, each species will have a mean phenotype that gives $\partial W_i / \partial z_i = 0$ as either a fitness maximum or fitness minimum on its adaptive surface that balances the contributions of the various fitness components to overall fitness (Abrams et al. 1993, Abrams and Matsuda 1997a). Our usual notion that natural selection moves populations uphill on the fitness surface implies a fitness maximum, but frequency-dependent selection generated by species interactions can cause stable fitness minima in some cases. Abrams et al. (1993) have illustrated the conditions where a stable fitness minimum for a species will result. A system may also express stable limit cycles in which both population abundances and trait values fluctuate through time (Abrams and Matsuda 1997b, Yoshida et al. 2003), as I will illustrate below.

As we will see, this framework clearly exposes the underlying drivers of the *ecological dynamics of natural selection*. The average fitness of the population changes as the mean trait value in the population alters, but average fitness also changes because the species' abundance as well as the traits and abundances of all the species with which it interacts also all change. Consequently, the fitness topography against which each species evolves may change continuously as species coevolve. Models that focus on genetic dynamics attempt to capture these fitness dynamics in formulations of density and frequency dependence, and various flavors of hard and soft selection (Levene 1953, Dempster 1955, Christiansen 1975, Nagylaki 1992, Charlesworth 1994). However, I think translating the processes to be considered here into those terms only obscures the ecological processes that produce natural selection. Previous analyses that utilize this more mechanistic approach to natural selection and coevolution have illustrated how fitness surfaces change as selection proceeds, but most have focused primarily on the evolutionary outcomes (e.g., Taper and Case 1985; Abrams et al. 1993; Abrams 2000; Abrams and Chen 2002; Abrams 2006; Price and Kirkpatrick 2009; Abrams and Fung 2010a, 2010b). In what follows, I will consider coevolution in many different types of community modules, and I will focus as much on the underlying causes of the ecological dynamics of natural selection as on the ultimate outcome.

Thus, the ecological opportunities available to species will change not only as the overall community structure changes through species additions and deletions (i.e., chapter 2), but also as the phenotypes and abundances of the species filling various community roles change. *An ecological opportunity represents both an ecological role to fill in a community and an evolutionary outcome of adapting to the community.*

TYPES OF TRAITS

The linkage between ecological and evolutionary dynamics is specified by how the traits of an individual interact with its environment to determine its overall fitness. Therefore, this framework also needs a mechanistic description of how the traits of individuals influence the various components of their absolute fitness. Many phenotypic traits of an organism may simultaneously influence its survival, growth, and fecundity. Some traits may be almost universally important; body size comes to mind as one such trait. However, even body size is not so important in every facet of demography and life history for every species (Harmon et al. 2010). Because the ecologically important traits contribute to determining the absolute fitness of the organism, small changes in the value of any one would result in a change in overall fitness. Mathematically, this means that in equation (3.7), $\partial f_{ij}(z_i, \bar{z}_j)/\partial z_i \neq 0$ over much of the possible range of trait values. All aspects of the morphology, physiology, and behavior of an organism can potentially influence its demographic performance, and each ecologically important trait may influence absolute fitness through its simultaneous effects on multiple fitness components.

The effect of a phenotypic trait on a particular fitness component can take many functional forms, depending on the mechanism of the interactions between individuals and populations. However, all can be categorized by two general properties. The first is whether the fitness component changes in the same direction with changes in the trait over its entire range. If a phenotypic change in one direction increases a fitness component over the entire phenotypic range, I will refer to it as a *unidirectional* trait (following Abrams 2000). This trait is experiencing *directional selection* (i.e., $\partial f_{ij}(z_i, \bar{z}_j)/\partial z_i > 0$ or $\partial f_{ij}(z_i, z_j)/\partial z_i < 0$ for all z_i) for this fitness component over its entire phenotypic range, although the *strength of the selection gradient* (i.e., the magnitude of change in the fitness component with a unit change in the trait) may vary. Many different interactions are governed by unidirectional traits. For example, increasing the amount of time spent in its burrow should always decrease the probability of a rabbit being killed by a fox, and the more time spent hunting by the fox should increase the number of rabbits it catches. Escape speed is also a common example; if the prey can run or swim faster than the pursuing predator, the prey will have a greater chance of escape, but if the predator can run faster than the prey, the prey will likely be caught when attacked. Other examples are interactions in which the consumer is gape limited so that it cannot eat a resource above a certain size. Prey morphological defenses such as spines, slime, armor, and shells are also unidirectional traits in interactions with predators.

Alternatively, a trait may have a reversal in the directionality of change in some fitness component with trait change over different ranges of the phenotype; that is, $\partial f_{ij}(z_i, \bar{z}_j)/\partial z_i > 0$ over some phenotypic range, but $\partial f_{ij}(z_i, \bar{z}_j)/\partial z_i < 0$ over another range. Such a trait must, therefore, have either a fitness component maximum or minimum at some trait value (i.e., where $\partial f_{ij}(z_i, \bar{z}_j)/\partial z_i = 0$). Such traits have been termed *bidirectional*, because the fitness component increases with increasing trait values over one range but decreases with increasing trait value over another range (Abrams 2000). If the population's trait distribution includes a fitness component maximum, the population experiences *stabilizing selection* from that component; whereas if the population's trait distribution includes a fitness component minimum, the population experiences *disruptive selection* from that component. If the population's trait distribution does not include the fitness maximum or minimum, it would experience directional selection. For example, consider a bird population feeding on the seeds of a plant that vary in size. Very small and very large seeds will have higher survival than seeds that closely match the sizes that are best manipulated and eaten using the bird's bill. The gill rakers of fish that are used to strain food particles from the water are also most efficient on a particular size of prey. In these cases, the consumer's feeding structure is most efficient on a particular size of resource, and the consumer is less efficient at feeding on resources that are both smaller and larger than this optimal size.

Traits also differ in whether their effect in determining the value of some fitness component does or does not depend on the trait value of another species; these are *dependent* or *independent* traits, respectively. For example, the contribution of swimming speed to determining a damselfly's survival under dragonfly predation cannot be determined without knowing how fast the dragonfly can strike and chase. Likewise, the contribution of size to determining a seed's survival under bird predation is unknown until one also knows the birds' bill sizes. These would both be dependent traits with respect to survival under predation. The contribution to determining the value of some fitness component by an independent trait does not depend on the trait values of other species. Increasing the time spend in a burrow will proportionally increase a rabbit's survival to a similar degree, regardless of the fox's phenotype. This does not mean that the fox's phenotype will have no influence on the rabbit's survival; it only means that the contributions of the rabbit and fox phenotypes to the rabbit's survival can be conceptually and mathematically partitioned in this trait. For dependent traits, this conceptual and mathematical partitioning cannot be done, because the contribution to the fitness component depends on the difference or ratio of the phenotypes of the interacting species (when measured on appropriate scales). Thus, we might expect frequency-dependent selection to be much more likely when dependent traits underlie a species interaction.

These categorizations highlight the dynamical nature of natural selection affecting the traits of interacting species. The relationship between fitness and the phenotype (i.e., the fitness surface as defined by equation (3.6)) is not a static feature of the environment, but rather has a dynamic that depends on both the abundances and traits of other interacting species. When dragonflies are rare, the fitness surface experienced by a damselfly population may have the same fundamental shape as when dragonflies are common; however, the height of the surface will be different in these two cases, because the rate at which dragonflies are attacking damselflies will change with dragonfly abundance. The rate at which fitness changes at a given difference in swimming speed (i.e., the strength of the selection gradient) will also increase with the number of foraging dragonflies. Additionally, the shape of the fitness surface will differ between damselfly populations that face slow versus fast dragonflies. The dynamics of this relationship drive *coevolution* between species; this occurs when an evolutionary response in one species changes the form and intensity of selection on its interaction partner and thereby causes an evolutionary response, which in turn changes the form and intensity of selection on the first species, and so on (Thompson 1994).

Any particular trait may also influence the values of multiple fitness components. Size may affect the survival of a seed in the face of predators, but seed size may also influence the probability that the resulting plant survives the seedling stage of its life history. Increasing speed to chase down fleeing prey may decrease other components of fitness in the dragonfly. These multiple fitness effects may produce *synergies* (change in the trait causes two fitness components to increase or decrease) or *trade-offs* (change in the trait causes one fitness component to increase and the other to decrease) among various fitness components as they contribute to determining the shape of the overall fitness surface (Arnold and Wade 1984b). Also, the categories in which a trait falls (i.e., independent or dependent, unidirectional, or bidirectional) will typically differ among the fitness components it influences. For example, seed size may be a bidirectional-dependent trait with respect to seed predation, but a unidirectional-independent trait with respect to seedling survival.

DYNAMICS OF NATURAL SELECTION
IN A VERY SIMPLE COMMUNITY

With this conceptual framework completed, we now need to actualize the simultaneous dynamics of abundances and traits that result from species interactions. Here, the focus will be on the interactions among consumers and resources to build on the purely ecological analyses presented in chapter 2. Let us begin by considering the interaction between one resource and one consumer. Also, assume that only one trait is ecologically important for each species (z_R for the resource and z_N for the consumer), and these traits influence both their per capita birth and death rates. Furthermore, their coevolutionary dynamics result from the functional response of their interaction, which is influenced by the traits of both species simultaneously. These per capita birth and death rates for each species are the separate components of absolute fitness that will define their evolutionary dynamics.

Using the basic models developed in chapter 2 to describe their population dynamics, the absolute fitnesses of individuals with specified trait values for an interacting consumer and resource species, respectively, are

$$\ln(W_N(z_N)) = \frac{dN_{z_N}}{Ndt} = \frac{ba(\bar{z}_R, z_N)R}{1 + a(\bar{z}_R, z_N)hR} - f(z_N) - gN$$

$$\ln(W_R(z_R)) = \frac{dR_{z_R}}{Rdt} = c(z_R) - dR - \frac{a(z_R, \bar{z}_N)N}{1 + a(z_R, \bar{z}_N)hR} \tag{3.8}$$

(Because only one species is present per trophic level, I will forego subscripts to identify species in this chapter.) In this formulation, the parameters of the model are functions of the traits. Note that the denominator of the resource's functional response is a function of the average trait values of both species; this is because the consumer species overall is satiated primarily by resource individuals with the average trait value. In contrast, the denominator of the consumer's functional response depends on the average trait value of the resource species but the actual trait value of the consumer individual; this is because a consumer individual's level of satiation is based on that individual's trait value (Abrams 2000). When evaluated at the current average trait value for each species, equations (3.8) govern the population dynamics of these species:

$$\frac{dN}{dt} = N\left(\frac{ba(\bar{z}_R, \bar{z}_N)R}{1 + a(\bar{z}_R, \bar{z}_N)hR} - f(\bar{z}_N) - gN\right)$$

$$\frac{dR}{dt} = R\left(c(\bar{z}_R) - dR - \frac{a(\bar{z}_R, \bar{z}_N)N}{1 + a(\bar{z}_R, \bar{z}_N)hR}\right) \tag{3.9}$$

Equations (3.9) also define the landscapes of average fitness against mean trait values and species abundances for these two species when expressed in their per capita forms (e.g., dN/Ndt). The equations governing trait dynamics are then given by substituting equations (3.8) into (3.7):

$$\frac{d\bar{z}_N}{dt} = V_{z_N}\left(\left.\frac{\partial\left(\frac{ba(\bar{z}_R, z_N)R}{1 + a(\bar{z}_R, z_N)hR}\right)}{\partial z_N}\right|_{z_N = \bar{z}_N} + \left.\frac{\partial(-f(z_N) - gN)}{\partial z_N}\right|_{z_N = \bar{z}_N}\right)$$

$$\frac{d\bar{z}_R}{dt} = V_{z_R}\left(\left.\frac{\partial(c(z_R) - dR)}{\partial z_R}\right|_{z_R = \bar{z}_R} + \left.\frac{\partial\left(-\frac{a(z_R, \bar{z}_N)N}{1 + a(z_R, \bar{z}_N)hR}\right)}{\partial z_R}\right|_{z_R = \bar{z}_R}\right) \tag{3.10}$$

The first term in each equation is the strength of the selection gradient on the respective species' birth fitness components, and the second term in each equation is the strength of the selection gradient on their death fitness components. These yield

$$
\begin{aligned}
\frac{d\bar{z}_N}{dt} &= V_{z_N} \left(\frac{bR\dfrac{\partial a(\bar{z}_R, z_N)}{\partial z_N}}{(1+a(\bar{z}_R, z_N)hR)^2}\Bigg|_{z_N = \bar{z}_N} - \frac{\partial f(z_N)}{\partial z_N}\Bigg|_{z_N = \bar{z}_N} \right) \\
\frac{d\bar{z}_R}{dt} &= V_{z_R} \left(\frac{\partial c(z_R)}{\partial z_R}\Bigg|_{z_R = \bar{z}_R} + \frac{N\dfrac{\partial a(z_R, \bar{z}_N)}{\partial z_R}}{(1+a(\bar{z}_R, \bar{z}_N)hR)}\Bigg|_{z_R = \bar{z}_R} \right)
\end{aligned}
\tag{3.11}
$$

Given these equations governing changes in abundances and trait means, all that is left is to specify the functional forms for how the parameters in the model depend on the traits of the species. As intuition would suggest, the different categories of traits have somewhat different effects on abundance and trait dynamics and different capabilities for how species may respond to different types of interactions. Even within a trait category, many different functional forms may be appropriate for different types of traits influencing different fitness components. Moreover, traits may be tied to various combinations of fitness components in different ways. An exhaustive analysis of various trait combinations is impossible to present. Here, I focus on a smaller set of trait combinations and functional forms, highlighting the resulting differences between different trait types to illustrate the major features of adaptive evolution that occurs as a result of species interactions.

Throughout this analysis, I will assume the resource's intrinsic birth rate, $c(z_R)$, and the consumer's intrinsic death rate, $f(z_N)$, are bidirectional-independent traits (fig. 3.1A and B, respectively).

I will use a quadratic function for the intrinsic birth rate of the resource:

$$
c(z_R) = c_0 \left(1 - \gamma(z_R - \tilde{z}_R^c)^2\right).
\tag{3.12}
$$

In this equation, the resource's intrinsic per capita birth rate has a maximum value of c_0 at its intrinsic birth rate optimum of $z_R = \tilde{z}_R^c$ and declines with larger deviations of the trait value from this optimum (fig. 3.1A). The parameter γ mediates the underlying strength of selection on z_R due to the birth fitness component—the rate at which the intrinsic birth rate declines away from $\tilde{z}_R^c$ with change in z_R. Therefore, z_R experiences overall stabilizing selection from the birth fitness component, and the strength of this stabilizing selection increases with increasing γ.

In analogous fashion, the intrinsic per capita death rate of the consumer is assumed to follow a quadratic function,

$$
f(z_N) = f_0 \left(1 + \theta(z_N - \tilde{z}_N^f)^2\right),
\tag{3.13}
$$

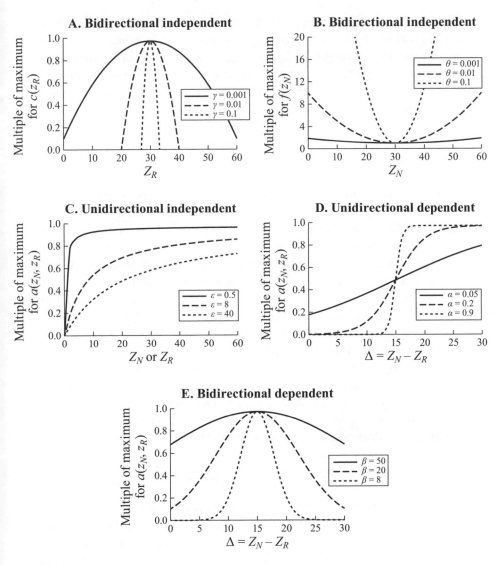

FIGURE 3.1. Functional forms used for the relationships between species trait values and parameters in the models. In each panel, the shape of the function for three different values of the tuning parameter that defines the underlying selection strength is shown. The functions are for the following trait types: (*A*) bidirectional-independent trait used for the resource intrinsic birth rate, (*B*) bidirectional-independent trait used for the consumer and predator intrinsic death rates, (*C*) unidirectional-independent traits defining the attack coefficients, (*D*) unidirectional-dependent traits defining the attack coefficient, and (*E*) bidirectional-dependent traits defining the attack coefficients.

where $z_N = \tilde{z}_N^f$ is the phenotype that minimizes the consumer's intrinsic death rate (i.e., its intrinsic death rate optimum), f_0 is the value at this minimum, and θ mediates the underlying strength of selection on z_N due to the death fitness component—the rate at which predator death rate increases away from $\tilde{z}_N^f$ with change in z_N (fig. 3.1*B*). Therefore, z_N also experiences stabilizing selection from the death fitness component and the strength of stabilizing selection increases with increasing θ.

The attack coefficient describing the per capita rate at which the consumer kills resource individuals is what generates the coevolutionary dynamics in this model. The attack coefficient can assume any of the various types of traits, depending on the specific traits involved in the consumer-resource interaction. In this analysis I consider three of the possible types. I neglect bidirectional-independent traits as a basis for the attack coefficient because trade-offs among fitness components cause the attack coefficient to rarely settle at the optimal value for the attack coefficient, and so the dynamics are in the end quite comparable to unidirectional-independent traits. Therefore, in the independent category, I present results only for unidirectional-independent traits defining attack coefficients.

First, many traits of interacting species have unidirectional-independent effects on their fitnesses and thus on the parameters in the model, and they can take myriad functional forms. For the analyses discussed here, imagine that z_R is a trait such as prey activity that affects the resource's exposure or conspicuousness to its consumers, and z_N is also a trait such as activity in which greater movement in the consumer increases its exposure to resources. Furthermore, assume that the attack coefficient is zero when $z_R = z_N = 0$ and increases asymptotically to a maximum value of a_0 as z_R and z_N increase. An equation with this general form is the Michaelis-Menten equation (Michaelis and Menten 1913),

$$a(z_R, z_N) = a_0 \frac{z_R}{(\varepsilon_R + z_R)} \frac{z_N}{(\varepsilon_N + z_N)}, \tag{3.14}$$

where a_0 is the asymptotic maximum, $z_R/(\varepsilon_R + z_R)$ is the independent effect that the resource trait has on the attack coefficient, and $z_N/(\varepsilon_N + z_N)$ is the independent effect of the consumer's trait on the attack coefficient (fig. 3.1*C*). The parameter ε_R is the underlying strength of selection on the attack coefficient via how fast the attack rate increases with z_R; ε_N does the same for z_N (fig. 3.1*C*).

When the consumer–resource interaction is influenced by traits such as evasion speed in the resource and pursuit speed in the consumer, the resource and consumer have unidirectional-dependent traits as their effects on the attack coefficient. Traits such as these are unidirectional because changes in only one direction increase the fitness contribution of this component for each species, but the difference between the trait values of the interacting species determine the fitness contribution to each. To model this type of trait combination determining the attack coefficient, the difference between the consumer and resource trait values are $\Delta = z_N - z_R$, and I use a logistic function to describe the attack coefficient

$$a(z_R, z_N) = \frac{a_0}{1 + e^{-\alpha\Delta}}, \tag{3.15}$$

where a_0 is again the asymptotic maximum and α defines the steepness of the transition from low to high values around $\Delta = 0$ (fig. 3.1D). In the case of pursuit and evasion speeds, if the resource can run much faster than the consumer, Δ will be a large negative number, and so the attack coefficient will be near zero. In contrast, if the consumer can run substantially faster than the resource, Δ will be a large positive number, and the attack coefficient will be near a_0. For such traits, this interaction will continually apply directional selection for larger trait values to the death fitness component of the resource and to the birth fitness component of the consumer, and the strength of selection will increase with increasing α.

Finally, consider the consumer-resource interaction when the attack coefficient is influenced by *bidirectional-dependent* traits, such as bird bill size and seed size. In this case, the maximal attack coefficient occurs when the traits exactly match one another to give $\Delta = 0$ (when measured on appropriate scales), and declines away from this point in both directions. I use a Gaussian function to model the attack coefficient for these traits of

$$a(z_R, z_N) = a_0\, e^{-\left(\frac{\Delta}{\beta}\right)^2}, \tag{3.16}$$

where a_0 is again the maximum, and β controls the steepness of decline away from $\Delta = 0$ in both directions (i.e., the strength of selection around $\Delta = 0$); see figure 3.1E. The attack coefficient will be near zero if the resource trait is very large or very small relative to the consumer trait (i.e., $|\Delta| \gg 0$), and will be near a_0 when the traits of the two species closely match (i.e., $\Delta \approx 0$). The form of selection experienced by the traits will also differ depending on the magnitude of the difference between them. When the species are similar (i.e., $\Delta \approx 0$), the consumer trait will experience stabilizing selection on its birth fitness component, and the resource trait will experience disruptive selection on its death fitness component. However, if the difference in trait values of the two species is large, they will experience directional selection in the same direction from this interaction on their respective fitness components.

The isoclines for R, N, $\bar{z}_R$, and $\bar{z}_N$ are complicated functions of population abundances and trait values, and so analytical solutions to the various models considered here are not possible. The following results are, therefore, based on extensive numerical simulations. The Matlab® code used for these simulations is available at http://enallagma.com/EvolutionaryCommunityEcology.

Throughout, I will illustrate the underlying mechanics of these models in features that are familiar to empirical ecologists and evolutionary biologists—namely. isoclines in phase portraits that drive abundance dynamics (e.g., fig. 2.2), and fitness surfaces that illustrate the relationship between fitness and phenotype. I do not intend this presentation as an exhaustive exploration of parameter space, but rather I will highlight the key dynamics and the drivers of those dynamics to expose the importance of the interplay between ecological and evolutionary dynamics.

I hope those who are interested in mutualisms will forgive me if I do not explicitly consider the evolution of mutualistic interactions here. Because of the length of this presentation, I am unable to consider that here. However, as I hope chapter 2 illustrated, many mutualistic interactions are really no different than consumer-resource interactions in understanding the properties of invasibility and coexistence. Preliminary analyses indicate that this is true for their evolution as well. I look forward to future analyses of the evolution of mutualisms with explicit trait mechanisms included.

THE DYNAMICS OF COEVOLUTION

When only one resource and one consumer are interacting, the fundamentals of coevolutionary dynamics are similar for all types of traits—independent or dependent, unidirectional or bidirectional—that define the nature of their interaction. Therefore, I will expound the central features of coevolution in this simplest of communities with a bidirectional-dependent trait defining the attack coefficient, as described by equation (3.16). The joint ecological and evolutionary dynamics of the system are then given by the following set of equations:

$$
\begin{aligned}
\frac{dN}{dt} &= N\left(\frac{ba_0 e^{-\left(\frac{\bar{\Delta}}{\beta}\right)^2} R}{1 + a_0 e^{-\left(\frac{\bar{\Delta}}{\beta}\right)^2} hR} - f_0\left(1 + \theta(\bar{z}_N - \tilde{z}_N^f)^2\right) - gN \right) \\
\frac{dR}{dt} &= R\left(c_0\left(1 - \gamma(\bar{z}_R - \tilde{z}_R^c)^2\right) - dR - \frac{a_0 e^{-\left(\frac{\bar{\Delta}}{\beta}\right)^2} N}{1 + a_0 e^{-\left(\frac{\bar{\Delta}}{\beta}\right)^2} hR} \right) \\
\frac{d\bar{z}_N}{dt} &= V_{z_N}\left(-\frac{2ba_0 \bar{\Delta} e^{-\left(\frac{\bar{\Delta}}{\beta}\right)^2} R}{\beta^2\left(1 + a_0 e^{-\left(\frac{\bar{\Delta}}{\beta}\right)^2} hR\right)^2} - 2f_0\theta(\bar{z}_N - \tilde{z}_N^f) \right) \\
\frac{d\bar{z}_R}{dt} &= V_{z_R}\left(-2c_0\gamma(\bar{z}_R - \tilde{z}_R^c) - \frac{2a_0 \bar{\Delta} e^{-\left(\frac{\bar{\Delta}}{\beta}\right)^2} N}{\beta^2\left(1 + a_0 e^{-\left(\frac{\bar{\Delta}}{\beta}\right)^2} hR\right)} \right)
\end{aligned}
\tag{3.17}
$$

This system is too complex to evaluate analytically. The results presented below summarize a thorough exploration of parameter space using numerical simulations.

To develop deeper insights on the interplay between ecological and evolutionary dynamics, first consider the following very simple case. Imagine an island inhabited by a plant species (resource), and a seed-eating bird species (consumer, perhaps a Darwin's finch) invades and feeds on this plant's seeds. Initially, the resource species is adapted to a community lacking the consumer (i.e., it begins with its phenotype at its intrinsic birth optimum, which is $\tilde{z}_R^c = 10.0$ in this case) and is at its demographic equilibrium abundance for this trait value (i.e., $R^* = c(\tilde{z}_R^c)/d = 50$ for the parameters considered). Then the consumer invades this community at low abundance, and both species respond demographically and evolutionarily to one another. The consumer can only invade and establish in the community if it can initially increase in abundance or adapt sufficiently to have a positive population growth rate before it goes extinct. In the scenario considered here, the consumer has a positive population growth rate when it invades, and both species evolve to a stable equilibrium for both abundances and traits (figs. 3.2 and 3.3). (I will return to the second scenario when I discuss speciation in chapter 4.) An animation of the changes in selection on the various fitness components and overall fitness for both species is provided at http://press.princeton.edu/titles /11175.html; figure 3.3 presents snapshots from this animation.

At the start, the resource experiences no selection pressure from the consumer because the consumer's abundance is too low: the resource's death fitness component experiences no selection gradient imposed by the consumer, and so the shape of its overall fitness topography is completely defined by the shape of the fitness topography of its birth fitness component (fig. 3.3A). Thus, initially the resource does not evolve, because the consumer's abundance is too low to inflict enough mortality to impose phenotypic selection on the resource. In contrast, the consumer immediately begins to evolve a higher trait value because of the selection gradient associated with its birth fitness component from eating resource individuals (fig. 3.3B); this causes the attack coefficient to increase, because this decreases the difference in the trait values (i.e., $\bar{\Delta} = \bar{z}_N - \bar{z}_R$) between the two species (i.e., to the left of the leftmost vertical dashed line in fig. 3.2C).

The resource only begins to evolve a higher trait value when the consumer's abundance increases to a level that inflicts substantial mortality (i.e., it creates a significant selection gradient on the resource's death fitness component, as shown in fig. 3.3B) and thus causes the resource's abundance to begin to decline (i.e., at the leftmost vertical dashed line in fig. 3.2A). At this point, the resource evolves rapidly to diverge from the consumer, which causes the former's intrinsic birth rate to decrease and the realized attack coefficient to also decrease rapidly (i.e., between the two vertical dashed lines in fig. 3.2C). The resource's abundance

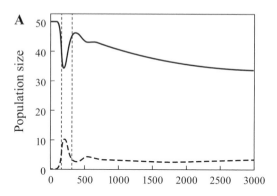

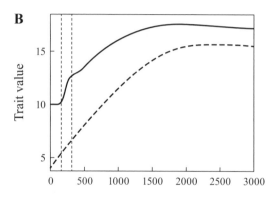

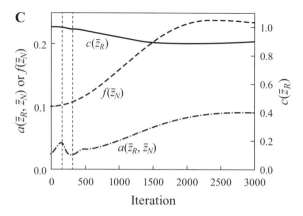

rebounds once its own evolution causes the realized attack coefficient to decrease sufficiently relative to its intrinsic birth rate.

Once this rapid phase of divergence is complete (i.e., the right vertical dashed line in fig. 3.2), the resource's and consumer's trait values continue to increase at a decelerating rate until they reach a stable equilibrium point in both abundances and trait values. Their evolution causes the resource's realized intrinsic birth rate $(c(\bar{z}_R))$ to decline, because the resource evolves away from its intrinsic birth optimum $(\bar{z}_R^c)$, and also causes the consumer's realized intrinsic death rate $(f(\bar{z}_N))$ to increase, because the consumer evolves away from its intrinsic birth optimum $(\bar{z}_N^f)$ (fig. 3.2C). The attack coefficient $(a(\bar{z}_R, \bar{z}_N))$ also increases as a result of this coevolution because the difference between the consumer's and resource's trait values decreases (fig. 3.2C). The positions of the abundance isoclines change as the realized parameters of the system evolve (fig. 3.2C), and the system eventually has abundance isoclines as in figure 2.2F at this stable equilibrium.

At the equilibrium, the consumer is at a local fitness maximum, but in this case the resource is at a local fitness minimum (fig. 3.3C). Because both species are at their demographic equilibria, the values of the overall fitness curves at the average phenotypes are both zero (remember that in this modeling framework, average fitness is measured on a log scale; fig. 3.3C) because their birth and death fitness components are equal in magnitude but opposite in sign (i.e., the values of the fitness component curves at the average phenotype sum to zero for each species; fig. 3.3C). In addition, each species is at an overall fitness optimum because the selection gradients on their underlying fitness components balance. The resource experiences directional selection for decreasing its trait value due to selection on its birth fitness component, but this is balanced by directional selection of the same magnitude to increase its trait value via its death fitness component (fig. 3.3C). Likewise, the consumer experiences directional selection for increasing its trait value due to selection on its birth fitness component, but this is balanced by directional selection of the same magnitude to decrease its trait value due to selection on its death fitness component (fig. 3.3C).

FIGURE 3.2. Abundance and trait dynamics of a consumer-resource coevolution. In this example, the resource is initially adapted to an environment lacking the consumer, and the consumer is then introduced at low abundance. The panels show the dynamics of (A) population sizes, (B) trait values, and (C) realized values of the attack coefficient (a, dot-dash line), resource intrinsic birth rate (c, solid line) and consumer intrinsic death rate (f, dashed line). In panels A and B, the solid lines identify the resource values, and the dashed lines identify the consumer values. The parameters used for this example are $c_0 = 1.0, d = 0.02, a_0 = 0.1, b = 0.1, h = 0.1, f_0 = 0.1, g = 0.0, \beta = 5.0, \gamma = 0.002, \theta = 0.01, \bar{z}_R^c = 10.0, \bar{z}_N^f = 4.0$, and $V_{z_R} = V_{z_N} = 0.2$. (This figure is redrawn from figure 2 of McPeek 2017, with permission.)

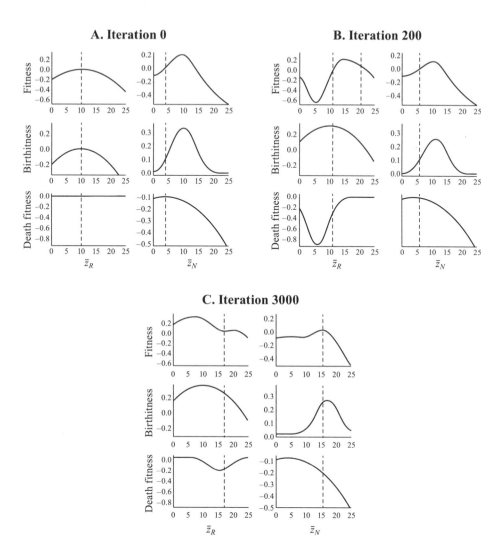

FIGURE 3.3. Fitness surfaces for the resource and consumer at various points in the scenario illustrated in figure 3.2. The final equilibrium for the example is given in figure 3.2. The groups of panels show the overall fitness surface, and the birth and death fitness component surfaces, for each species at iterations (*A*) 0, (*B*) 200, and (*C*) 3000. The *vertical dashed lines* in each panel identify the trait value of that species at that point. Note that the consumer is at a stable fitness maximum and the resource is at a stable fitness minimum at the equilibrium in iteration 3000. The parameters are given in figure 3.2. (This figure is redrawn from figure 2 of McPeek 2017, with permission.)

The resource is maintained at this local fitness minimum because if its average trait value moves in either direction, changes in the consumer's abundance will alter the selection gradient on the resource's death fitness component to move it back to this trait optimum (see also Abrams et al. 1993, Abrams and Matsuda 1997a). If the resource's trait mean is perturbed to a lower value away from this equilibrium, the consumer's abundance immediately increases because of the resulting increase in the attack coefficient, which increases the steepness of the selection gradient associated with the resource's death fitness component from predation; this then alters the shape of the resource's overall fitness surface to increase its trait value. In contrast, if the resource's trait mean is perturbed to a higher value, the consumer's abundance immediately decreases because of the resulting decrease in the attack coefficient, which reduces the steepness of the selection gradient on the resource's death fitness component, and thus changes the shape of the resource's overall fitness surface to decrease its trait value. Although the resource is experiencing disruptive selection at this minimum fitness optimum, any evolutionary response by the resource away from this equilibrium will alter its overall fitness surface to bring the population back to the equilibrium due to the ecological response of consumer abundance. Likewise, opposing directional selection pressures impinging on the consumer's various fitness components maintain it at its fitness optimum (fig. 3.3C).

Remember that the selection gradient associated with a fitness component is the slope of the line tangent to the fitness surface at the average phenotype. Thus, the slopes of the tangent lines on the birth and death fitness components (i.e., the terms inside the parentheses in equation (3.10)) are equal in magnitude and opposite in sign for each species at this evolutionary equilibrium. In this case, as probably in most cases in nature, a fitness optimum on the overall fitness surface results from a balancing of various underlying fitness components (Travis 1989). Also, note in equations (3.17) that the selection gradient of the resource's death fitness component depends on the abundance and average trait value of the consumer, and the selection gradient of the consumer's birth fitness component depends on the abundance and average trait value of the resource. *The fitness components of each species change in response to abundance and trait changes in the other species, and both the various fitness components and the various selection gradients must balance in each species at the demographic and evolutionary equilibrium.*

Throughout the course of this coevolution, the fitness surfaces for the two species are not static—their shapes change because of changes in both abundances and traits—and the traits of the two species closely follow the changing positions of the optimal phenotypes on the overall fitness surfaces (fig. 3.3; see also the animation of the fitness component surfaces in this figure at http://press .princeton.edu/titles/11175.html). Changes in the optimal phenotype for both species are driven by how changes in their abundances and traits determine the mortality inflicted on the resource by the consumer (the resource's death fitness component) and how this mortality translates into consumer births (the consumer's birth fitness component). Also, the same phenotype $(\tilde{z}_R^c)$ gives the highest birth fitness component value to the resource throughout, but the fitness value at this fitness component optimum changes because of density dependence in birth rate, and the same is true for the death fitness component of the consumer.

Trait Cycling

Another interesting situation that illuminates the ecological drivers of trait evolution is when the entire system enters a stable limit cycle or chaotic cycle in which both abundances and traits change continuously over time (Hochberg and Holt 1995, Abrams and Matsuda 1997b, Abrams 2000, Yoshida et al. 2003). Such a situation is illustrated in figure 3.4 when bidirectional-dependent traits define the attack coefficient. Trait cycling occurs in areas of parameter space with a number of specific features: the resource's maximum intrinsic birth rate (c_0) and the maximum attack coefficient (a_0) are relatively high, the underlying strengths of selection gradients on the resource's intrinsic birth rate (γ) and consumer's intrinsic death rate (θ) are weak relative to the underlying selection strength on the attack coefficient (β), and the handling time of the consumer (h) is zero or relatively low. In other words, cycling occurs when the interaction is a greater determinant of the shape of the overall fitness surfaces of both species relative to other selection pressures, and when the demographic response of the consumer is not substantially damped by satiation when the resources are abundant.

In contrast to trait cycling, the species' abundances cycle in areas of parameter space where the abundance isoclines cross in ways that cause dynamic instability (see chapter 2). These areas of parameter space are generally characterized by the consumer having such a high handling time that the consumer isocline crosses the resource isocline to the left of its apex (Rosenzweig and MacArthur 1963, Rosenzweig 1969), or by particular community module configurations (Holt and Polis 1997, Křivan and Diehl 2005, Tanabe and Namba 2005). The species' traits may evolve into areas where abundance limit cycles occur, but once there the traits do not evolve appreciably. Consequently, the cycles are driven exclusively by the dynamical properties of abundance regulation given the traits of the interacting species and the resulting parameters.

Conversely, if traits cycle, abundances always cycle as well. The dynamics of this system is governed by a set of four isoclines in the four-dimensional joint abundance/trait space of equation (3.17). The shapes of the four isoclines near where they cross to form a local equilibrium is what fosters the cycling. Because the trajectory of the cycle causes the system's position to change on all four axes at once, the abundance isoclines (e.g., fig. 2.2) appear to change shape when considered in isolation, as do the trait isoclines. The trait isoclines have interesting shapes, but because we almost exclusively consider the dynamics of evolution in the context of the shapes of the fitness surfaces, I will not consider the trait isocline shapes here.

When trait cycling occurs with bidirectional-dependent traits defining the attack coefficient, the resource continually cycles through evolving away from the consumer and then evolving toward the consumer, while the consumer continually chases the resource in trait space (fig. 3.4B). This coevolutionary chase can be understood by the continual changes in the two species' fitness surfaces (fig. 3.4D–G). As a result, their abundances also cycle (fig. 3.4A) because the realized demographic parameters are continually changing (fig. 3.4C). (In this description, I will focus on the evolutionary aspects of the cycle. I will leave it for the reader to discern how this evolution drives the population dynamics seen in figure 3.4A.)

The most important evolutionary insights are made by considering selection associated with the various fitness components of the two species through one of these trait cycles. Figure 3.4D–G shows the fitness surfaces for the two species at four different points in the trait cycle. (An animation of figure 3.4 is also available at http://press.princeton.edu/titles/11175.html.) At iteration 2500 (fig. 3.4D), the resource is near its maximum trait value in the cycle, and the consumer is evolving toward the resource (fig. 3.4B). Here, the resource is very near the local adaptive peak in overall fitness at the high trait value because the strengths of the selection gradients on its birth and death fitness components are strong and nearly balance. Note that the resource's overall fitness surface at this point has a second adaptive peak that is substantially below its intrinsic birth optimum, and the bottom of the fitness valley between the two peaks is very near the consumer's trait value. At this point, the consumer continues to evolve toward the resource because the positive selection gradient on its birth fitness component is much higher than the negative selection gradient on its death fitness component.

As the consumer continues to evolve toward the resource (e.g., iteration 2650; fig. 3.4E), the height of the adaptive peak on which the resource resides declines to the point where this is no longer an adaptive peak at all. This is caused by the shift in the balance of selection gradients on the resource's birth and death fitness components. Here, a very interesting evolutionary dynamic has happening: as the consumer evolves toward the resource, the magnitude of the attack coefficient increases as a result, and thus the ecological interaction strength between them (the strength of the selection gradient on the resource's death fitness component

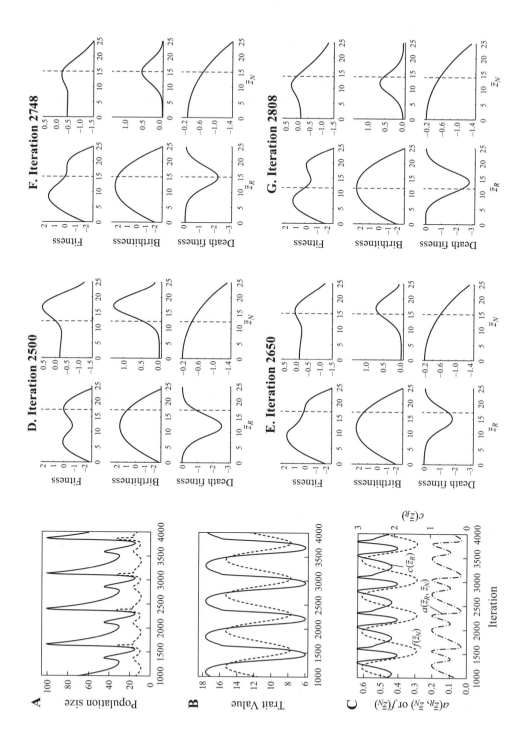

caused by predation from the consumer) actually decreases. Stronger selection now impinges on the resource's birth fitness component, and the fitness peak on which it resided has now disappeared. Consequently, the resource reverses its evolutionary course and begins to evolve toward the consumer, even though this increases the realized attack coefficient and the consumer's abundance still further, and so increases its death rate due to predation (fig. 3.4C).

At iteration 2748, the resource's and consumer's trait values match (fig. 3.4F). Now the selection gradients resulting from the interaction between the species are zero; even though the attack coefficient is at its maximum value and the consumer is at its maximal abundance, and so is imposing maximal death rate due to predation (see equation (3.16) and fig. 3.4C), no selection impinges on either species because of the trophic interaction between them. Overall selection to decrease the resource's mean trait value is due solely to the negative selection gradient to increase its intrinsic birth rate, and overall selection to decrease the consumer's mean trait value is due solely to the negative selection gradient to decrease its intrinsic death rate.

As the resource's mean trait value passes the consumer's, natural selection due to predation reappears and strengthens again on both species to decrease both their trait values, whereas the selection gradients on their respective intrinsic birth and death rates decrease in magnitude. At iteration 2808, the resource reaches its intrinsic birth optimum $(\tilde{z}^c_R)$, and so its intrinsic birth rate is maximized (see equation (3.12) and fig. 3.4C), but the selection gradient on its birth fitness component is zero (fig. 3.4G). By this time, the resource's overall fitness surface again has two fitness peaks, but overall selection is now pushing the resource toward the lower peak. As the resource's trait value passes $\tilde{z}^c_R$ and continues to decrease, the selection gradient on its birth fitness component increases to pull it back toward $\tilde{z}^c_R$, and the selection gradient on its death fitness component decreases but remains larger than the birth selection gradient, which pushes the resource to lower trait values. As a result, the height of the fitness peak to which it is evolving begins to decline;

FIGURE 3.4. An example of trait cycling when bidirectional-dependent traits define the attack coefficient between the consumer and resource. The panels show the dynamics of (A) population sizes, (B) trait values, and (C) realized values of the attack coefficient (a, dot-dash line), resource intrinsic birth rate (c, solid line) and consumer intrinsic death rate (f, dashed line). In panels A and B, the solid lines identify the resource values, and the dashed lines identify the consumer values. Panels D–G show these overall fitness surfaces and the birth and death fitness component surfaces at specific iterations in the simulation. The parameters used for this example are $c_0 = 3.0$, $d = 0.02$, $a_0 = 0.2$, $b = 0.1$, $h = 0.0$, $f_0 = 0.2$, $g = 0.0$, $\beta = 5.0$, $\gamma = 0.01$, $\theta = 0.01$, $\tilde{z}^c_R = 12.0$, $\tilde{z}^f_N = 1.0$, and $V_{z_R} = V_{z_N} = 0.2$. (This figure is redrawn from figure 2 of McPeek 2017, with permission.)

it continues to decline as the consumer continues to evolve to chase the resource in phenotype space. The resource's trait value reaches its lowest value when the birth and death selection gradients again balance. At this point, the lower fitness peak has decreased in height to such a degree that it is no longer a peak (analogous but reversed to the overall fitness surface in iteration 2650), and overall selection then causes the resource's trait value to increase and again evolve toward that of the consumer. The other half of the trait cycle is then simply this entire series of relationships in reverse. Note that the resource evolves a more extreme phenotype than the consumer on both extremes of the cycle (fig. 3.4B).

With this cycling, the resource is undergoing continual shifts between two adaptive peaks, while the consumer is chasing a continuously moving single fitness peak (fig. 3.4 and http://press.princeton.edu/titles/11175.html). These peak shifts occur exclusively because of the dynamics of the overall fitness surfaces for the two species caused by the changing strengths of selection gradients associated with their various fitness components. The resource is cycling between extreme phenotypes that give it relatively low birth and death rates, but it must pass through a period of high birth and death rates to traverse from one to the other. The magnitudes of selection gradients on its two fitness components follow a countervailing cycle, both being steep at the trait extremes of the cycle and relatively shallow between.

When unidirectional-dependent traits define the attack coefficient, trait cycling occurs over a smaller total area of parameter space but under similar parameter relationships. Trait cycling in this case also takes a quite different form. For this situation, equation (3.15) is used for the attack coefficient in equations (3.9) and (3.10) to give

$$
\begin{aligned}
\frac{dN}{dt} &= N\left(\frac{\dfrac{ba_0 R}{1+e^{-a\bar{\Delta}}}}{1+\dfrac{a_0 hR}{1+e^{-a\bar{\Delta}}}} - f_0\left(1+\theta(\bar{z}_N - \tilde{z}_N^f)^2\right) - gN \right) \\[2mm]
\frac{dR}{dt} &= R\left(c_0\left(1-\gamma(\bar{z}_R - \tilde{z}_R^c)^2\right) - dR - \dfrac{\dfrac{a_0 N}{1+e^{-a\bar{\Delta}}}}{1+\dfrac{a_0 hR}{1+e^{-a\bar{\Delta}}}} \right) \\[2mm]
\frac{d\bar{z}_N}{dt} &= V_{z_N}\left(\dfrac{ba_0\alpha e^{-a\bar{\Delta}}R}{(1+e^{-a\bar{\Delta}})^2\left(1+\dfrac{a_0 hR}{1+e^{-a\bar{\Delta}}}\right)^2} - 2f_0\theta(\bar{z}_N - \tilde{z}_N^f) \right) \\[2mm]
\frac{d\bar{z}_R}{dt} &= V_{z_R}\left(-2c_0\gamma(\bar{z}_R - \tilde{z}_R^c) + \dfrac{2a_0\alpha e^{-a\bar{\Delta}}N}{(1+e^{-a\bar{\Delta}})\left(1+\dfrac{a_0 hR}{1+e^{-a\bar{\Delta}}}\right)} \right)
\end{aligned}
\qquad (3.18)
$$

Generally, when trait cycling occurs with unidirectional-dependent traits defining the interaction between the two species, the consumer evolves to have a higher trait value than the resource (fig. 3.5B). The resource is again undergoing shifts between two alternative adaptive peaks, and the consumer is chasing a single moving peak. (An animation of figure 3.5 is also available at http://press .princeton.edu/titles/11175.html.) For much of the cycle, the selection gradients on the resource's death fitness component and the consumer's birth fitness component are nearly zero because the consumer's trait value is much higher than the resource's trait value (fig. 3.5). This means that overall selection on the resource favors trait values very near its intrinsic birth optimum ($\tilde{z}_R^c$) (fig. 3.5D).

As the predator evolves lower trait values, the steep transition zone of change in the attack coefficient (fig. 3.1D) and thus in the resource's death fitness component moves toward the resource's trait value (fig. 3.5D–E). Once the consumer evolves to a low enough value to increase the selection gradient on the resource's death fitness component, the overall fitness peak on which the resource resides temporarily disappears in favor of higher trait values in the resource, and thus it rapidly transitions to the alternative adaptive peak (fig. 3.5E). This evolutionary response that decreases the attack coefficient causes a spike in the resource's abundance even though its intrinsic birth rate declines sharply (fig. 3.5A–C). The consumer then rapidly reverses its evolutionary course to evolve higher trait values due to the change in the selection gradient on its birth fitness component (fig. 3.5E–F). In addition, this response to selection in the consumer causes its abundance to decline because of the increase in its intrinsic death rate (fig. 3.5A–C). As the consumer evolves higher trait values, the selection gradient on the resource's birth fitness component transitions back to the flat part of the fitness component surface, and overall selection then favors the resource to evolve back toward the reformed lower adaptive peak in overall fitness near its intrinsic birth optimum (fig. 3.5F–G).

Although the resource is again shifting between alternative adaptive peaks, the two peaks in this case represent very different demographics. One fitness peak is very near its intrinsic birth optimum; the resource has its highest birth rate but also a high death rate at this peak, and the selection gradients on both fitness components are relatively weak. The other fitness peak is far from its intrinsic birth optimum; here it has its lowest birth rate and lowest death rate, but the selection gradients on both fitness components are steepest. Thus, trait cycling with unidirectional-dependent traits defining the attack coefficient moves the resource between two extreme phenotypes that represent a trade-off of fitness components. In contrast, remember that cycling with bidirectional-dependent traits defining the attack coefficient moves the resource between alternative phenotypes that both have low birth and death rates.

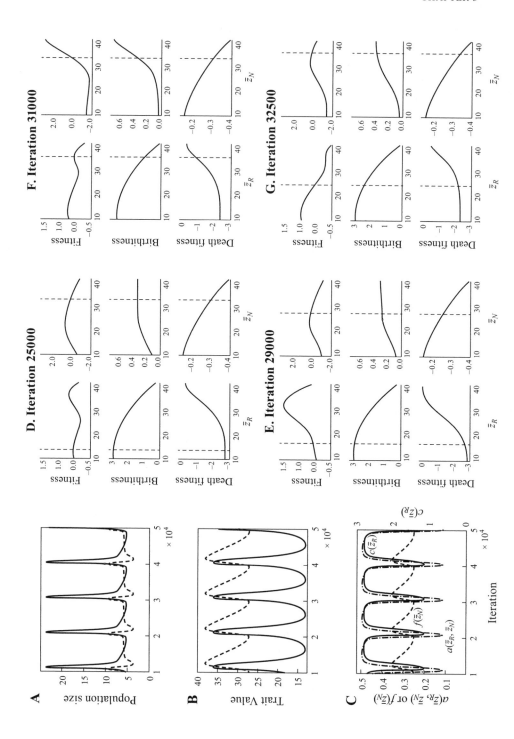

Trait cycling again typically occurs only in areas of parameter space where the underlying selection strengths on the resource's intrinsic birth rate and the consumer's intrinsic death rate are weak relative to the underlying selection strength on the attack coefficient (α). For example, in this case with unidirectional-dependent traits underlying the attack coefficient, trait cycling does not occur if the underlying selection strength on the attack coefficient is weak (i.e., small α) relative to the underlying selection strengths on the intrinsic resource birth rates (i.e., large γ) or the intrinsic consumer death rate (i.e., large θ) (cf. fig 3.6A–C to 3.6D–F). Likewise, if γ and θ are increased relative to α, cycling is also stopped (Fig. 3.6G–I). Note that in this latter case, the trait values favored by selection are closer to the intrinsic birth and death optima for the respective species than in the case where the attack coefficient is the weaker factor (cf. figs. 3.6B and H). However, if α is then increased again relative to γ and θ, trait cycling occurs again (fig. 3.6J–L).

If the traits of the two species have unidirectional-independent effects on the attack coefficient, as when defined by equation (3.14), trait cycling occurs only in a very narrow range of parameter space, and it occurs differently from dependent traits. Substituting equation (3.14) into equations (3.9) and (3.10) yields

$$\frac{dN}{dt} = N\left(\frac{\dfrac{ba_0\,\bar{z}_R\,\bar{z}_N\,R}{(\varepsilon_R+\bar{z}_R)(\varepsilon_N+\bar{z}_N)}}{1+\dfrac{a_0\,\bar{z}_R\,\bar{z}_N\,hR}{(\varepsilon_R+\bar{z}_R)(\varepsilon_N+\bar{z}_N)}} - f_0\left(1+\theta(\bar{z}_N-\bar{z}_N^f)^2\right) - gN \right)$$

$$\frac{dR}{dt} = R\left(c_0\left(1-\gamma(\bar{z}_R-\bar{z}_R^c)^2\right) - dR - \frac{\dfrac{a_0\,\bar{z}_R\,\bar{z}_N\,N}{(\varepsilon_R+\bar{z}_R)(\varepsilon_N+\bar{z}_N)}}{1+\dfrac{a_0\,\bar{z}_R\,\bar{z}_N\,hR}{(\varepsilon_R+\bar{z}_R)(\varepsilon_N+\bar{z}_N)}} \right)$$

$$\frac{d\bar{z}_N}{dt} = V_{z_N}\left(\frac{\dfrac{ba_0\,\varepsilon_N\,\bar{z}_R\,R}{(\varepsilon_R+\bar{z}_R)(\varepsilon_N+\bar{z}_N)^2}}{\left(1+\dfrac{a_0\,\bar{z}_R\,\bar{z}_N\,hR}{(\varepsilon_R+\bar{z}_R)(\varepsilon_N+\bar{z}_N)}\right)^2} - 2f_0\theta(\bar{z}_N-\bar{z}_N^f) \right) \qquad (3.19)$$

$$\frac{d\bar{z}_R}{dt} = V_{z_R}\left(-2c_0\gamma(\bar{z}_R-\bar{z}_R^c) + \frac{\dfrac{a_0\,\varepsilon_R\,\bar{z}_N\,N}{(\varepsilon_R+\bar{z}_R)^2(\varepsilon_N+\bar{z}_N)}}{1+\dfrac{a_0\,\bar{z}_R\,\bar{z}_N\,hR}{(\varepsilon_R+\bar{z}_R)(\varepsilon_N+\bar{z}_N)}} \right)$$

FIGURE 3.5. An example of trait cycling when unidirectional-dependent traits define the attack coefficient between the consumer and resource. The panels are as described in figure 3.4. The parameters used for this example are $c_0 = 3.0$, $d = 0.02$, $a_0 = 0.5$, $b = 0.1$, $h = 0.0$, $f_0 = 0.15$, $g = 0.0$, $\alpha = 0.25$, $\gamma = 0.001$, $\theta = 0.001$, $\bar{z}_R^c = 12.0$, $\bar{z}_N^f = 1.0$, and $V_{z_R} = V_{z_N} = 0.2$. (This figure is redrawn from figure 5 of McPeek 2017, with permission.)

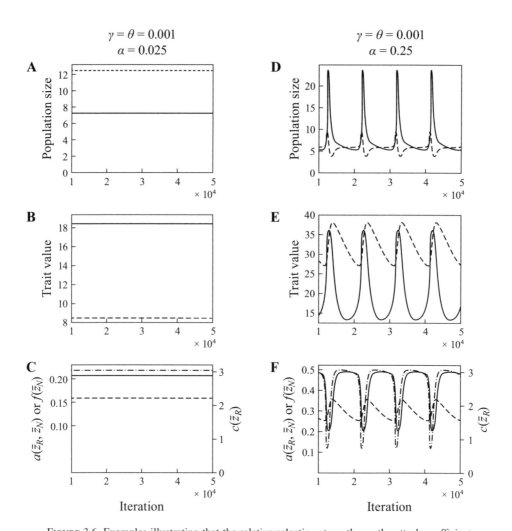

FIGURE 3.6. Examples illustrating that the relative selection strengths on the attack coefficient and the intrinsic resource birth and consumer death rates influence the prevalence of trait cycling. Each column of panels shows the results of a simulation with specific values for the parameters defining the underlying selection strengths on these three fitness components (specific values are given *above* each column). The *top row* of panels gives the dynamics of population sizes, and the *middle row* of panels gives the dynamics of the trait values. In these panels, the *solid lines*

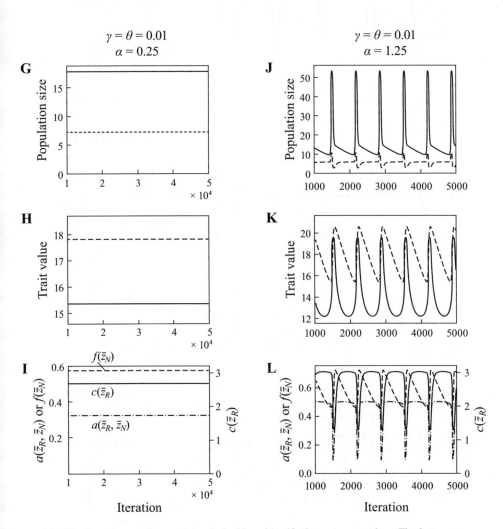

identify the resource values, and the *dashed lines* identify the consumer values. The *bottom row* of panels gives the dynamics of the realized values of the attack coefficient (*a, dot-dash* line), resource intrinsic birth rate (*c, solid* line), and consumer intrinsic death rate (*f, dashed* line). All other parameters are as given in figure 3.5. (This figure is redrawn from figure 6 of McPeek 2017, with permission.)

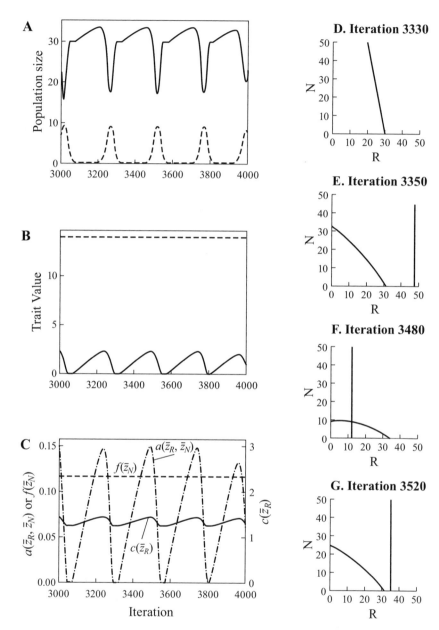

FIGURE 3.7. An example of trait cycling when unidirectional-independent traits define the attack coefficient between the consumer and resource. Panels *A–C* are as described in figure 3.4. Panels *D–G* show the abundance isoclines at specific iterations of the simulation. The parameters used for this example are $c_0 = 2.0$, $d = 0.04$, $a_0 = 3.5$, $b = 0.1$, $h = 0.3$, $f_0 = 0.1$, $g = 0.0$, $\varepsilon_R = 20.0$, $\varepsilon_N = 20.0$, $\gamma = 0.001$, $\theta = 0.001$, $\bar{z}_R^c = 20.0$, $\bar{z}_N^f = 1.0$, and $V_{z_R} = V_{z_N} = 0.2$.

The first striking difference in trait cycling when independent traits are involved is that only the trait of the resource cycles (fig. 3.7B). (An animation of figure 3.7 is also available at http://press.princeton.edu/titles/11175.html.) Moreover, what drives the resource's trait cycles is the cycling in the abundance of the consumer. However, the consumer's abundance cycles are caused by changes in the positions of the abundance isoclines due to the evolution of the resource, and not from inherent instability caused by the shapes of the abundance isoclines.

When the resource is at its lowest trait value in the cycle, the attack coefficient is at a minimum. In the cycling system depicted in figure 3.7, the minimum trait value of the resource in the cycle is zero, which in this case means the attack coefficient is zero. At this point in the cycle, with $a(\bar{z}_R, \bar{z}_N) = 0$ (fig. 3.7C), the resource's abundance isocline is parallel to the N-axis, and the consumer's abundance isocline is at $+\infty$ on the R-axis. Thus, the consumer is at that instant being driven extinct. As a result, the resource increases to its equilibrium abundance and evolves higher trait values to move toward its intrinsic birth optimum, because mortality from the consumer is very low. However, this increase in the resource's trait value increases the attack coefficient, which moves the consumer's isocline to lower values of the R-axis (fig. 3.7E). As the resource continues to evolve higher trait values, eventually the consumer's isocline moves to lower values on the R-axis so that it crosses the resource isocline, and the consumer then increases in abundance (fig. 3.7F). When the resource reaches its maximum trait value in the cycle, the attack coefficient is at its maximum, and consequently the predator isocline is at its lowest value on the R-axis (fig. 3.7F). At this point, the consumer has increased sufficiently in abundance to alter the overall fitness surface of the resource to favor lower trait values. The attack coefficient plummets as a result (fig. 3.7C), the consumer's abundance isocline then moves toward $+\infty$ on the R-axis, its abundance plummets, and the cycle begins again. All this merely causes a reversal in the direction of selection on the resource's trait, and not shifts between alternative adaptive peaks.

COEVOLUTION IN A SIMPLE ECOLOGICAL SYSTEM

Trait cycles show strikingly how the balance among selection gradients on underlying fitness components determine the directionality of trait change in both species, the dynamic nature of fitness and natural selection, and the coupling of abundance dynamics and coevolution for the two species. However, in much of parameter space, stable equilibria for the entire system result. The same types of dynamics occur as the system proceeds to a stable equilibrium, and the same balances are struck at these equilibria.

The outcome of coevolution between these two species depends on the ecological context in which their interaction takes place and on the underlying performance relationships between the traits and the various fitness components. For example, these two species may interact in various locations across the landscape (i.e., in different communities) that differ in many ways. One location may have high nutrient concentrations and good water availability in the soil, whereas another location may have nutrient-poor soils and low water availability. A plant species in these different areas would have a high intrinsic birth rate (c_0) in the community in the first location, but a low intrinsic birth rate in the second. The locations may also differ in conditions (e.g., temperature) that cause different levels of stress for the consumer or differ in the availability of other essential resources (e.g., water) that the consumer needs, and these differences may cause differences in the consumer's minimum intrinsic death rate (f_0) among locations. The maximum value of the attack coefficient (a_0) may also differ among communities because of environmental differences that cause prey to be more easily recognized or captured. For example, differences in turbidity would influence the ability of aquatic predators to see their prey, and different lakes may thus have different maximum possible values of attack coefficient. Locations may differ in structural complexity, and this also affects the ability of predators to capture prey. Alternatively, the maximum attack coefficient may take on different values because of other traits of the consumers and resources not being modeled here (e.g., visual acuity of the consumer); these also influence the likelihood of a prey being captured. These basic parameters reflect the abiotic features of the environment in which this species interaction occurs and other properties of the species' phenotypes not being modeled.

In this section, I explore how differences in these parameters, which will reflect differences in the ecological background of the community and other intrinsic properties of the species, affect the outcome of coevolution for these two species. I primarily present results for bidirectional-dependent traits underlying the attack coefficient, since parameter effects are generally the same across the various types of traits. However, I point out where discrepancies arise with other trait types. This is also not meant to be an exhaustive analysis of parameter space, but rather highlights the major trends.

First, consider the outcome of coevolution in communities that differ in productivity to cause differences in the resource's maximum intrinsic birth rate (c_0); see figure 3.8A–B. In a community where the c_0 is very low (e.g., <2.8 for the parameters considered in figure 3.8A–B), the resource is not abundant enough to support a consumer population, regardless of whether the consumer can evolve or not. In other words, the consumer is incapable of evolving to satisfy its invasibility criterion (i.e., equation (2.3)). As a result, the resource evolves to its intrinsic birth optimum (i.e., $\tilde{z}_R^c = 12.0$ in fig. 3.8A–B), and its abundance is $R^* = c(\tilde{z}_R^c)/d$.

Bidirectional dependent traits

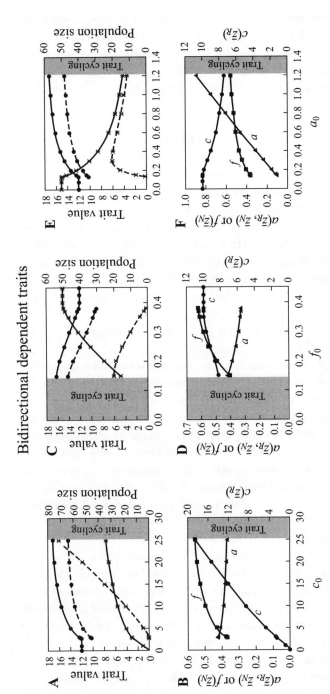

FIGURE 3.8. Effects of different values of (A–B) the resource's maximum intrinsic birth rate (c_0), (A–B) the resource's maximum intrinsic birth rate (f_0), and (E–F) the maximum attack coefficient (a_0) on the dynamics of coevolution between a resource and consumer with bidirectional-dependent traits defining the attack coefficients. Symbols identify equilibrium values for specific parameter combinations. The *top row* of panels shows the trait values (*filled circles*) and population sizes (×) for the resource (*solid lines*) and consumer (*dashed lines*). The *bottom row* of panels shows the realized values of the attack coefficient (*a, triangles*), resource intrinsic birth rate (*c, circles*) and consumer intrinsic death rate (*f, squares*). Parameter areas where trait cycling and population cycling occur are identified. Other parameters used in all these simulations are $c_0 = 10.0$, $d = 0.2$, $a_0 = 0.5$, $b = 0.1$, $h = 0.1$, $f_0 = 0.2$, $g = 0.0$, $\beta = 5.0$, $\gamma = 0.01$, $\theta = 0.001$, $\tilde{z}_R^c = 12.0$, $\tilde{z}_N^c = 1.0$, and $V_{z_R} = V_{z_N} = 0.2$. (Panels A, C, and E are redrawn from figure 3 of McPeek 2017, with permission).

In communities in which the resource has a maximum intrinsic birth rate above this critical value (e.g., >2.8 for the parameters considered in fig. 3.8A–B), the consumer can evolve to satisfy its invasibility criterion and thus support a local population. In this parameter range, in communities with higher values of c_0, the resource evolves to a trait value farther from its intrinsic birth optimum, and the consumer follows and then equilibrates at a trait value farther from its intrinsic death optimum (fig. 3.8A–B).

This latter result is caused by the ecological feedbacks between these two species. Larger values of c_0 caused the consumer to equilibrate at a higher abundance (chapter 2), which in turn causes a steeper selection gradient on the resource's death fitness component. As a result, the resource strikes the balance between the selection gradients of its two fitness components farther from its intrinsic birth optimum ($\tilde{z}_R^c$). In addition, the higher resource productivity increases the consumer's realized birth rate, which allows it to support a population at a higher realized death rate (fig. 3.8B); this in turn permits it to strike the balance between the selection gradients of its fitness components farther from its intrinsic death optimum ($\tilde{z}_N^f$). Thus, the productivity of the environment affecting the resource only indirectly influences which trait value is favored for it by determining the consumer's abundance. Yet again, we cannot understand the outcome of coevolution without understanding the drivers of species abundances.

A similar line of explanation holds when comparing communities that develop in locations with different environmental conditions that cause differences in the consumer's minimum intrinsic death rate (f_0) (fig. 3.8C–D). If f_0 is too high in a particular location, again the consumer is incapable of evolving to support a population (e.g., $f_0 > 0.39$ for the parameters considered in fig. 3.8C–D). Below this point, the consumer will evolve farther from its intrinsic death optimum in communities in which it has a lower f_0 value, which also forces the resource to evolve farther from its intrinsic birth optimum. A lower value of f_0 results in a lower intrinsic death rate for the consumer at a given trait value (fig. 3.8D), which results in a larger consumer abundance (fig 3.8C); this in turn causes a steeper selection gradient on the resource's death fitness component.

Likewise, the outcome of coevolution will differ between communities with environmental differences that would affect the maximum value of the attack coefficient (a_0) (fig. 3.8E–F). Here again, if the maximum attack coefficient is too low, the consumer will not be able to support a population, even if it can evolve (e.g., $a_0 < 0.135$ for the parameters considered in fig. 3.8E–F). Because a higher value of a_0 will directly create steeper selection gradients on both the resource's death fitness component and the consumer's birth fitness component, both species will evolve to trait values that are farther from their respective intrinsic optimal phenotypes in communities where a_0 is larger (fig. 3.8E).

These results suggest a general set of predictions that can be easily tested. For example, they predict that selection gradients on fitness components will be steeper in species up and down the food web in communities (1) with higher basal productivity (results for larger c_0), (2) with more benign conditions for the consumer (lower f_0), and (3) where environmental conditions permit the consumer to catch resources at higher maximum rates (higher a_0). Moreover, if one interprets the distance from the intrinsic birth optimum as a measure of the elaboration of an antipredator defense (e.g., swimming faster, producing more of a noxious chemical, lengthening spines, making exceedingly large or small seeds), prey should also evolve more elaborate defenses in communities with these same environmental conditions (i.e., higher basal productivity, more benign conditions for consumers, and higher maximal capture rates). Also, changes in species abundances across these environmental conditions match what one would expect purely based on ecological considerations—namely, abundances of both species should increase with productivity, the resource should increase with increasing death rate of the consumer, and the resource should decline and the abundance of the consumer should first increase and then decrease with an increasing attack coefficient (cf. fig. 3.8 with results in chapter 2). Therefore, species coevolution should reinforce ecological patterns of community structure along environmental gradients.

The steepness of the underlying selection gradient on the various fitness components (i.e., the shapes of the relationships in fig. 3.1 controlled by γ, θ, β, α, or ε_R, and ε_N) may also differ among communities, or among different consumer-resource interaction pairs within a community. For example, in an environment in which the distance to safety (e.g., thick bushes for cover) is very short, a resource species many only need to run slightly faster than its consumer to effectively evade capture; but in an environment in which the distance to safety is far, the resource species may need to be substantially faster than the consumer to have the same chances of escaping capture. These environments would cause a difference in α for the unidirectional-dependent traits of escape and pursuit speed (fig. 3.1D), and this difference will influence how the consumer and resource coevolve, because the positions of evolutionary equilibria are set by the balance of selection gradients on different fitness components.

Because the relative strengths of the selection gradients acting on different fitness components determine the shape of the overall fitness surface against which the species evolve (Arnold and Wade 1984b, McPeek 1996b), differences in the underlying selection gradients will also shape the outcome of coevolution. For example, compare the result of coevolution in communities where the coefficient scaling the change in the attack coefficient with bidirectional-dependent traits (i.e., where a smaller β value makes the gradient steeper). If the realized attack coefficient changes very slowly with a change in the difference in species' trait

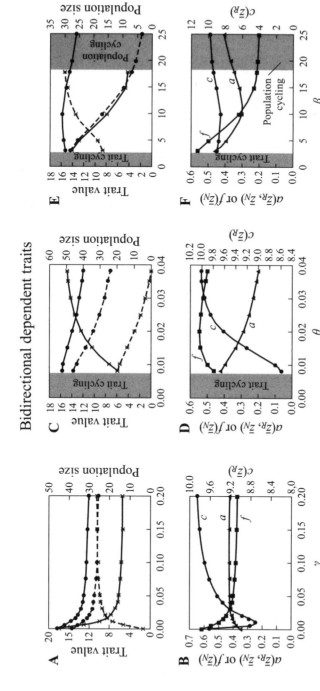

FIGURE 3.9. Effects of different values of (A–B) the underlying selection strength for the resource intrinsic birth rate and the minimum value (γ), (C–D) the underlying selection strength for the consumer intrinsic death rate (θ), and (E–F) the underlying selection strength for the attack coefficient (β) on the dynamics of coevolution between a resource and consumer with bidirectional-dependent traits defining the attack coefficients. Symbols identify equilibrium values for specific parameter combinations. Panels and symbols are as given in figure 3.8. Parameter areas where trait cycling occurs are identified. Parameters in these simulations are the same as those used for the results presented in figure 3.8. (Panels A, C, and E are redrawn from figure 3 of McPeek 2017, with permission).

values (i.e., large value of β), the consumer does not have to evolve far from its intrinsic death optimum to capture substantial quantities of resources (fig. 3.9E–F). Making the intrinsic gradient on the attack coefficient steeper (i.e., smaller β) causes the consumer to be closer to the resource's trait value, which moves it farther from its intrinsic death optimum, and thus causes the consumer's abundance to decrease and the resource's abundance to increase. Interestingly, differences in β have little effect on the resource's trait value; the realized selection gradient on the resource's death fitness component changes little because the increase in the attack coefficient is offset by the decrease in the consumer's abundance.

Altering the intrinsic gradients on the resource's intrinsic birth rate (i.e., γ in equation (3.12)) or the consumer's intrinsic death rate (i.e., θ in equation (3.13)) also change the outcome of coevolution (fig. 3.9A–D). A steeper intrinsic gradient on the resource's intrinsic birth rate (increasing γ) forces the resource to evolve a trait value closer to its intrinsic birth optimum (fig. 3.9A–B), and the analogous result is true for the consumer (fig. 3.9C–D). For example, higher values of θ cause the consumer to evolve with have a trait value closer to $\tilde{z}_N^f$, which decreases the realized attack coefficient (fig. 3.9C–D). As a consequence of the decreased realized attack coefficient, the resource will evolve to have a trait value closer to $\tilde{z}_R^c$. If the consumer cannot support a population in the system at $\bar{z}_N = \tilde{z}_N^f$ (as for the parameters considered in fig. 3.9C–D), high values of θ (i.e., strong intrinsic selection gradient on the consumer's death fitness component) will prevent the consumer from existing in the system, even with adaptation. These considerations illustrate the critical importance of quantifying the strengths of selection gradients on various fitness components to understand the overall form and outcome of natural selection.

Finally, the distance between the intrinsic trait optima, $\tilde{z}_R^c$ and $\tilde{z}_N^f$, of the consumer and resource also strongly shapes the ecological system that will evolve, and the importance of the absolute distance between them depends on the underlying selection gradient of the attack coefficient. I show results for each trait type underlying the attack coefficients here because the effect of increasing $\tilde{z}_R^c$ relative to $\tilde{z}_N^f$ depends on the attack coefficient traits. The consequences of altering the distance between $\tilde{z}_N^f$ and $\tilde{z}_R^c$ illustrate the main difference between the three trait types. Since the consumer does not have to match the resource in phenotype in any particular way when independent traits underlie the attack coefficient, altering $\tilde{z}_R^c$ relative to $\tilde{z}_N^f$ has little effect on the abundances of the two species over much of this range; the prey is then the primary species to evolve, regardless of the selection strength (fig. 3.10A–D). In addition, if the resource's optimum is near the value where the realized attack coefficient would be very low, the consumer may not be able to support a population (fig. 3.10A–D).

In contrast, increasing the distance between $\tilde{z}_R^c$ and $\tilde{z}_N^f$ for unidirectional and bidirectional-dependent traits can eventually drive the consumer extinct as it

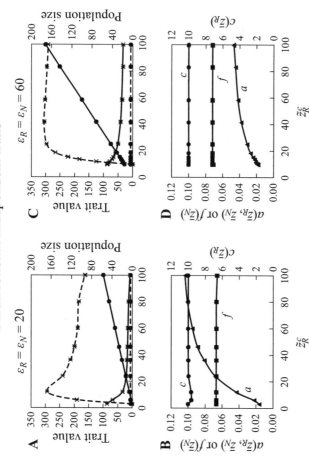

FIGURE 3.10. Effects of different values of the resource's intrinsic birth optimum $\tilde{z}_R^c$ for various levels of the underlying selection strengths for the attack coefficients when (A–D) unidirectional-independent traits, (E–J) unidirectional-dependent traits, and (K–P) bidirectional-dependent traits on the dynamics of coevolution between a resource and consumer. Panels and symbols are as given in figure 3.8. The underlying selection strength parameters used in simulations are as given above the column of panels. Other parameters used in simulations are as follows: unidirectional-independent traits (panels A–D) $c_0 = 10.0$, $d = 0.2$, $a_0 = 0.5$, $b = 0.1$, $h = 0.3$, $f_0 = 0.05$, $g = 0.0$, $\gamma = 0.0$, $\theta = 0.01$, $\gamma = 0.01$, $V_{z_R} = V_{z_N} = 1.0$, $\tilde{z}_N^f = 1.0$, $V_{z_R} = V_{z_N} = 0.2$; unidirectional-dependent traits (panels E–J) $c_0 = 10.0$, $d = 0.2$, $a_0 = 0.5$, $b = 0.1$, $h = 0.3$, $f_0 = 0.05$, $g = 0.0$, $\gamma = 0.01$, $\theta = 0.01$, $\tilde{z}_N^f = 0.01$, $V_{z_N} = 1.0$, $V_{z_R} = V_{z_N} = 0.2$; and bidirectional-dependent traits (panels K–P) $c_0 = 10.0$, $d = 0.2$, $a_0 = 0.5$, $b = 0.1$, $h = 0.1$, $f_0 = 0.2$, $g = 0.0$, $\gamma = 0.01$, $h = 0.1$, $f_0 = 0.2$, $g = 0.0$, $\gamma = 0.01$, $\theta = 0.01$, $\tilde{z}_R^c = 12.0$, $\tilde{z}_N^f = 1.0$, $V_{z_R} = V_{z_N} = 0.2$.

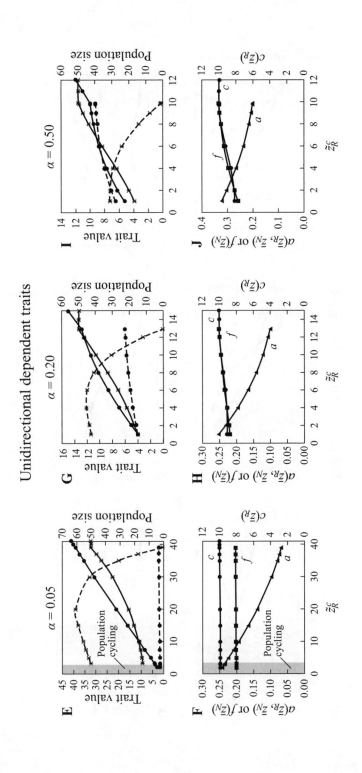

Bidirectional dependent traits

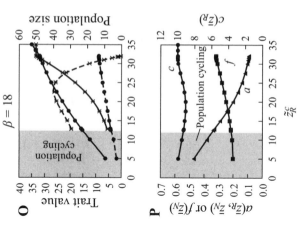

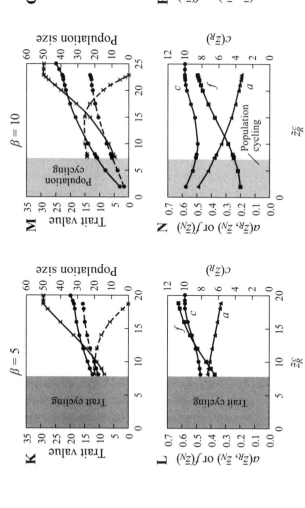

coevolves away from its intrinsic death rate optimum to maintain an adequate capture rate of the resource (fig. 3.10E–P). Stronger selection on the attack coefficient (i.e., larger values of α or lower values of β) also causes the consumer to evolve to extinction at lower values of $\bar{z}_R^c$ relative to $\bar{z}_N^f$ (fig. 3.10E–P).

ADD A TROPHIC LEVEL

Now consider how this simple consumer-resource system evolves when a predator that feeds on the consumer (i.e., a third trophic level), and potentially the resource as well (i.e., an intraguild predator), is added to the system. The predator feeds on the consumer, and possibly the resource, both with a saturating functional response (i.e., intraguild predation/omnivory) and experiences its own density-dependent death rate. Adding this predator to the system elaborates the dynamical abundance and trait system to the following general form:

$$\frac{dP}{dt} = P\ln(\bar{W}_P) = P\left(\frac{wv(\bar{z}_R,\bar{z}_p)R + nm(\bar{z}_R,\bar{z}_p)N}{1+v(\bar{z}_R,\bar{z}_p)uR + m(\bar{z}_R,\bar{z}_p)lN} - x(\bar{z}_P) - yP\right)$$

$$\frac{dN}{dt} = N\ln(\bar{W}_N) = N\left(\frac{ba(\bar{z}_R,\bar{z}_N)R}{1+a(\bar{z}_R,\bar{z}_N)hR} - \frac{m(\bar{z}_R,\bar{z}_p)P}{1+v(\bar{z}_R,\bar{z}_p)uR + m(\bar{z}_R,\bar{z}_p)lN} - f(\bar{z}_N) - gN\right)$$

$$\frac{dR}{dt} = R\ln(\bar{W}_R) = R\left(c(\bar{z}_R) - dR - \frac{a(\bar{z}_R,\bar{z}_N)N}{1+a(\bar{z}_R,\bar{z}_N)hR} - \frac{wv(\bar{z}_R,\bar{z}_p)P}{1+v(\bar{z}_R,\bar{z}_p)uR + m(\bar{z}_R,\bar{z}_p)lN}\right)$$

$$\frac{d\bar{z}_p}{dt} = V_{z_p}\left(\frac{wR\dfrac{\partial v(\bar{z}_R,z_p)}{\partial z_p} + nN\dfrac{\partial m(\bar{z}_N,z_p)}{\partial z_p} + RN(nu-wl)\left(v(\bar{z}_R,z_p)\dfrac{\partial m(\bar{z}_N,z_p)}{\partial z_p} - m(\bar{z}_N,z_p)\dfrac{\partial v(\bar{z}_R,z_p)}{\partial z_p}\right)}{(1+v(\bar{z}_R,z_p)uR + m(\bar{z}_R,z_p)lN)^2}\right.$$
$$\left.\left. - \frac{\partial x(z_p)}{\partial z_p}\right|_{z_p=\bar{z}_p}\right)_{z_p=\bar{z}_p}.$$

$$\frac{d\bar{z}_N}{dt} = V_{z_N}\left(\frac{bR\dfrac{\partial a(\bar{z}_R,z_N)}{\partial z_N}}{(1+a(\bar{z}_R,z_N)hR)^2}\right|_{z_N=\bar{z}_N} - \frac{P\dfrac{\partial m(z_N,\bar{z}_p)}{\partial z_N}}{1+v(\bar{z}_R,\bar{z}_p)uR + m(\bar{z}_R,\bar{z}_p)lN}\Bigg|_{z_N=\bar{z}_N} - \frac{\partial f(z_N)}{\partial z_N}\Bigg|_{z_N=\bar{z}_N}\right)$$

$$\frac{d\bar{z}_R}{dt} = V_{z_R}\left(\frac{\partial c(z_R)}{\partial z_R}\right|_{z_R=\bar{z}_R} - \frac{N\dfrac{\partial a(z_R,\bar{z}_N)}{\partial z_R}}{(1+a(\bar{z}_R,\bar{z}_N)hR)}\Bigg|_{z_R=\bar{z}_R} - \frac{P\dfrac{\partial v(z_R,\bar{z}_p)}{\partial z_R}}{1+v(\bar{z}_R,\bar{z}_p)uR + m(\bar{z}_R,\bar{z}_p)lN}\Bigg|_{z_N=\bar{z}_N}\right)$$

$$(3.20)$$

In the added functional responses, $m(\bar{z}_N,\bar{z}_P)$ is the attack coefficient, n is the conversion efficiency, and l is the handling time for the predator feeding on the consumer; $v(\bar{z}_R,\bar{z}_P)$ is the attack coefficient, w is the conversion efficiency, and u is the handling time for the predator feeding on the resource.

The predator also has an intrinsic death rate given by $x(\bar{z}_P)$ and a density-dependent increase in death rate defined by y. The functions describing the relationships between trait and parameter values follow the same form as those for the consumer. The functions defining the two attack coefficients of the predator for the different trait types are given in box 3.1. The intrinsic death rate is given by

$$x(z_p) = x_0 \left(1 + \delta(z_p - \tilde{z}_P^x)^2\right),\tag{3.21}$$

where $\tilde{z}_P^x$ is the trait value that minimizes the predator's intrinsic death rate, x_0 is the minimum intrinsic death rate at this optimum when $z_p = \tilde{z}_P^x$, and δ mediates the underlying selection gradient on z_p due to the death fitness component.

First, consider the evolution of this system in the absence of intraguild predation (i.e., $v(\tilde{z}_R, \tilde{z}_p) = 0$). The first general feature to emerge when a third trophic level is added to create a food chain is that the basal resource generally evolves trait values that are closer to its own intrinsic birth optimum as compared to when the predator is not feeding on the consumer (fig. 3.11). If the predator can invade, adapt, and ultimately support a population, it reduces the consumer's abundance and alters the trait value favored in the consumer, which in turn changes the attack coefficient of the consumer feeding on the resource. Thus, the presence of the predator alters both the ecological (i.e., consumer's abundance) and evolutionary (i.e., consumer's trait value) conditions experienced by the resource. When the attack coefficient between the predators and their prey are determined by unidirectional-dependent (fig. 3.11E–H) or bidirectional-dependent traits (fig. 3.11I–L), the realized attack coefficient between the consumer and resource increases in the presence of the predator because of changes in both the consumer's and resource's phenotypes. Despite the increase in the attack coefficient on the resource, the reduction in the consumer's abundance reduces the selection gradient sufficiently on the resource's death fitness component to favor a trait value closer to $\tilde{z}_R^c$ in the presence of the predator.

If the attack coefficients are determined by unidirectional-independent traits (fig. 3.11A–D), the consumer evolves a lower trait value that lowers the attack coefficient on itself from the predator, which concomitantly reduces its own attack coefficient on the resource. Here, the resource evolves to have a phenotype closer to its intrinsic birth optimum, because of both an evolutionary reduction in the attack coefficient on it and an ecological reduction in the consumer's abundance. Thus, the types of traits involved in the species interactions determine how the ecological structure and interaction strengths among species change as new species are added to a community.

Comparing communities with different maximum attack coefficients of the predator on the consumer, m_0 has little effect on the trait value that is favored in the predator (fig.3.11C, G, and K). However, communities with higher values of m_0 have lower abundances of both the predator and the consumer, which causes the resource to evolve a trait value closer to its intrinsic birth optimum. Different

---

BOX 3.1. FUNCTIONS FOR THREE DIFFERENT TRAIT TYPES
DEFINING THE PREDATOR ATTACK COEFFICIENTS

When a top predator is added to a simple system containing a basal resource and an intermediate consumer, two additional functional responses must be defined. These are given in equations (3.20). Thus, an attack coefficient for the predator feeding on each must also be defined that incorporates the effects of the phenotypic traits of the interacting species to determining the realized value of each in the model. Thus, $m(z_N, z_P)$ is the attack coefficient of the predator feeding on the consumer, and $v(z_R, z_P)$ is the attack coefficient of the predator feeding on the resource.

When unidirectional-independent traits define each, the functional forms are

$$m(z_N, z_P) = \frac{m_0 z_N z_P}{(\eta_N + z_N)(\eta_P + z_P)} \quad \& \quad v(z_R, z_P) = \frac{v_0 z_R z_P}{(\kappa_R + z_N)(\kappa_P + z_P)}.$$

As with the analogous equation (i.e., equation [3.14]) for the consumer feeding on the resource, m_0 and v_0 are the asymptotic maxima for each, and η_x and κ_x are scaling parameters that define the underlying selection strength for the respective species traits.

When unidirectional-dependent traits define these attack coefficients, the functional forms are

$$m(z_N, z_P) = \frac{m_0}{1 + e^{-\rho\Omega}} \quad \& \quad v(z_R, z_P) = \frac{v_0}{1 + e^{-\tau\Sigma}},$$

where $\Omega = z_P - z_N$ and $\Sigma = z_P - z_R$, and ρ and τ are scaling parameters that define the underlying selection strengths on the respective attack coefficients in this case.

When bidirectional-dependent traits define these attack coefficients, the functional forms are

$$m(z_N, z_P) = m_0\, e^{-\left(\frac{\Omega}{\phi}\right)^2} \quad \& \quad v(z_R, z_P) = v_0\, e^{-\left(\frac{\Sigma}{\psi}\right)^2},$$

where ϕ and ψ are scaling parameters that define the underlying selection strengths on the respective attack coefficients in this case.

---

values of the underlying selection strength on $m(\bar{z}_N, \bar{z}_P)$ cause different trait values that evolve in the predator, giving a higher realized value of $m(\bar{z}_N, \bar{z}_P)$ with stronger underlying selection (fig. 3.11D, H, and L).

Communities with higher productivity of the basal resource (i.e., larger values of c_0) have higher abundances of all three species regardless of the trait

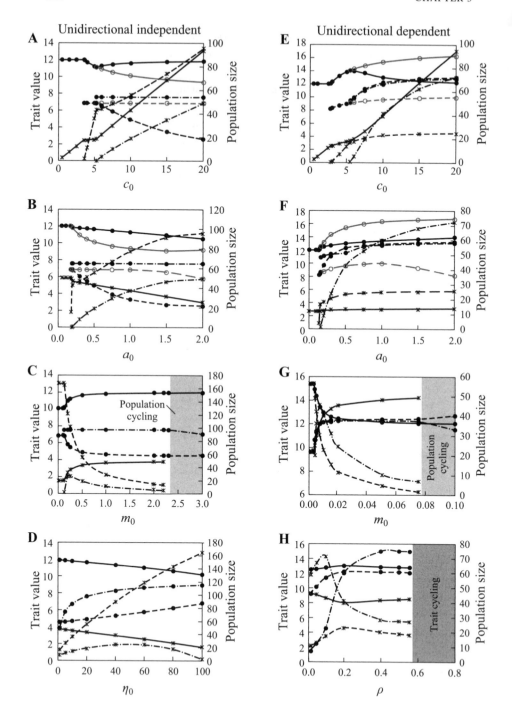

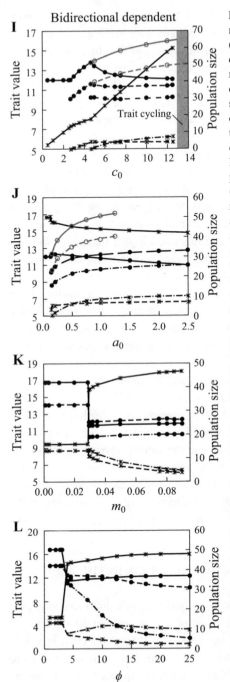

Bidirectional dependent

FIGURE 3.11. Effects of different values of the maximum intrinsic birth rate of the resource (c_0), the maximum attack coefficient of the consumer on the resource (a_0), the maximum attack coefficient of the predator on the consumer (m_0), and the various underlying selection strengths (η_P, ρ, ϕ) on the attack coefficient of the predator on the consumer; these are for the various trait types on the dynamics of coevolution in a three trophic level food chain with one species at each trophic level. Each panel shows the trait values *(filled circles)* and population sizes (×) for the resource *(solid lines)*, consumer *(dashed lines)*, and predator *(dot-dashed lines)*. Results from simulations in which the predator is absent are shown in *gray*, so that the effects of adding the top predator can be compared. The *left column* shows results for simulations where all attack coefficients are defined by unidirectional-independent traits, the *middle column* for unidirectional-dependent traits, and the *right column* for bidirectional-dependent traits. Parameter areas where trait cycling and population cycling occur are identified. Parameters other than the gradient parameter used in simulations are as follows: unidirectional-independent traits (panels A–D) $c_0 = 10.0$, $d = 0.2, a_0 = 0.5, b = 0.1, h = 0.3, m_0 = 0.05$, $n = 0.1, l = 0.3, f_0 = 0.05, g = 0.0, x = 0.01$, $y = 0.0, \varepsilon_R = 20.0, \varepsilon_N = 20.0, \gamma = 0.01, \eta_N = 20.0, \eta_P = 20.0, \theta = 0.01, \delta = 0.01, \tilde{z}_R^c = 20.0, \tilde{z}_N^f = 1.0, \tilde{z}_P^x = 1.0, V_{z_R} = V_{z_N} = V_{z_P} = 0.2$; unidirectional-dependent traits (panels E–H) $c_0 = 10.0, d = 0.2, a_0 = 0.5, b = 0.1, h = 0.2$, $m_0 = 0.01, n = 0.1, l = 0.2, f_0 = 0.2, g = 0.0$, $x = 0.01, y = 0.0, \alpha = 0.1, \rho = 0.1, \gamma = 0.01$, $\theta = 0.01, \delta = 0.01, \tilde{z}_R^c = 12.0, \tilde{z}_N^f = 1.0, \tilde{z}_P^x = 1.0, V_{z_R} = V_{z_N} = V_{z_P} = 0.2$; and bidirectional-dependent traits (panels I–L) $c_0 = 10.0, d = 0.2, a_0 = 1.0, b = 0.1, h = 0.1, m_0 = 0.05$, $n = 0.1, l = 0.1, f_0 = 0.2, g = 0.0, x = 0.01$, $y = 0.0, \beta = 5.0, \phi = 5.0, \gamma = 0.01, \theta = 0.01, \delta = 0.02, \tilde{z}_N^c = 12.0, \tilde{z}_N^f = 1.0, \tilde{z}_P^x = 1.0$, $V_{z_R} = V_{z_N} = V_{z_P} = 0.2$.

types underlying the attack coefficients (fig. 3.11A, E, and I). The evolutionary responses of the three species when c_0 is greater are more heterogeneous. Higher values of c_0 cause the resource to evolve closer to its intrinsic birth optimum for all trait types underlying the attack coefficients, but the evolutionary responses in the consumer and predator depend on the trait types. The phenotype of the predator responds to different levels of productivity of the resource substantially only when unidirectional-dependent traits determined the attack coefficients (fig. 3.11E), whereas with higher values of c_0, the consumer evolves closer to its intrinsic death optimum with unidirectional-independent traits (fig. 3.11A) and away from its intrinsic death optimum with unidirectional-dependent traits (fig. 3.11E).

The evolutionary responses of species at the three trophic levels when the top predator is also an intraguild predator differs depending on the trait types defining the attack coefficients (fig. 3.12). With both independent- and dependent-unidirectional traits defining the attack coefficients, a higher maximum value of the attack coefficient for the predator feeding on the resource (v_0) causes smooth transitions in trait values for all species across communities. In communities with higher values of v_0, the resource evolves to a trait value closer to its intrinsic birth rate optimum, the consumer evolves farther from its intrinsic death rate optimum, and the predator evolves closer to its intrinsic birth optimum (fig. 3.12A–B). With high values of v_0 at which the consumer cannot persist, the system returns to two trophic levels.

In contrast, when bidirectional-dependent traits define the attack coefficients, higher values of v_0 result in more graded transitions (fig. 3.12C). Over a range of low values of v_0 (0.0001–0.0015 in fig. 3.12C), changing its value has little effect on the traits that evolve in the three species or their abundances, with the predator having a trait value between the resource and consumer. Above this range, the predator evolves to feed more heavily on the resource by shifting its trait value below that of the resource and thus closer to its intrinsic death rate optimum. Here again, the outcome of coevolution by natural selection critically depends on the structure of the community and on the trait types defining the species interactions.

IMPLICATIONS FOR MEASURING SELECTION IN THE WILD

This theoretical exploration of the ecological dynamics of natural selection provides important guidance into what measures of natural and sexual selection in wild populations quantify. First, few if any studies of natural selection in a wild population actually measure overall fitness. The studies that come the closest are those that follow marked populations of large mammals over multiple generations (e.g., Ozgul et al. 2009). However, most studies take a snapshot of selection within

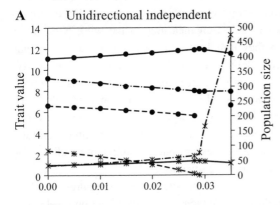

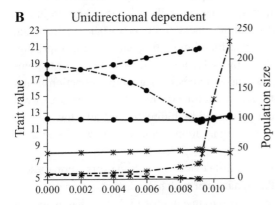

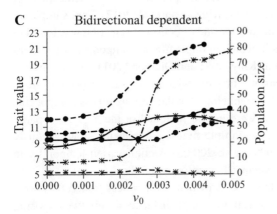

FIGURE 3.12. Effects of different values of the predator's maximum attack coefficient on the resource (v_0) (i.e., the strength of intraguild predation) for the various trait types on the dynamics of coevolution in a three-trophic-level food chain with one species at each trophic level. (*A*, unidirectional-independent traits; *B*, unidirectional-dependent traits; and *C*, bidirectional-dependent traits.) Symbols are as specified in figure 3.11. Parameters other than the gradient parameter used in simulations are as follows: (*A*) unidirectional-independent traits $c_0 = 10.0$, $d = 0.2$, $a_0 = 0.5$, $b = 0.1$, $h = 0.1$, $m_0 = 0.1$, $n = 0.1$, $l = 0.1$, $w = 0.1$, $u = 0.1$, $f_0 = 0.05$, $g = 0.0$, $x = 0.01$, $y = 0.0$, $\varepsilon_R = 20.0$, $\varepsilon_N = 20.0$, $\eta_N = 20.0$, $\eta_P = 20.0$, $\kappa_R = 20.0$, $\gamma = 0.01$, $\theta = 0.01$, $\delta = 0.01$, $\tilde{z}_R^c = 20.0$, $\tilde{z}_N^f = 1.0$, $\tilde{z}_P^x = 1.0$, $V_{z_R} = V_{z_N} = V_{z_P} = 0.2$; (*B*) unidirectional-dependent traits $c_0 = 10.0$, $d = 0.2$, $a_0 = 0.5$, $b = 0.2$, $h = 0.1$, $m_0 = 0.1$, $n = 0.1$, $l = 0.1$, $w = 0.1$, $u = 0.1$, $f_0 = 0.2$, $g = 0.0$, $x = 0.01$, $y = 0.0$, $\alpha = 0.2$, $\rho = 0.2$, $\tau = 0.2$, $\gamma = 0.01$, $\theta = 0.01$, $\delta = 0.01$, $\tilde{z}_R^c = 12.0$, $\tilde{z}_N^f = 1.0$, $\tilde{z}_P^x = 1.0$, $V_{z_R} = V_{z_N} = V_{z_P} = 0.2$; and (*C*) bidirectional-dependent traits $c_0 = 1.0$, $d = 0.02$, $a_0 = 0.75$, $b = 0.1$, $h = 0.05$, $m_0 = 0.1$, $n = 0.1$, $l = 0.05$, $w = 0.1$, $u = 0.05$, $f_0 = 0.005$, $g = 0.0$, $x = 0.005$, $y = 0.0$, $\beta = 5.0$, $\phi = 5.0$, $\psi = 5.0$, $\gamma = 0.02$, $\theta = 0.02$, $\delta = 0.02$, $\tilde{z}_R^c = 12.0$, $\tilde{z}_N^f = 1.0$, $\tilde{z}_P^x = 1.0$, $V_{z_R} = V_{z_N} = V_{z_P} = 0.2$.

one generation, and typically only on a subset of the entire life cycle of the organism being studied. Most also consider only one fitness component. For example, in their comprehensive analysis of selection measures in the wild, Kingsolver et al. (2001) divided the studies they found into three categories based on what measure of fitness was being considered: survival, mating success, or fecundity. Not surprisingly, they found that linear (i.e., directional) selection was common, but quadratic selection was not (see also Hendry and Kinnison 1999, Hoekstra et al. 2001, Rieseberg et al. 2002, Kingsolver and Diamond 2011, Kingsolver et al. 2012). Quadratic selection is a measure of the curvature of the fitness surface experienced by the population and is a necessary but not sufficient condition for the identification of stabilizing or disruptive selection.

The theoretical investigations presented here suggest that measures of selection on fitness components should most frequently identify directional selection as important, since the evolution of a species typically reaches an equilibrium at which the selection gradients of different fitness components balance (e.g., fig. 3.3C). Thus, directional selection being found more frequently is not surprising, especially since stabilizing and disruptive selection should also be harder to identify (Haller and Hendry 2014), and most studies of phenotypic selection in the wild measure fitness components and not lifetime overall fitness. Moreover, even if multiple fitness components are influenced by bidirectional traits, the balance required will typically mean that few if any fitness components will be at their optimal trait values.

However, these analyses also found little evidence for trade-offs among selection pressures, which should be the signature of such balancing selection gradients (i.e., the selection gradients summing to zero in equation (3.7)); see Kingsolver et al. 2001, Kingsolver and Diamond 2011, Kingsolver et al. 2012). For example, in one compilation of studies, the magnitude and direction of selection was found to balance in only 2% of the cases in which multiple fitness components were measured in association with one trait (Kingsolver and Diamond 2011). These authors offer a number of plausible reasons for why evidence of such expected fitness trade-offs is sparse among the data, including that the various fitness components or traits involved in the trade-off were not all included in the analyses, spatial environmental variation influences multiple fitness components simultaneously (e.g., Rausher 1992, Stinchecombe et al. 2002), selection varied temporally in direction, and indirect selection occurred on correlated traits.

The most likely culprit for the lack of evidence for fitness trade-offs is that optimizing selection rarely results from the simple balance of only two opposing selection gradients operating on one trait in real organisms. The models considered in this chapter only included two overall fitness components (birth and death rates) acting on one trait. However, even for this simple scenario, many more than two selection gradients may act on a single trait—particularly when a demo-graphic rate is influenced by multiple selection gradients, as when the species interacts with more than one species. We will see much more of this in chapter 5, but even in the models considered in this chapter, the resource evolves to balance three fitness components when a consumer and an intraguild predator are present (equation (3.20)). One would conclude that the resource's fitness components balance only when all three are measured simultaneously. To properly see the balancing in this case, the resource's survival would have to also be separated into the components due to predation by the intraguild prey and intraguild predator. If interactions with many more species are also considered, as well as multiple traits influencing various fitness components, and the fact that one must measure fitness components over the entire life cycle of an organism, then the task for actually identifying where the trade-offs lie becomes daunting, if not impossible.

Given the ubiquity of directional selection that is apparent in these summary analyses, one must ask also why rapid changes in the phenotypes of species are not occurring (Merilä et al. 2001). This is a particularly troubling question given the consistency in the direction of selection over multiple generations measured in many species (Siepielski et al. 2009, Kingsolver and Diamond 2011, Siepielski et al. 2011a, Siepielski et al. 2013). Many plausible explanations exist for how consistent selection gradients that a population experiences generation after generation may still result in little if any phenotypic change over time, particularly since heritabilities of many traits under selection seems to be adequate to permit responses (Mousseau and Roff 1987). These include strong antagonistic genetic correlations that prevent response to selection, unmeasured countervailing selection pressures, and greater environmental phenotypic variation in natural populations (see Merilä et al. 2001 for an excellent discussion of these).

The theoretical analyses presented here suggest an additional explanation that is actually embodied in all of these, but one that will be difficult to document for the reasons given in the previous paragraph. This explanation is that many species are at evolutionary equilibria. The direction of selection is consistent across years and generations in many species, whereas the magnitudes of the selection gradients do fluctuate (Siepielski et al. 2009, Kingsolver and Diamond 2011, Siepielski et al. 2011a). These conditions would be expected for species at or near the point where the selection gradients they experience are relatively balanced, but short-term fluctuations in the abundances of interacting species cause the magnitudes to vary. This short-term variability is probably caused by short-term fluctuations in abiotic conditions that affect the magnitudes of species' intrinsic birth and death rates. Large and changing types of selection should be expected most when species are first adapting to a new selection regime (e.g., figs. 3.2 and 3.3), as when they are introduced into a new environment (Hendry and Kinnison 1999).

THE BALANCE THAT IS STRUCK

The outcome of coevolutionary interactions among species is determined by the balances that are struck among both the fitness components and selection gradients experienced by each species. Also, the outcomes of these coevolutionary interactions influence both what evolves in each species and the abundance dynamics that results in the community.

If the community reaches a stable equilibrium point in both abundances and traits, the balancing fitness components in each species are what generates stability of the abundances. At equilibrium, the overall realized birth and death rates sum to zero for each species, with these abundances being defined by the position where their abundance isoclines intersect (chapter 2). I hope it was clear that everything considered in chapter 2 applies when we consider the community's evolutionary aspects. The only difference with evolutionary dynamics is that the abundance isoclines for the species change shape and position as their traits change in evolutionary response to one another.

The evolutionary balance requires that the selection gradients for each species also sum to zero. The selection gradients are defined by the underlying relationships of fitness components with the traits experiencing selection; and the relative magnitudes of these selection gradients for the various fitness components will define the outcome of that selective process. In other words, if the selection gradient associated with one fitness component is disproportionately strong relative to all others, the overall fitness surface's shape will be defined primarily by this selection gradient, and the species will consequently evolve mostly in response to this selection gradient (McPeek 1996a). For example, when the underlying selection strength on the resource's intrinsic birth rate is high (i.e., large γ) relative to the other fitness components, the resource evolves a trait value very close to its intrinsic birth optimum (fig. 3.9A). However, when the underlying selection gradient on the attack coefficient is relatively high (i.e., small β), the resource evolves farther from its intrinsic birth optimum if that will decrease predation, and the consumer will evolve farther from its intrinsic death optimum (fig. 3.9E).

The best theoretical illustrations of the importance of this balance are when trait cycling occurs. Trait cycling takes place when the balance among selection gradients continually shift back and forth. In the models studied here, prey will periodically evolve to become more vulnerable to their predators when the selection gradients on other fitness components are stronger. This cycling only occurs in areas of parameter space where the underlying selection strengths are relatively equal (fig. 3.6).

The magnitudes of the fitness components also influence the balance struck among the various selection gradients impinging on each species. For example, increasing its maximum intrinsic birth rate permits the resource to evolve farther from its intrinsic birth optimum; similarly decreasing the consumer's minimum intrinsic death rate permits it to evolve farther from its intrinsic death optimum. Moreover, because many of the fitness components are influenced by the abundances of themselves or other species, alterations of these parameters in one species can shift these balances merely by changing features of the system that alter species abundances (e.g., the strength of density dependence). In other words, one cannot determine what trait value will be favored in one species without knowing the abundances of all species with which it interacts in the community.

All these considerations imply that a much richer ecological and evolutionary set of information is needed if we are to actually understand various patterns of natural selection in the wild. Moreover, this richer set of considerations makes the study of natural selection a *predictive endeavor*. Given the last few paragraphs, one would predict that a resource species in a productive environment would adapt primarily by elaborating defenses to thwart its enemies (e.g., predators, diseases), whereas the same resource species in an unproductive environment would adapt primarily by adapting to maximize its birth rate at the expense of defenses against enemies (fig. 3.8). However, in the productive environment one would measure a strong selection gradient for its birth rate (because the species is far from $\tilde{z}_R^c$), whereas in the unproductive environment the selection gradient on its birth rate would be weak (because the species is near $\tilde{z}_R^c$). This simple example also suggests that many of these predictions may be quite counterintuitive if one were only considering evolutionary features in isolation.

In addition, a critical feature for determining what is favored by natural selection is the differences in strengths of selection gradients experienced by different fitness components (McPeek 1996a). Comparable sets of predictions can be made for the same species in different environmental settings or when comparing different sets of trophic interactors based on the relationships for other model parameters considered in figures 3.8–3.12. A full explanation for patterns of natural selection will require not only quantifying the shapes of fitness surfaces, but also the magnitudes of the associated fitness components and selection gradients on those fitness components, and the abundances and phenotypes of the species that influence those fitness components.

These analyses also highlight that adaptive evolution does not act to perpetuate a species. Species may be incapable of evolving the phenotypes that would be necessary to prevent their extinction. Moreover, adaptive evolution can in some cases favor species to actively move to trait values that will ensure their extinction

(Webb 2003, Parvinen 2005). Adaptive evolution does not act "for the good of the species" overall, and cannot act to make species better at all things. The richness of the study of evolution is embedded in understanding the conflicting ecological and evolutionary demands that any species faces in adapting to the community in which it lives.

UNIFYING FRAMEWORK

The theory pioneered by Lande (1982), Iwasa et al. (1991), and Abrams et al. (1993) for understanding the joint dynamics of species' abundances and traits in a community context (e.g., equations (3.9) and (3.10)) provides a unifying framework for understanding the joint ecological and evolutionary dynamics of a community. The underlying currency is the fitnesses of individuals and how these depend on the abiotic environment in which those individuals find themselves, and the abundances and traits of all the species in the community. The average of these individual fitnesses for each population is the explicit currency that defines the ecological and evolutionary trajectories of all the populations.

In a population dynamics context, we speak of average fitness as the overall per capita demographic rate of the population: dN/Ndt. The average fitness of the population determines whether the abundance will increase, decrease, or not change in the next instant of time.

In an evolutionary context, we speak of average fitness as the fundamental metric of natural selection: $\ln(\bar{W}_N)$. How this will change with a change in the distribution of phenotypes of the population determines whether natural selection can potentially shift that phenotypic distribution (Lande 1982).

Because the symbology of population dynamics and natural selection are different, we typically do not realize that the fundamental basis of both is the same thing; that is, dN/Ndt and $\ln(\bar{W}_N)$ are the same quantity (equation (3.4)), and changes in the abundance of a population and its average trait value are governed by the same basic metric (Charlesworth 1994; Lande 2007, 2008)!

We can see the relationship better by specifying the overall average fitness landscape for a species embedded in a community. The average fitness of a population (i.e., $dN/Ndt = \ln(\bar{W}_N)$) maps onto a system of axes describing the abiotic factors (one subset of axes) for that site, and the abundances (a second subset) and mean traits (a third subset) of this species and all the other species in the community. In this work, I have not explicitly modeled abiotic factors, but rather subsumed their effects into the parameters of the models. In many cases, one would want to also model the dynamics of abiotic factors explicitly, particularly when they have dynamics themselves (e.g., species utilizing resources that can be depleted, such as water, nitrogen, phosphorus, silica, light). Regardless of whether they can be altered by the actions of the biotic community, the average fitness of a population may change along an abiotic factor axis. As this chapter makes clear for each species in the community, the traits and abundances of all species are also important axes in this system.

To make this discussion concrete and to illustrate visually how ecology and evolution fit together, first consider an extremely simple community composed of a single resource species by itself at a site. Assuming the same relationships for this resource as for all other resources discussed in this chapter, its average fitness is given by

$$\frac{dR}{Rdt} = \ln(\bar{W}_R) = c_0 \left(1 - \gamma(z_R - \tilde{z}_R^c)^2\right) - dR, \tag{3.22}$$

with a single bidirectional-independent trait that influences the value of its maximum birth rate, and a linear density-dependent decrease in fitness with increasing abundance (i.e., logistic population growth). Because I am limited to three dimensions, figure 3.13. does not illustrate the abiotic axes that also influence this species' average fitness; because they are not considered explicitly, the fitness surface would change shape as the values of the abiotic factors change. In addition, I am limited to considering only one trait; a full representation would require as many trait axes for this species as it has ecologically important traits. More than one abundance axis would also be needed if this species possessed a complex life cycle with multiple stages. In this illustration, the intrinsic birth optimum is at $\tilde{z}_R^c = 10$. At any point in time, the population of this species is a point in the mean trait-abundance plane, and height of the topography above this point is the average fitness of the individuals in the population. The ecological and evolutionary dynamics of the population move this point as we have described until it eventually reaches the stable equilibrium point of $[R^*, z_R^*] = [50, 10]$.

I show two slicing planes through this three-dimensional fitness topography to illustrate the ecological and evolutionary features driving the dynamics of the population. The first slicing plane runs parallel to the mean trait axis and crosses the abundance axis at $R = 50$; the upper left panel shows the intersection of the fitness surface with this slicing plane (fig. 3.13). This is the relationship between fitness and the mean trait at this abundance, and thus describes phenotypic selection gradients acting on the population for different mean trait values when the abundance of the population is 50. The dynamics of the fitness topography that we explored in this chapter and that we will continue to explore in chapter 5 are simply determined by sliding this slicing plane along the abundance axis as the species' numbers change. Obviously, in more complex communities, the position of the corresponding slicing plane would be moving simultaneously along multiple trait and abundance axes for all the species, including the species of interest and the abiotic factor axes.

The other slicing plane runs parallel to the abundance axis and crosses the mean trait axis at 10 (fig. 3.13). The intersection of the fitness surface with this slicing plane demarcates where the population abundance would equilibrate given the mean trait value in the population. The equilibrium population abundance, given the current mean trait value, is given by the abundance having $\ln\left(\bar{W}_{X_i}\right) = 0$ in the upper right panel; this is a stable equilibrium in this case. If the population abundance is below this value, the abundance will increase, and if above it, population abundance will decrease. It is easy to see why the population would equilibrate at different abundances if the mean trait value in the population changes.

Without boggling the mind too much, now extrapolate this simple picture to the analogous relationship for a multispecies community. The axis system for a community with more interacting species would simply have more mean trait axes and more abundance axes (and don't forget about the abiotic factor axes as well), but we can conceptualize the dynamics of the system in exactly the same way. In fact, I have been doing this throughout this chapter. The per capita population growth forms of equations (3.17)–(3.20) all represent more elaborate multidimensional forms of the figure depicted in figure 3.13C, each with more abundance and trait axes. Each species in the community will have a different fitness topography associated with this same multidimensional axis system of abiotic factors and species' traits and abundances, and each species will respond to its own fitness surface associated with this same axis system. A joint abundance and trait equilibrium for the community occurs at each point in these multidimensional spaces where for every species $\ln(\bar{W}) = 0$, and the fitness topography of each and every species at this point has $\partial\ln(W)/\partial z\big|_{z=\bar{z}} = 0$ for all of its own trait axes. The locations of these equilibrium points will define domains of attraction, and within each domain of attraction the species' abundances and traits would either approach the equilibrium (i.e., stable equilibrium) or enter into a limit cycle or chaos of only abundances or of both traits and abundances.

These are the two orientations on which we focus without realizing that they are two perspectives of the same feature. Ecologists take the population regulation perspective (i.e., dR/Rdt vs. abundance in fig. 3.13B) and tend to ignore that species' traits are also critical to population dynamics. Likewise, evolutionary biologists take the natural selection perspective (i.e., $\ln(\bar{W}_R)$ vs. mean trait in fig. 3.13A) and tend to ignore that the fitnesses of individuals depend on the abundances of all the species in the community. The changing shapes of isoclines and of fitness topographies that we have explored here and will explore further in chapter 5 are simply the result of moving these slicing planes along the axes not considered in each perspective. The shapes of isoclines change as the traits of

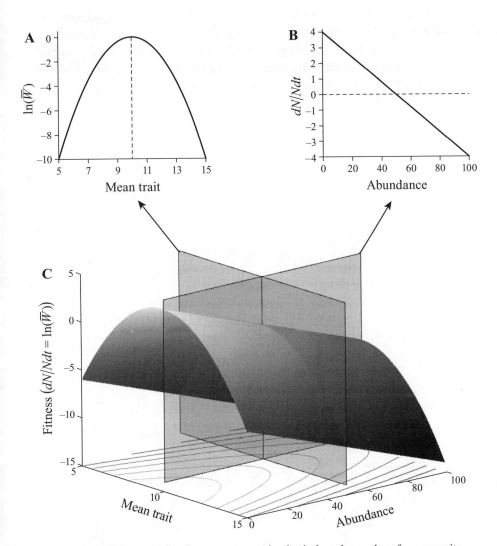

FIGURE 3.13. The fitness surface for a resource species that is the only member of a community, whose average individual fitness is determined by equation (3.22). For this species, the parameters of the model are $c_0 = 4.0$, $d = 0.08$, $\gamma = 0.1$, and $\bar{z}^c_{x_i} = 10.0$. The *lower* panel shows the fitness surface as a function both the mean trait value in the population and the species' abundance. The *upper left* panel shows the intersection of the fitness surface with the slicing plane at $R = 50$, which is the fitness topography that determines phenotypic selection on this trait at this abundance. The *upper right* panel shows the intersection of the fitness surface with the slicing plane at $z_R = 10.0$, which is the fitness relationship that governs population dynamics at this trait value. (This figure is redrawn from figure 1 of McPeek 2017, with permission.)

the interacting species change (e.g., fig. 3.7), and the shapes of fitness landscapes change as the abundances of all interacting species and the traits of other species change (e.g., fig. 3.3). This conceptual framework plainly reveals that both perspectives are simultaneously essential to understanding either the population dynamics of interacting species or the natural selection of any one species in the community.

CHAPTER FOUR

New Species for the Community

As you walk on a trail through the woods, have you ever pondered where all the various species came from and what this plot of ground may have looked like a thousand years ago, ten thousand years ago, or a million years ago? How long have these various tree species been growing here, and how long have these salamander species been hiding under these rocks? When did these fish species show up in this lake, and what types of fish were here before them?

We know that life on Earth is in constant flux, at least on a geological time scale. The only remnants of the huge dinosaurs that once dominated the landscape are the birds that fly above our heads. The only remaining sarcopterygian fishes are the coelacanths and lungfishes, unless you also count all the amphibians, reptiles, birds, and mammals running around as being related to them. The equisetums, ferns, and cycads have lost their dominant places in the organization of the forests as well and are now minor components of the understory. In fact, even just 18,000 years ago, there were no plants or animals at all on the spot where I am presently typing these words; at that time, this spot was covered by a kilometer-thick glacier.

Thus, we know that a community found at any location on Earth is a dynamic entity with changes in composition over some time frame. What drives this change? Extinction is simple to explain; species go extinct when the environment changes enough to make it demographically impossible to support populations anywhere. The puzzle is the source of new species. Where do new species come from, and could that source influence the structure of the resulting community?

Based on pioneering laboratory experiments in the 1930s and 1940s (Gause 1934, Gause et al. 1934, Park et al. 1941, Crombie 1947, Park 1948), ecologists have generally taken as axiomatic that for the long-term ecological equilibrium for a community, only one species can occupy a given niche. In this context, a niche is equivalent to an ecological opportunity. Under this "one species per niche" axiom, the types of species found in any local community reflect the diversity of available niches, and when two or more species occupy the same niche in one local community, competition should quickly drive the poorer competitors extinct (Hutchinson 1958, 1959; MacArthur 1965, 1972). A number of ecological mechanisms have been identified that can work in opposition to competitive exclusion, and thereby permit more co-occurring species than available niches—most notably various forms of disturbance and predation (Payne 1966, Dayton 1971, Paine 1974, Connell 1978, Sousa 1979, Paine and Levin 1981, Pickett and White 1985, Chesson and Huntly 1997). However, the "one species per niche" axiom remains the dominant theoretical touchstone of ecology.

Although questions of coexistence on a local scale dominate current investigations of biodiversity, macroevolutionary and biogeographic processes operating on much larger spatial, and much longer temporal, scales are the ultimate sources of species that interact in local communities (Ricklefs 1987, Cornell and Lawton 1992, Cornell and Karlson 1996). Understanding patterns of biodiversity on any scale of analysis must, therefore, reconcile the mechanisms by which new species come into being and are dispersed across the landscape with mechanisms governing local coexistence (Ricklefs 1987). This is the subject of this chapter.

In general, many ecologists have directly extended the "one species per niche" axiom into the realm of speciation. Many ecologists consider speciation to be a process by which lineage diversification accompanies ecological diversification, because speciation events are thought to be driven primarily by populations adapting to fill different niches (e.g., Rosenzweig 1978, Pimm 1979). In effect, speciation from this perspective is considered to be simply a *by-product* of adaptive ecological differentiation. Therefore, speciation need only be considered in explaining diversity patterns when all the available niches are not filled:

> If the areas being compared are not saturated with species, an historical answer involving rates of speciation and length of time available will be appropriate; if the areas are saturated with species then the answer must be expressed in terms of the size of the niche space . . . and the limiting similarity of coexisting species (MacArthur 1965, p. 510).

To translate this quote of MacArthur's into the framework of chapter 2, only coexisting species are present in a "saturated" community, and all possible functional groups that could invade and coexist are present. From this standpoint, speciation is seen as a process that neatly packs one species into each niche and can be ignored if the community is "saturated" with coexisting species. Moreover, speciation is viewed as never acting in opposition to the ecological processes enforcing the "one species per niche" rule. Additionally, in his stimulating book on macroecology, Brown (1995, p. 166) makes the following statement: "In order for a lineage to split and for at least two descendant species to survive, it is necessary that ecological and/or biogeographic, as well as genetic, differentiation occur. If the new species are sympatric, they must differ sufficiently in their ecological requirements so as to coexist." Ecological differentiation associated with speciation is an unquestioned assumption for many ecologists. If this view was accurate, the mode of speciation could be ignored, and only the speciation rate would be relevant to ecological discussions. However, as we will see here, this assumption is both theoretically not necessary and empirically not demonstrable in many mechanisms of speciation.

Speciation occurs by myriad processes, many of which involve no necessary adaptive ecological differentiation at all, as we will see in this chapter. We must therefore consider explicitly the consequences of the many and varied modes of speciation in a community ecology context. These considerations illustrate that the tempo *and* mode of speciation can have profoundly important consequences for patterns of biodiversity and the structure of local communities. We must also consider how all of the various mechanisms might introduce species of varying degrees of differentiation from existing community members. New actors enter the evolutionary play from innumerable sources, but whether they can perform on the ecological stage is a separate question.

WHAT IS A SPECIES?

A species is a fundamental unit of biological organization, and what a species is plays a central role in the theory of many disciplines, including community ecology. Some of the fundamental issues that community ecologists address are the causes of patterns of *species richness* (i.e., the number of species in a defined area) and *species diversity* (i.e., metrics influenced by both the number and relative frequencies of species in a defined area). Community assembly over both the short and long term involves the introduction of new species via the process of speciation within an area and migration from other areas. Therefore, we must consider what a species is, and we must consider explicitly how new species arise and are introduced into a community.

To know what a species is, we need a definition. As anyone who has spent at least 10 minutes on the subject knows, the best way to start a fight among biologists is to ask two or more of them to define species. Dozens of species definitions can be found in the literature (Wilkins 2009a, 2009b; Hausdorf 2011), and most involve either explicitly or implicitly the concept of reproductive isolation between distinct *lineages*. A lineage is a group of individuals with ancestor-descendent relationships (fig.4.1; Simpson 1951, 1961). The key feature making an ancestor-descendant relationship important is that genetic materials—alleles at genes, epigenetic modifications in DNA, or cultural information—are passed among those individuals. Because genetic material is passed, those individuals with ancestor-descendant relationships will show resemblance in both genotypes and phenotypes to one another. This intergenerational continuity creates the additive genetic variation on which the various mechanisms of evolution depend and so allows the mechanisms of evolution to operate (e.g., equation (3.7)). In fact,

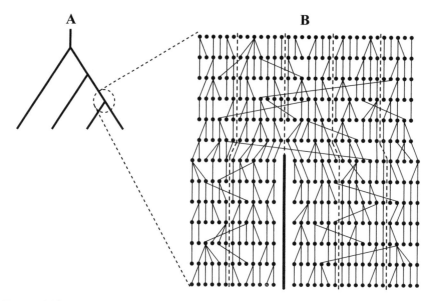

FIGURE 4.1. Ancestor-descendent relationships that define the speciation of one lineage into two different lineages. Panel *A* shows a phylogeny of four species, and panel *B* illustrates a speciation event in this clade. Each *circle* in panel *B* represents an individual, and *lines between circles* represent ancestor/descendent relationships. *Dashed lines* delineate populations and the *thick black line* identifies the reproductive barrier that causes speciation. This figure is a representation that illustrates the same ideas as a number of previous illustrations in the literature; for example, see figure 7.2 in Barton 1988, and figures 1.13 and 2.1 in Avise 2000.

Darwin (1859) defined what we now call evolution by natural selection as "descent with modification," and passing genetic material from ancestors to descendants establishes the "descent" component of the process.

Darwin (1871, Vol. 1, pp. 214–215) stated, "Even a slight degree of sterility between any two forms when first crossed, or in their offspring, is generally considered as a decisive test of their specific distinctness; and their continued persistence without blending within the same area, is usually accepted as sufficient evidence, either of some degree of mutual sterility, or in the case of animals of some repugnance to mutual pairing." His definition presaged the most durable species definition—the *biological species* concept. The biological species concept derives from a description by Dobzhansky (1937b) that emphasized the process through which species come into existence—namely, the stage of differentiation when reproductive isolation is established between two or more groups of organisms. Mayr (1942) criticized Dobzhansky's species definition, arguing "a species is not a stage of a process, but the result of a process." However, Mayr's (1942) definition of a biological species is essentially the same as Dobzhansky's: "Species are groups of actually or potentially interbreeding natural populations, which are reproductively isolated from other such groups" (also see discussion in Mayr 1940). In other words, a species forms an independent inheritance unit constituted by ancestor-descendent relationships.

Complete reproductive isolation between lineages means that genetic material in one cannot move into another. The biological species concept is usually considered to not apply to the Bacteria and Archaea because of substantial lateral gene transfer. However, some genetic material does sometimes pass between what seem to be "good species," even in groups where the biological species concept is considered the standard. Therefore, these barriers are not completely impermeable (Mallet 2008). Most modifications, refinements, and alternatives have been attempts to make the biological species concept operational for different types of information that can be gleaned about species other than their reproductive isolation status (Simpson 1951, Paterson 1985, Templeton 1989, Mayden 1997, de Queiroz 1998, Sites and Marshall 2004, de Queiroz 2007, Mallet 2008, Hausdorf 2011). However, the inability to freely and consistently pass substantial amounts of genetic material from one lineage to another is the key criterion underlying most of these alternative species definitions.

Thus, speciation is the establishment of reproductive isolation among lineages and so is caused by the myriad mechanisms that reduce or prevent successful mating between individuals in those lineages. Dobzhansky (1937a) and Mayr (1942) provided the first comprehensive catalogs of these mechanisms (see also Coyne and Orr 2004). Some directly involve ecological differentiation of lineages, with reproductive isolation created as a by-product. However, many generate reproductive isolation without regard to the ecological context of the species and frequently with little or no change in the ecological character of the lineage.

As you might imagine, the properties of new species will have important and disparate ecological consequences for the overall structures of communities. These consequences are determined by the differences in ecologically important phenotypes that are created at the time of speciation. However, those ecological consequences are also downstream; they become relevant only after reproductive isolation is established. I think many people make the mistake of bypassing these critical stages that follow the establishment of reproductive isolation, and simply jumping to what they assume will be the outcome of introducing new species into a community—a set of wholly coexisting species. For example, adaptive dynamics approaches to modeling speciation assume that speciation occurs only when a "branching point" is reached (i.e., two close trait values that have a fitness valley between them). As a result, this approach can only introduce ecologically differentiated and thus coexisting species into a community (Meszéna et al. 1997,

Geritz et al. 1998, Doebeli and Dieckmann 2000, Doebeli 2011). *This is a mistake when it comes to understanding how community structure develops.* As we will see below, many speciation mechanisms produce either ecologically differentiated but potentially maladapted species or new species that are not ecologically differentiated at all from their progenitors.

Speciation mechanisms are also frequently categorized by the geographical relationships among the progenitor and daughter species (e.g., allopatric, vicariant, parapatric, peripatric, microallopatric, sympatric) at the time of speciation (White 1978, Losos and Glor 2003, Coyne and Orr 2004, Gavrilets 2004). This geography is critically important because it delineates whether new species are entering new communities at their inception or are not community invaders at all. Additionally, the geography influences whether daughter species must co-occur with their progenitors from the start. Alternatively, they would only interact with their progenitor if they expanded their ranges to invade the communities inhabited by those immediate ancestors. Because many of the various mechanisms generating reproductive isolation tend to occur with a specific geographic configuration, I do not consider geography independent of the mechanisms generating reproductive isolation but instead explore them together.

Ignoring the dynamics that ensue immediately after new species are introduced by various mechanisms is a critical omission to understanding the full development of community structure. Different speciation mechanisms have disparate ecological consequences and so the mode of speciation cannot be ignored. Melding the rich and varied mechanisms of speciation with the subsequent dynamics that result is critical to explaining why many co-occurring but not coexisting species (i.e., neutral, sink and walking-dead species) may be present in local communities. This chapter explicitly considers the phenotypic and geographic patterns of species introductions via speciation that are produced by the various isolating mechanisms. In chapter 5, I will consider what happens after these various types of species are introduced to a local community.

MECHANISMS GENERATING REPRODUCTIVE ISOLATION

Speciation is not a single process. Myriad mechanisms can produce reproductive isolation among lineages, and sometimes multiple processes may operate simultaneously. Dobzhansky (1937a) and Mayr (1942) provided the first comprehensive catalogs of these mechanisms. The summary later provided by Coyne and Orr (2004) shows that very few if any have been added since those original compilations. Following these compilations over the years, I group these isolating mechanisms here into four major categories (table 4.1).

TABLE 4.1. Various speciation modes and their properties

Speciation Mode	Generations Required	Phenotypic Differentiation of Ecologically Important Traits	Coexistence as a By-product	Geographical Structure*	Examples
Ecological differentiation	Few to millions	Large	Yes	Sympatric, microallopatric, allopatric	Host shifts (herbivores among plants), habitat shifts, local adaptation in allopatry
Genomic and genetic incompatibilities	Thousands to millions	None to small	No	Allopatric primarily	Chromosomal rearrangements, Bateson-Dobzhansky-Muller incompatibilities
Hybridization & polyploidization	1	Small to large	No	Sympatric, microallopatric	Allo- and autopolyploidy
Sexual selection and mate recognition	Few to millions	None to small	No	Microallopatric, allopatric	Prezygotic mate recognition differentiation in courtship songs, pheromones, mechanical contacts, color, behavior

*Sympatric—same community; microallopatric—different spatial locations (e.g., adjacent host plants, adjacent soil patches, etc.) in same community connected by link species; allopatric—different communities in different regions that result in very low dispersal rates between the communities.

In what follows, I focus on the outcome of generating reproductive isolation and not on what initiates the evolution of reproductive isolation or the genetic causes and consequences of speciation. Specifically, I concentrate on (1) how various mechanisms generating reproductive isolation shape phenotypic differences among progenitor and daughter species, (2) what types of traits are affected to generate reproductive isolation relative to the ecologically important traits favored by natural selection in the communities in which these species are embedded, and (3) the geographic relationships among progenitor and daughter species and what these mean for changing species diversity within local communities (i.e., α-diversity) and among communities across the landscape (i.e., β-diversity). Readers interested in the conditions that favor the operation of different speciation mechanisms and the specific mechanisms that enforce reproductive isolation should pursue the references cited throughout in more depth.

Ecological Speciation

When ecologists think of speciation, we typically think of ecological speciation; we are, after all, very chauvinistic. Fundamentally, ecological speciation occurs when natural selection causes lineages to diverge from one another in traits that determine ecological performance, and reproductive isolation occurs as a by-product of, or evolves in concert with, this differentiation (Schluter 2000, Coyne and Orr 2004, Rundle and Nosil 2005, van Doorn et al. 2009, Nosil 2012). Ecological differentiation may be driven by alternative selection pressures existing over a range of alternative geographic scenarios; the geography is important in determining the effects of this differentiation on local and regional community structure immediately following the speciation event and whether other traits must also be involved in generating reproductive isolation. Consequently, this section is organized to consider ecological speciation mechanisms operating over large to small geographic scales.

Mayr's (1942, p. 160) "dumbbell" model of *allopatric speciation* is typically what many have in mind when they conjure up the idea of large-scale allopatric speciation—that is, a widely distributed species with its range dissected by some large-scale barrier to gene flow (e.g., a mountain range, a wide river, or other type of inhospitable habitat) that persists over many, many generations as the populations on either side diverge from one another (fig. 4.2A; Mayr 1942, Simpson 1944, Barton 1988, Coyne and Orr 2004). If the ecological selection regimes differ on either side of the barrier, the two areas will ecologically diverge from one another (e.g., the areas will represent different points along the ecological gradients considered in chapter 3). Reproductive isolation may develop in a number of ways in concert with ecological differentiation in the various areas. Allopatry by

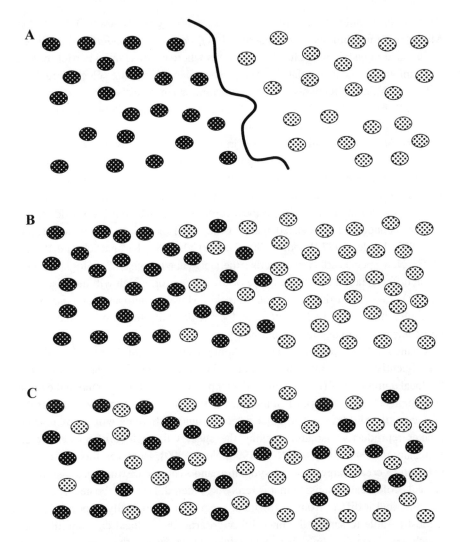

FIGURE 4.2. Various habitat patch configurations that generate ecological gradients leading to speciation. Panel *A* illustrates a vicariant barrier that reduces or eliminates movement of individuals between two geographic areas that differ in their ecological conditions. Panel *B* illustrates such an ecological gradient across the landscape that lacks a dispersal barrier at the transition zone. Panel *C* shows a landscape where habitat patches are interspersed throughout the landscape.

itself is a barrier to reproduction among the types if dispersal between the areas is prevented. However, were the types in the different areas to interbreed, they may still be incompatible. Reproductive isolation may result directly from the differentiation because of the fitness shortcomings of hybrids in both areas. Alternatively, reproductive isolation may develop because other genetic incompatibilities develop in parallel to ecological differentiation (e.g., Bateson-Dobzhansky-Muller incompatibilities, as discussed below). *Parapatric speciation* (i.e., differentiation along ecological clines or gradients where no substantial barrier to gene flow among adjacent populations exists other than a changing selection regime) will also produce effectively the same outcome when speciation is accomplished (fig 4.2*B*; Fisher 1930, Endler 1977, Lande 1980, Gavrilets 2000b, Doebeli and Dieckmann 2003, Coyne and Orr 2004).

None of these speciation scenarios in which an initially widespread progenitor species is dissected into some number of smaller but allopatric or parapatric daughter species do anything to alter patterns of local community structure. Each population adapts to the selection regime defined by the abiotic conditions and existing species interactions in the local community in which it is embedded, and this local adaptation is balanced against gene flow from nearby populations (Slatkin 1973, Endler 1977, Lande 1980, Pease et al. 1989, Kirkpatrick and Barton 1997). New species are created across the landscape because of ecological population differentiation across the range of the progenitor, which will increase β-diversity across the barrier or ecotone transition. However, allopatric and parapatric speciation have no immediate effect on local α-diversity since no new species are being introduced to any local community.

Change to local species richness will only occur if one of the resulting species subsequently expands after differentiation into the range of another by invading the local communities (i.e., having both sympatric and syntopic [found in the same community] distributions). Most analyses of such scenarios focus on whether the species will hybridize and whether selection against hybrids will promote even greater reproductive isolation among the species through *reinforcement* (e.g., Dobzhansky 1940, Kelly and Noor 1996, Noor 1999). However, this subsequent invasion to create large areas of sympatric and syntopic distributions also has a critical ecological dimension that is rarely considered. Because the new species is invading and it differs from the resident species in ecologically important traits, it must be able to satisfy the invasibility criteria for the local communities into which it is expanding; only then can it advance very far into the range of the other (chapter 2). If the invader cannot coexist, it will at best only be able to establish as a sink species in local communities at the edge of its range (Kirkpatrick and Barton 1997).

Considering these ecological consequences also suggests insights regarding hybrid zones between species, which are also typically considered only from a genetic perspective. Most examples of these hybrid zones occur at such ecotones that were thought to be the barriers that originally fostered speciation (Barton and Hewitt 1985, Harrison and Rand 1989, Harrison 1992). One or both hybridizing species may thus be sink species in the communities of the hybrid zones, which would just as easily explain why introgression of species is stopped and the positions of the hybrid zones are relatively stable.

Ecological speciation is not limited to differentiation across broad geographic areas. In fact, many of the best examples of ecological speciation imply that adaptation of one or a few local populations to a novel ecological environment is a powerful engine of speciation (Schluter 2000, Rundle and Nosil 2005, Nosil 2012). The populations may be at the periphery of a progenitor species range. This is peripatric speciation; imagine figure 4.2B shows the boundary of a species range where the dark-stippled patches support communities in which the species of interest resides, and the light-stippled patches support communities in which the species of interest would be a walking-dead species were it to invade. Otherwise, the populations may be embedded within the progenitor's range (fig 4.2C). The mechanism fostering ecological speciation in both these geographic cases is the same, and therefore I make no distinction here. I include in this category both shifts onto new resource hosts by herbivores or pathogens, and habitat shifts of other species types from one local community into another.

Speciation via a *habitat/host shift* (for simplicity I will refer to both as a habitat shift) is initiated by a founding population being established in the community of a habitat where this species is initially a walking-dead species. The ecological and evolutionary dynamics of this founding population probably most closely follows the scenario considered by Gomulkiewicz and Holt (1995) in their analysis of a population undergoing *evolutionary rescue* (although this is not strictly "rescue" in this case); see also Iwasa et al. 2004, Orr and Unckless 2008, Chevin and Lande 2009, Loverdo and Lloyd-Smith 2013, Uecker et al. 2014. They also follow the dynamics like that described in figures 3.2–3.3, except that the introduced species begins as a walking-dead species. As a result, the synergy of ecological and evolutionary dynamics of the species and the community are critical to the success or failure of a habitat shift.

Imagine a landscape as pictured in figure 4.2B or C in which a progenitor species inhabits the communities found in the dark-stippled habitat patches, but is absent from communities in the light-stippled patches. On rare occasions, individuals of this progenitor establish small populations in the community found in a light-stippled patch. Because it is continually being excluded from these patches when these founder populations are established, this is a walking-dead species in the local community that is typically extirpated. For example, in my own work on *Enallagma* damselflies (Odonata), one set of species is found as larvae only in lakes with fish as the top predators. Females of these species sometimes make mistakes and oviposit in lakes that have large, active dragonflies as the top predators (McPeek 1989). However, larvae of these species are found only very rarely in lakes where fish are absent and large dragonflies are the top

A

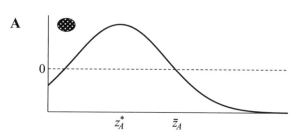

B

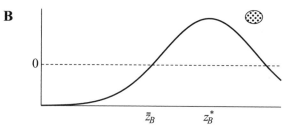

C

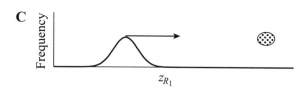

D

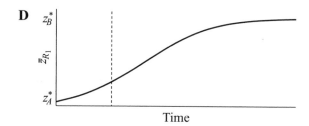

E

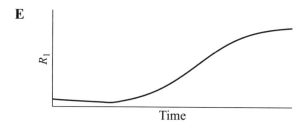

predators. Experiments have shown that predation by the large dragonflies is primarily responsible for excluding the species found in fish lakes from inhabiting these dragonfly lakes (McPeek 1990b). Fish and dragonfly lakes are intermingled across the landscape just as in figure 4.2C, and so these founder populations are probably being established continuously and then extirpated.

These founder populations, however, also experience natural selection in this new habitat. When a founder population of the species that is adapted to the community found in the dark-stippled patches is established in one of the communities in the light-stippled habitat, imagine that it moves from the selection regime depicted in figure 4.3A to the one in figure 4.3B. This founder population is now under selection to increase the value of the trait. For example, when they are experimentally transplanted to dragonfly lakes, the founder fish-lake *Enallagma* populations experience strong phenotypic selection pressures from dragonfly predation (Strobbe et al. 2009, Strobbe et al. 2010). As this founder population adapts to this new selection regime, the average fitness of the population will increase (chapter 3), which means that the population is declining at an ever slower rate as it adapts (Gomulkiewicz and Holt 1995). However, until this population's average trait value changes to the point where the population's average fitness is above replacement (i.e., the value of $\bar{z}_{R_1}$ giving $\ln\left(\bar{W}^B_{R_1}\right) > 0$ in fig. 4.3B), the population is a walking-dead species in this community. If the founder population can adapt past this point to change itself from a walking-dead species into a coexisting species ($\ln\left(\bar{W}^B_{R_1}\right) > 0$ when $R_1 \approx 0$), it will have successfully invaded this new habitat (fig. 4.3; see Gomulkiewicz and Holt 1995). In other words, can this founder population adapt enough before going extinct to satisfy the *invasibility criterion* of the local community module in which it now finds itself? Because of the inherent race between extirpation and adaptation, this initial stage of speciation via habitat shifts must occur very rapidly, probably over, at most, tens of generations.

FIGURE 4.3. The ecological and evolutionary response of a founder population making a successful habitat shift to a new environment (as in Gomulkiewicz and Holt 1995). Panel *A* illustrates the fitness surface for the ancestral habitat of this population (e.g., the *dark-stippled* habitat patches in fig. 4.2C), and panel *B* gives that fitness surface for the habitat being invaded (e.g., the *light-stippled* habitat patches in fig. 4.2C). The parameters z^*_A and z^*_B identify the optimal phenotypes in the two habitats. Panel *C* shows the initial phenotypic distribution when the population is founded in the new habitat type, and the arrow shows the direction of adaptive evolution (i.e., uphill on the fitness surface in panel *B*). Panel *D* shows the change in the average phenotype in this population over time as it adapts to the ecological conditions in this new habitat. Panel *E* shows the change in abundance for this population over time. The population declines in abundance until the mean phenotype passes the value at which $\ln(\bar{W}_B) = 0$. This is the point where adaptation has converted this from a sink to a source population.

Once the founder population has adapted enough to be a coexisting species in the local community, dispersers from this population can then colonize other habitat patches in the area that support similar communities (e.g., light-stippled habitat patches in fig. 4.2*B* or *C*, to continue our hypothetical scenario). Because they now satisfy the invasibility criterion for these communities, only the rate of dispersal among these patches should limit this expansion (we'll get to this in chapter 6).

A critical evolutionary consequence of this invasion is that the populations adapted to these two different communities are now spatially segregated, and so the probability of breeding between the two is greatly reduced—a reproductive barrier is thus established. Any individuals that moved to the other community would have low fitness themselves, and if they survived long enough to mate, their hybrid offspring would have low fitness in both communities (e.g., Hatfield and Schluter 1999, Nosil et al. 2005). This level of ecological differentiation to reduce hybrid viability in both ecological contexts can be established in tens to hundreds of generations (Hendry et al. 2007). Other genetic incompatibilities may develop over much longer periods of time as well, but reproductive isolation will initially be determined by allopatry plus hybrid inviability due to the difference in phenotypes required to perform well in the two community types.

Thus, such habitat shifts come with a price; in successfully invading and adapting to this new community, the new species has greatly decreased its average fitness in the ancestral community (e.g., figs. 4.3*A* and *B*). Members of this new adapted population are now restricted to coexisting only in this new community and would be a walking-dead species in the ancestral community, just as the progenitor species is restricted to coexisting in only the ancestral community. Again, using my own empirical work as an example, another set of *Enallagma* species are found only in dragonfly lakes. When these are experimentally transplanted into fish lakes, they are extirpated by fish predation (McPeek 1990b). The differences in predator susceptibility that enforce these habitat distributions result from large phenotypic differences between fish-lake and dragonfly-lake species (e.g., McPeek 1990a, 1995; McPeek et al. 1996; McPeek 1999, 2000; Stoks et al. 2003; Stoks and McPeek 2006). Phylogenetic studies have shown that fish lakes are the ancestral habitat for *Enallagma*. The dragonfly-lake species are the result of three independent habitat shifts in which these lineages adapted to this new environment to coexist with dragonfly predators, but in so doing, lost the ability to coexist with fish predators (McPeek and Brown 2000, Turgeon et al. 2005, Stoks and McPeek 2006). Comparisons of selection gradients measured in the field also suggest that the phenotypic differences between fish-lake and dragonfly-lake *Enallagma* species would require only tens of generations to evolve (McPeek 1995, Strobbe et al. 2010).

Given the race between extirpation and adaptation, a number of factors will influence the probability of a successful habitat shift (Gomulkiewicz and Holt 1995). The first is the phenotypic distance from the initial average phenotype of the founding population to the nearest adaptive peak on the new fitness landscape. Founder populations that start closer to the peak will have a higher probability of success, because they will start with less of a fitness disadvantage and will have less distance to adapt to make them source populations. The importance of this can be seen in the phylogenetic patterns of habitat shifts across the *Enallagma* phylogeny. The genus has two primary subclades with fish lakes being ancestral in both. Fish-lake species in one subclade are on average closer to the phenotypes that are favored in dragonfly lakes, and all three habitat shifts into dragonfly lakes occurred in this subclade (McPeek 2000). Second, greater additive genetic variation of the traits (e.g., V_{z_R} in equation (3.7)) under selection in the founder population will speed adaptation and thus increase the probability of a successful habitat shift. In fact, colonization of such new areas may permit the expression of cryptic genetic variation on which these new selection pressures can act (Gomulkiewicz et al. 1999, Ledón-Rettig et al. 2014). Finally, a larger size of the founding population will provide more time to adapt before the founder population is driven extinct. It is interesting to note that many examples of habitat shifts are associated with periods of climatic upheaval (e.g., Schluter and McPhail 1993, Rundle et al. 2000, Turgeon et al. 2005), which suggests that many of these issues (particularly initial population size) may be moot in many cases for reasons I will discuss later in this chapter.

This listing of what favors a successful habitat shift also highlights the primary reason why ecological opportunities in communities may be unfilled and why historical contingency plays a large role in community structure in general. A species with an appropriate phenotype must establish the founder population to accomplish the habitat shift. If such a species is not present in the region, such a founder population cannot be established. As Darwin (1859) first recognized, evolution can only modify what is available, and this carries down to the construction of community structure.

The reader may have noted that I have referred to the initial founder population as a walking-dead species in the community that it has invaded and not a sink species. This is because continual immigration of individuals from the populations adapted to the ancestral communities will frequently slow or prevent adaptation to the new selection regime (Kawecki 1995, Holt 1996, Holt and Gomulkiewicz 1997, Ronce and Kirkpatrick 2001, Tufto 2001). Phenotypic plasticity—the ability of individuals to express different phenotypes in response to different ecological conditions—may also facilitate habitat shifts by permitting colonizing populations to persist for a longer time (West-Eberhard 1989, Pfennig et al. 2010,

Scheiner et al. 2015). As a result, successful habitat shifts are expected to be rare; for example, only three that we know of occurred in the eight to nine-million-year history of the *Enallagma* clade (Turgeon et al. 2005, Callahan and McPeek 2016), but more may have successfully taken place over this time, followed by extinction. However, over the history of a clade, these rare events are what permit expansion of the clade's ecological breadth.

Such habitat shifts also represent many of our best empirical examples of ecological speciation. Shifts from marine to freshwater environments are common in many lineages, including sticklebacks (Schluter and McPhail 1993, McKinnon et al. 2004), copepods (Lee 1999, 2000), and amphipods (Hou et al. 2011). Sticklebacks (Schluter and McPhail 1993, Rundle et al. 2000) and whitefish (Lu and Bernatchez 1999, Østbye et al. 2006) have also diversified into benthic and limnetic forms within some freshwater lakes. Shifts among host plants have played an important role in the diversification of many herbivorous insect clades; examples are *Rhagoletis* flies (Feder et al. 2003), *Eurosta* gall-forming flies (Abrahamson and Weis 1997), and *Timema* walking sticks (Nosil et al. 2008). Assemblages of endemic plants on serpentine soil illustrate that habitat shifts can create very localized communities (Krukenberg 1951, Anacker 2014). In addition to *Enallagma* damselflies, *Lestes* damselflies (Stoks and McPeek 2006) and mosquitofish (Langerhans et al. 2007) have diversified into the various predation regimes found in different freshwater ponds and lakes. Mice and lizards have also diversified in pelage and body color to match environmental backgrounds in response to predation (Manceau et al. 2010). Caribbean lizards radiated to use different types of perches on and around trees (Losos 1990a, 1990b; Losos et al. 1998). I could go on, as the examples are abundant.

In all these cases, the habitat shifts increased both the α-diversity in the new communities receiving successful invaders and the β-diversity across the landscape. It is important to realize that such habitat shifts can occur at any position in a community, and consequently the shifts are like introducing coexisting species at random functional positions into a local community.

Speciation may also increase the number of functional positions and α-diversity in a community through the process of *sympatric ecological speciation*. Mayr (1963) dismissed the idea that speciation could be accomplished without geographic isolation, but the topic of sympatric speciation driven by ecological differentiation has remained a fruitful area for both conceptual and empirical inquiry. Little doubt exists that sympatric ecological speciation is possible and has occurred to create new species in some communities, but it seems to be very rare relative to ecological speciation mechanisms creating allopatric daughter species (Berlocher and Feder 2002, Coyne and Orr 2004, Gavrilets 2004, Bolnick and Fitzpatrick 2007). However, models of sympatric ecological speciation have sharpened insights into the various roles of ecology, coexistence, and reproductive isolation in creating biodiversity (Maynard Smith 1966, Kondrashov and Mina 1986, Dieckmann and Doebeli 1999, Kondrashov and Kondrashov 1999, Doebeli and Dieckmann 2000, Via 2001, Bolnick 2006, Gavrilets 2006).

Mechanisms of sympatric speciation via ecological differentiation all involve the breaking of a gene pool of a species in one community into multiple independent components that are pulled apart from one another by diversifying natural selection. I will consider the conditions that foster this ecological divergence in much greater detail in the next chapter. The central problem with sympatric speciation, and the reason it is probably rare, is the difficulty of generating reproductive isolation in concert with this disruptive selection (Berlocher and Feder 2002, Kirkpatrick and Ravigné 2002, Gavrilets 2004, Bolnick and Fitzpatrick 2007). Because the point of this chapter is not to explore when speciation will happen, I will not evaluate these various issues. However, what I do want to convey are the consequences to community structure of various speciation mechanisms, and these various mechanisms generating reproductive isolation in sympatric ecological speciation scenarios do have alternative implications for the resulting communities; I will explore these issues here.

Some modes of ecological differentiation within a community can directly generate reproductive isolation, and these operate largely like the parapatric speciation considered above because they typically involve some host or habitat shift. For example, the classic and best studied example of sympatric speciation via ecological differentiation is the shift of the tephritid fly *Rhagoletis pomonella* from feeding and living on the ancestral hawthorn fruits in eastern North America to feeding and living on apples (Bush 1969, Feder et al. 1994, Feder et al. 2003). Because the flies mate on the fruits on which they oviposit and on which their larvae will develop, reproductive isolation is established as a by-product of the ecological adaptation to the different hosts (Feder et al. 1994). In fact, most examples of sympatric ecological speciation are host shifts of herbivorous insects (reviewed in Berlocher and Feder 2002), or microhabitat differentiation over very small spatial scales; examples are trophic morphs of fish species that feed in different areas of a lake (Gíslason et al. 1999, Wilson et al. 2004) or differentiation of mole rats onto different soil types (Hadid et al. 2013). Such host/habitat shifts are creating a new species within the same community. For example, the different *Rhagoletis* species are developing on the fruits of different trees in the same stand, and parasitoids and predators that move among the trees in search of their prey will link the *Rhagoletis* species on different host plants into apparent competition modules within the community.

Reproductive isolation may also arise either as a pleiotropic effect because some of the traits involved in ecological differentiation also influence mating preferences; these have been termed "magic traits" by Gavrilets (2004). For example, many species have difficulty mating when males and females are too different in size, and if ecological differentiation results in differences in body size, reproductive isolation may occur as a pleiotropic consequence. Alternatively, mating preferences may be genetically independent of the differentiating ecological traits, and sympatric speciation is still possible (excellent reviews of the theoretical literature can be found in Berlocher and Feder 2002, Gavrilets 2004, Bolnick and Fitzpatrick 2007); reproductive isolation can evolve in parallel to ecological differentiation because assortative mating among the ecologically differentiating types is also favored (van Doorn et al. 2009). However, few if any examples of sympatric ecological speciation have been suggested for these mechanisms.

In all these cases, sympatric speciation via ecological differentiation is introducing new species that satisfy the invasibility criteria of the community in which differentiation is occurring, and thus directly increases the community's α-diversity. In general, sympatric ecological speciation should not fundamentally change the community structure in which the progenitor is embedded. Differentiation of a species into two may add a related functional group to change the module (e.g., differentiation of the intermediate-trophic-level consumer in a linear food chain to make a diamond module); alternatively, differentiation may simply add more coexisting species to the same functional group (e.g., another apparent competitor, another resource competitor, or another diamond-module intermediate consumer; see fig. 2.1).

Genomic and Genetic Incompatibilities

Many different genetic mechanisms may cause reproductive isolation among putative species. For example, the most obvious candidate is implied by the fact that many closely related species have different numbers of chromosomes, including *Vandiemenella* grasshoppers, Malagasy lemurs, *Equus* horses, *Clarkia* wildflowers, *Mimulus* monkeyflowers, *Heliantus* sunflowers, *Thomomys* and *Perognathus* pocket gophers, and *Mus* and *Peromyscus* mice, to name just a few (White 1969, King 1993, Coyne and Orr 2004, Livingstone and Rieseberg 2004, Brown and O'Neill 2010). In fact, chromosomal breaks, fusions, and inversions have occurred throughout the evolutionary history of primates (Dutrillaux 1979, Stanyon et al. 2008); these events include human chromosome 2, which is the fusion of two smaller chromosomes found in chimpanzees (Yunis and Prakash 1982, Ijdo et al. 1991, Consortium 2005). It is therefore natural to think that such fundamental differences among species may play a direct role in the speciation events leading to those species.

A number of different mechanisms have been proposed in which chromosomal differences can generate reproductive isolation by creating problems during meiosis in hybrid individuals (see reviews in White 1969, 1978; King 1993). However, these earlier theories have a flaw in logic, particularly when applied to animal populations. The basis of reproductive isolation is that hybrids, which would be heterozygous for chromosomal differences, are hypothesized to have low fitness relative to homozygotes. The offspring of any individual with a new chromosomal variant will be heterozygotes by definition, and consequently are at an immediate and great selective disadvantage. This means that chromosomal variants should only increase because of genetic drift acting in opposition to natural selection in very small populations (Futuyma and Mayer 1980, Brown and O'Neill 2010, Schumer et al. 2014). This criticism is not important in plants that routinely reproduce by asexual means and that frequently undergo polyploidization (Ramsey and Schemske 1998, Rieseberg 2001, Ramsey and Schemske 2002, Rieseberg and Willis 2007).

While chromosomal rearrangements may not directly generate reproductive isolation (particularly in animals), they may indirectly reduce recombination between chromosomal regions that harbor genetic differences fostering reproductive isolation (Rieseberg 2001, Navarro and Barton 2003, Brown and O'Neill 2010). We must then consider the consequences of these genetic differences on the geographic and ecological structures of the resulting species.

Genetic incompatibilities that cause reproductive isolation are typically separated into those that prevent the fusion of gametes to produce a viable zygote (postmating, prezygotic) and those that reduce the viability of the resulting zygote (postzygotic). In postmating, prezygotic mechanisms, either the egg or sperm have viability problems themselves during insemination (e.g., sperm is lost from a heterospecific female's reproductive tract, pollen fails to germinate on a heterospecific stigma), the sperm and egg do not recognize one another (e.g., recognition proteins are different and sperm and egg fusion does not occur), or homospecific sperm outcompetes heterospecific sperm to effect mating (e.g., pollen germinate and grow faster through homospecific styles) (see the review in Coyne and Orr 2004).

Genetic incompatibilities also arise that reduce the intrinsic postzygotic fitness of hybrids. These mechanisms are generally referred to as Bateson-Dobzhansky-Muller incompatibilities, because these scientists proposed how such incompatibilities can arise without the intermediates being disfavored by natural selection (Bateson 1909, Dobzhansky 1937a, Muller 1942). They involve the evolution of alleles at different loci that are incompatible with one another (Orr and Presgraves 2000, Orr and Turelli 2001, Coyne and Orr 2004, Gavrilets 2004). For example, imagine a two-locus genetic system with A and a alleles at one locus, and B and b alleles at the other, and assume that any individual that has at least one a allele and one *b* allele has reduced viability or fecundity. Our imaginary species begins as an *AABB* homozygote in every population across its range. A Bateson-Dobzhansky-Muller incompatibility arises if populations in one area of this species' range became fixed for the *a* allele (*AABB*, *AaBB*, and *aaBB* are all equally fit) so that all individuals are *aaBB*, while populations in allopatry or parapatry from these became fixed for the *b* allele (*AABB*, *AABb*, and *AAbb* are all equally fit) so that all individuals in this area are *AAbb*. All offspring from matings between individuals from these two areas would now have *a* and *b* alleles and would therefore be less fit than the offspring of matings between individuals that are both from the same area. Bateson-Dobzhansky-Muller incompatibilities are a common cause of reproductive isolation in animals and plants (Coyne and Orr 2004).

For present purposes, two features of all these genetic incompatibility mechanisms are important: (1) the differences that cause genetic incompatibility must arise in allopatric or parapatric areas of an existing species' range, and (2) the changes that cause incompatibilities do not affect the level of adaptation to their local communities (Coyne and Orr 2004, Gavrilets 2004). Consequently, no new species are introduced into local communities when such a speciation event occurs, and the evolution of these incompatibilities do not affect the processes or degree of local adaptation to the community in which the species resides. While the evolution of genetic incompatibilities will shape patterns and rates of clade-level diversification and increase species richness over large biogeographic areas, speciation caused by these forms of reproductive isolation should not affect the primary structure of communities. However, as we will see below, creating new species in different parts of the landscape does provide new species if these communities were to somehow later be mixed together.

Hybridization and Polyploidy

Hybridization and polyploidy, sometimes separately and sometimes together, have clearly been involved in giving rise to numerous species of animals and plants (little is known about their importance in other eukaryotes). Hybridization among species sometimes occurs in animals and is common in plants (Mallet 2005), and polyploidization (doubling the number of chromosomes) can occur through many different genetic routes (Ramsey and Schemske 1998). Thus, the seeds for these speciation mechanisms are common in many taxa, but new species are formed by these mechanisms much more commonly in plants than in animals (Muller 1925, Orr 1990, Mable 2004). For example, Otto and Whitton (2000) estimate that approximately 7% of speciation events in ferns and approximately 2–4% of speciation events in flowering plants involve polyploidization. While much less common, clear examples of animals resulting from these processes do exist (Mable 2004, Gross and Rieseberg 2005, Mallet 2007, Otto 2007).

Whole-genome duplication events that double the number of chromosomes within a single species—termed *autopolyploidy*—were clearly important in the deep history of flowering plants (Bowers et al. 2003, Jiao et al. 2011, Soltis and Soltis 2016) and vertebrates (Dehal and Boore 2005, Donoghue et al. 2014). Auto-polyploidy continues to occur routinely in plants to produce new species (Stebbins 1940; Grant 1971; Ramsey and Schemske 1998, 2002; Rieseberg and Willis 2007, Soltis et al. 2007, Parisod et al. 2010). Estimates suggest that autotetraploids are produced among gametes in flowering plants per generation on the order of 10^{-5} (Ramsey and Schemske 1998). Polyploidization following a hybridization event between two species—termed *allopolyploidy*—is an even more common specia-tion mode in plants (Stebbins 1940; Grant 1971; Ramsey and Schemske 1998, 2002; Otto 2007; Rieseberg and Willis 2007; Wood et al. 2009). Estimates suggest that allotetraploids are produced among the gametes of a hybrid individual on the order of 10^{-2}–10^{-3} (Ramsey and Schemske 1998).

Polyploidization is, however, not essential to the process. The offspring of a hybridization event between two species may in some rare cases be reproductively isolated from their parents and thus constitute new species produced from a single mating event. This speciation process—termed *homoploid hybrid speciation*—is much rarer than speciation events involving polyploidization, but clear examples exist in both plants and animals (Rieseberg 1997, Buerkle et al. 2000, Gross and Rieseberg 2005, Mallet 2005, Schumer et al. 2014). In fact, one of the best studied examples of hybrid speciation involves homoploid hybrid speciation in *Heliantus* sunflowers (Rieseberg et al. 2003).

Hybrids are typically assumed to be intermediate in phenotype to the hybridiz-ing progenitors. However, reviews of the phenotypic and ecological properties of hybrids and polyploids relative to their progenitors suggest that the only general-ization to be made is that they are simply phenotypically different from their pro-genitors. The phenotypic distribution of hybrids spans the entire range between, and is inclusive of, the parental species' phenotypes, but frequently some hybrid phenotypes are outside the range of the parentals. Gene dosage effects, novel gene combinations, and epistatic interactions all have the potential to produce novel phenotypes not found in the parental species that can favor occupation of eco-logical conditions outside the scope of the parentals (Levin 1983, Rieseberg and Ellstrand 1993, Ramsey and Schemske 2002, Ramsey and Ramsey 2014). These phenotypic differences will make them ecologically different as well. For exam-ple, diploids and tetraploids of the saxifrage *Heuchera grossulariifolia* attract different pollinator species and are attacked by different herbivore species even when growing sympatrically (Thompson et al. 2004).

By definition, all of these are modes of sympatric speciation, because the new species are the direct offspring of a mating event. Thus, the new species are immediately members of the same community as their progenitors. Moreover, each newly introduced species begins at a small population abundance as an invader and may or may not satisfy its invasibility criterion (i.e., $\ln(\bar{W}) > 0$ or $\ln(\bar{W}) < 0$, respectively) because its initial phenotypic distribution is constructed without regard to the fitness topography it is experiencing. Polyploidy and hybridization as speciation mechanisms seem to be more prevalent in (but not exclusive to) taxa that have some degree of asexual reproduction, hermaphroditism, and selfing involved in the breeding systems, as well as a long-lived life cycle (Rieseberg 1997, Mallet 2007). All of these would permit a single polyploid individual to produce a small population on its own that could persist for a while, even with $\ln(\bar{W}) < 0$.

If the new species meets its invasibility criterion from the start (i.e., $\ln(\bar{W}) > 0$), it begins as a coexisting species in the community and will increase in abundance, and the rest of the community members will have to adjust accordingly. Its introduction may change the coexistence criteria for other species in the community, including its own progenitors. That is, do the progenitors still meet their own invasibility criteria once this daughter species is present? Also, other species may be extirpated locally. Along the way, the new species may have to also adapt to move to its local fitness peak. Its introduction may also alter the selection regimes of the other community members and cause adaptive adjustments in them as well.

However, the new species may initially not satisfy its invasibility criterion (i.e., $\ln(\bar{W}) < 0$) and then start as a walking-dead species. It may thus go extinct, which is probably the fate of most new species produced by these means. However, just as with habitat shifts, this initially walking-dead species may ultimately be successful if it can adapt sufficiently to produce $\ln(\bar{W}) > 0$ before it is driven extinct (Gomulkiewicz and Holt 1995, Chevin and Lande 2009).

In fact, hybrid and polyploid species may undergo habitat shifts away from their progenitors as well (e.g., shifting habitats as in figure 4.2C). For example, three *Helianthus* species resulted from three independent homoploid hybridization events between *H. annuus* and *H. petiolaris* (Rieseberg 1991, Rieseberg et al. 2003). The parental species have broad distributions, but the three new species have small geographic ranges and are found "in the most extreme habitats of any *Helianthus* species" (Rieseberg et al. 2003, p. 1213). These distributions imply that each new hybrid species was unable to coexist with the parentals in their habitats, but each accomplished a habitat shift into its more extreme habitat, either because its initial phenotype was partially or wholly adapted to this new environment or it was able to adapt sufficiently to maintain populations there.

Sexual Selection, Sexual Conflict, and Mate Recognition

Fundamentally for sexually reproducing species, speciation influences whether this female will mate with that male. If the individuals of one sex choose among individuals of the other to determine which are suitable and unsuitable mates, the choosing individuals are applying selection to the traits (Lande 1981; Sved 1981b, 1981a; Arnold 1983). These decisions can impose strong selection on these phenotypes because they influence the fitnesses of the chooser, the chosen, and the unchosen (Darwin 1871, Andersson 1994, Hoekstra et al. 2001, Kingsolver et al. 2001, Siepielski et al. 2011a).

If differences in mating preferences arise among populations in terms of what traits define suitable and unsuitable mates, speciation can result. Moreover, this can be a very effective means of generating new species, because differentiation directly affects mating preferences (Lande 1981, West-Eberhard 1983, Panhuis et al. 2001, Kirkpatrick and Ravigné 2002, Ritchie 2007). In fact, a number of comparative studies have shown that clades in which various types of sexual selection operate (e.g., those with dichromatic species, polyandrous mating systems, greater sexual ornamentation) tend to have more species than clades lacking such features in the sexual systems (reviewed in Barraclough et al. 1995, Owens et al. 1999, Panhuis et al. 2001, Ritchie 2007).

Individuals (usually females) may express mate choice preferences for a variety of reasons. Obviously, individuals that choose mates that maximize the number and quality of their offspring should be at a fitness advantage in a population, and the individuals being chosen would also have a mating advantage. Choices based on such criteria are typically characterized as choice for *good genes* that will be passed on to the chooser's offspring (Bateson 1983, Andersson 1994). Choice may be based on the values of ecologically important traits directly or traits that signal the potential ecological quality of a mate; examples are nuptial gifts, indicators of immunocompetence to fight parasite infections (Hamilton and Zuk 1982) or magic traits (Gavrilets 2004). If mate choice is based on these types of traits, sexual selection will tend to reinforce what is favored by natural selection. Sexual selection for traits that would directly increase fitness or that would indirectly serve as indicator traits can both facilitate ecological speciation by enhancing assortative mating among different ecological types (Lande 1981, Kondrashov and Kondrashov 1999, Kirkpatrick and Ravigné 2002, Gavrilets 2004, Ritchie 2007, van Doorn et al. 2009).

Fisher (1930) proposed that mate choice for any trait can confer an overall fitness advantage to the bearers of that trait, even if the trait has negative ecological consequences, and this can lead to a *Fisherian runaway sexual selection* process. The classic example presented is that of a bird in which females prefer males with longer tails, but these males have lower survival because of the ecological consequences of being poorer flyers. Various models have demonstrated that such scenarios are possible (O'Donald 1980, Lande 1981, Kirkpartick 1982). The evolutionary equilibrium is a balance that is struck between various fitness components (as in chapter 3, but with one fitness component being mating success), and the resulting species will be somewhat maladapted to its ecological environment. If mating preferences for different male traits evolve in different populations, this can lead to rapid allopatric speciation (Lande 1981). Other models have been proposed for how differentiation via sexual selection may also be accomplished in sympatry (Turner and Burrows 1995, Higashi et al. 1999), but the conditions to accomplish this are probably quite rare (Kondrashov and Mina 1986, Ritchie 2007)

Mate preferences may also be based on the inherent biases that exist in the sensory systems used in communication between males and females (Endler and Basolo 1998, Kokko et al. 2003, Fuller et al. 2005, Horth 2007). Particular females may be able to best see certain colors or hear certain frequencies because of their sensory systems, and these *sensory biases* can skew their mating probabilities toward males that match these biases (Ryan and Keddy-Hector 1992, Wilczynski et al. 1992, Boughman 2001, Seehausen et al. 2008). These biases may be directed toward all types of traits, both those that are ecologically important (e.g., body size for conspicuousness) and those that do not affect the ecological performance of the individual (e.g., mating calls and breeding coloration). These biases are often influenced by the environment in which the choices are expressed (e.g., fish show different biases for male colors with water depth because of color attenuation or turbidity); see Endler 1992, Endler and Basolo 1998, Boughman 2001, Seehausen et al. 2008, Scordato et al. 2014. Consequently, sensory biases are expected to drive primarily allopatric speciation between populations that differ in conditions favoring different signals (Boughman 2002). However, sympatric speciation driven by sensory biases in female preferences for male breeding coloration has been demonstrated for African cichlid species that breed at slightly different water depths (Seehausen et al. 2008).

Conflict between the sexes over mating can also promote speciation (Gavrilets 2000b, Gavrilets and Waxman 2002, Martin and Hosken 2003). *Sexual conflict* occurs when factors that enhance the reproductive success of one sex reduce the fitness of the other sex (Arnqvist and Rowe 2005). For example, in many species, male reproductive success increases with more matings, but female reproductive success plateaus or declines at high mating frequencies. Sexual conflict typically results in an arms race between the sexes, with females usually trying to reduce the number of matings in various ways, and males trying to increase the number. Consequently, sexual conflict is most likely expected to cause differentiation in traits that have little or no effect on ecological performance but are important for mating, insemination, or fertilization. Sexual conflict can drive speciation in both allopatry and sympatry (Gavrilets 2000a, Gavrilets and Waxman 2002).

The mating decisions involved in sexual selection can increase an individual's fitness. Selection also favors individuals that can discriminate among potential mates on the basis of whether they are partially or fully reproductively isolated from the choosing individual; that is, discriminating among conspecifics and heterospecifics can minimize the chances of producing no or inferior offspring. These types of mating decisions are typically identified under the rubric of *mate recognition* (Paterson 1985, Ryan and Rand 1993, Phelps et al. 2006, Mendelson and Shaw 2012). The traits involved in mate recognition are the species-specific identifiers that are often used by taxonomists to place individuals into a species (Eberhard 1985, Huber 2003). These include the unique and distinctive color patterns seen in the (primarily) males during the breeding season (Price 1998, Seehausen and van Alphen 1998, van Alphen et al. 2004, Price 2008, Salzburger 2009); unique mating calls (Otte 1989, Wells and Henry 1992, Ryan and Rand 1993, Shaw 2000, Gerhardt 2005); volatile and contact pheromones that mates use to attract or discriminate among suitors (Lofstedt 1993, Wyatt 2003, Bickford et al. 2007, Johansson and Jones 2007, Higgie and Blows 2008); and morphologically distinct structures used during courtship and mating (e.g., genitalia, claspers and other secondary sexual structures) that differ among species (Dufour 1844, Eberhard 1985, Shapiro and Porter 1989, Arnqvist 1998, Hosken and Stockley 2004, McPeek et al. 2008, Simmons 2014).

Individuals of one sex (typically males) display these traits, and individuals of the other sex (typically females) have preferences for mating with individuals displaying particular values of these traits; the *mate recognition system* of a species is the combined interaction of that species' identifier traits expressed in one sex and the preference functions for those traits expressed in the other sex. These identifier traits usually are merely badges of species identity to potential mates, and the particular features of such traits are typically ecologically unimportant, as are the preference functions of the other sex. While the general expression of these traits may have ecological consequences, the particular mating call, pheromone blend, or shape of a secondary sexual structure that a male expresses does not influence its ability to interact with predators or mutualists, to garner resources, to fight off diseases, and so forth. For example, calling male frogs may generally attract predators, but predators do not typically key to specific features of one species' call.

Clearly, differentiation of the traits involved in mate recognition will directly result in speciation (Hoskin and Higgie 2010). If the selection pressure that would favor such differentiation involves discriminating among conspecific and heterospecific mates, how can differentiation be generated within a species? Imagine two species that co-occur in a local community, and these two species have similar but not identical mate recognition systems (e.g., two frog species with similar but not identical mating calls). These two species are also partially or completely reproductively isolated from one another: for example, they have accumulated Bateson-Dobzhansky-Muller incompatibilities with one another, and are ecologically capable of supporting populations in the same community. However, males of both species promiscuously court females of both species. Females that sometimes make mistakes in their mating choices by choosing to mate with a heterospecific will have lower fitness, which can generate intense selection on the preferences to discriminate conspecifics from heterospecifics. Females that prefer conspecific males with traits that are the most different from heterospecific males will be favored, because they will make the fewest mating mistakes (McPeek and Gavrilets 2006, Pfennig and Ryan 2006). Selection should, therefore, cause the average female mating preference in each species to move away from the heterospecific male trait distribution. This evolution of female preference in turn cause the male traits of the two species to diverge as a by-product (Lande 1981; Sved 1981b, 1981a). This divergence should occur in only a few to ten or so generations (Higgie et al. 2000, McPeek and Gavrilets 2006, Pfennig and Ryan 2006). Thus, in a single community, these two species diverge in their mate recognition systems to ensure that females do not make mating mistakes with heterospecifics: this is frequently termed reproductive character displacement.

However, across the range of one (or both) of these species, some populations may co-occur with the other species and some may not (fig. 4.4). For example, imagine two species coming into secondary contact at their range boundaries, either at an ecotone (fig. 4.4A) or simply expanding their ranges in the same type of habitat (fig. 4.4B). In populations where the two species are sympatric, they first will differentiate from the other in its mate recognition system, but in populations where the other species is absent, no differentiation will occur. Thus, a by-product of reproductive character displacement will be that populations interacting with this other species will also differentiate from populations of its own species that do not interact with this other species (Howard 1993, McPeek and Gavrilets 2006, Pfennig and Ryan 2006, Hoskin and Higgie 2010). If the resulting differentiation in the mate recognition system is large enough, populations that are sympatric and allopatric with this other species may differentiate enough to make individuals from these different populations not recognize one another as potential mates. In other words, local reproductive character displacement from one species can generate reproductive isolation between different populations of the same species (Howard 1993, McPeek and Gavrilets 2006, Pfennig and Ryan 2006, Hoskin and Higgie 2010); Hoskin and Higgie (2010) termed this process *reproductive character displacement speciation.*

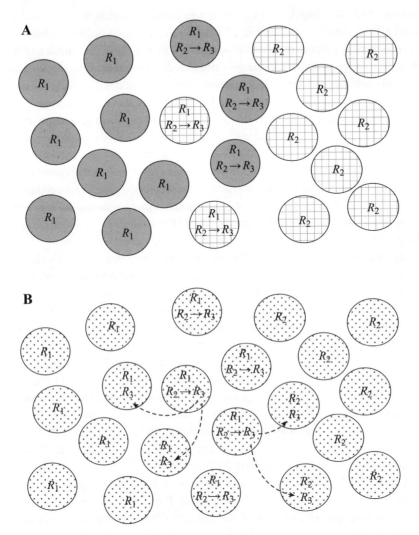

FIGURE 4.4. Two scenarios in which reproductive character displacement speciation occurs. In both, species R_2 differentiates its mate recognition system from that of R_1 in communities where the two are sympatric. McPeek and Gavrilets (2006) showed that this is most likely to occur when R_1's abundance is substantially greater than that of R_2, such as when R_2 colonizes a community in which R_1 is already at its equilibrium abundance. Panel A shows these two species meeting at the ecotone as defining the edges of the two progenitor species respective range boundaries. Panel B shows this occurring in the same types of habitat patches—for example, as these two species recolonize a deglaciated area—and the new (R_3) species then spreads to other local communities with the both progenitors.

Obviously, this implies differentiation in allopatry which does not add new species to the local community in which the differentiation occurs. In figure 4.4, the communities where differentiation occurs still only have two species in them at the end of the process. Moreover, since this differentiation only involves traits affecting mating preferences (e.g., mating calls, pheromones, breeding colors and displays), the ecological character of the community remains unchanged. Consequently, dispersal of R_3 back into its ancestral areas where R_2 already resides would be introducing a neutral species. This may be a prime source of neutral species in communities. I will elaborate on this shortly.

A number of examples of this process have now been studied experimentally. A well-studied example is the contact zone between two partially reproductively isolated lineages of the green-eyed tree frog, *Litoria genimaculata*, in northeast Queensland, Australia (Hoskin et al. 2005). Hybridization between these two lineages was much more detrimental to one lineage (identified as the "S" lineage in the paper). Populations of the S lineage that were sympatric with the other lineage were differentiated in both male mating calls and female preferences for these calls from populations that were allopatric with the other lineage. In fact, experimental mate choice tests showed "complete premating isolation" between the S-lineages in sympatry and allopatry with the other lineage; females of each S-lineage group (i.e., sympatric or allopatric with the other lineage) chose to mate only with males from their own group (Hoskin et al. 2005). Other examples where reproductive character displacement from one species causes reduced mating success with populations not undergoing character displacement have been seen in a number of frogs (Littlejohn and Loftus-Hills 1968, Lemmon 2009, Rice and Pfennig 2010, Richards-Zawacki and Cummings 2011, Pfennig and Rice 2014), fish (Kozak et al. 2015), and insects (Higgie and Blows 2008, Ortiz-Barrientos et al. 2009, Porretta and Urbanelli 2012).

I have also hypothesized that this mechanism is one of the major drivers of speciation causing the radiation of *Enallagma* damselflies that produced 17 species from two progenitors within the past 250,000 years (McPeek and Brown 2000, Turgeon et al. 2005). While ecological speciation via habitat shifts clearly caused three of these speciation events (habitat shifts discussed above), eleven of the speciation events produced species that are ecologically indistinguishable from one another; they differ only in the claspers used by males to grasp females during mating, and the surfaces of females where these claspers grasp the females (McPeek et al. 2008, McPeek et al. 2011). Females use the tactile cues from the shape of the claspers as they are grasped to discriminate males as conspecifics or heterospecifics (Paulson 1974, Robertson and Paterson 1982). We have genetic evidence for restricted hybridization in local areas between the distantly related species that appear to be the progenitors of these two radiations (Turgeon et al.

2005), suggesting the importance of selection pressures on this mate recognition system. We have hypothesized that these *Enallagma* lineages underwent a radiation driven primarily by reproductive character displacement speciation as they recolonized lakes that formed during repeated cycles of deglaciation (McPeek et al. 2008). The random nature of colonization would have thrown different combinations of species together, which could explain why many species would have been formed by this process simultaneously (see below also).

Accumulating examples of co-occurring species that differ only in mating system features suggest that reproductive character displacement speciation may be an important source of neutral species in communities. Up to three cryptic species of *Chrysoperla* lacewings that differ only in courtship songs can be found living together (Wells and Henry 1992: the authors also review other examples of co-occurring crickets that differ only by courtship songs). Otte (1989) discusses three radiations of crickets on Hawaii in separate colonization events (see also Shaw 2000). He describes many different areas where multiple species, based on call differentiation, occupy the same microhabitats: "species believed to be particularly closely related often have what appear to us to be virtually identical ecologies" (Otte 1989, p. 509). The same has been noted for the Hawaiian *Drosophila* radiation as well (Carson and Kaneshiro 1976). Price et al. (2000) postulated that the sexually dichromatic *Dendroica* warblers of North America are so ecologically similar because these speciation mechanisms drove their radiation about three million years ago. As many as 12 *Enallagma* species can be found co-occurring in lakes with fish as top predators in eastern North America. We have shown experimentally that two of these species whose last common ancestor forms the basal split in the genus eight to nine million years ago are ecologically indistinguishable (Siepielski et al. 2010), which implies that all the co-occurring species are similarly ecologically indistinguishable.

Mate recognition and sexual selection are not necessarily distinct processes but rather fall along a continuum (Ryan and Rand 1993, Boake et al. 1997, Pfennig 1998, Castellano and Cermelli 2006, Mendelson and Shaw 2012). Mate recognition decisions typically function to prevent low fitness whereas sexual selection decisions are typically associated with increasing the individual's fitness—two sides of the same coin. For present purposes, the critical issue for all speciation mechanisms involving the differentiation of mate choice all along this continuum is what traits are involved.

I hope this brief review makes clear that ecological differentiation is a prominent but not the only (and perhaps not even the most frequent) speciation mechanism. Ecological speciation will directly create coexisting species. Hybridization and polyploidy will produce new species with ecologically important traits that are different from their progenitors but that have no relationship to what is adaptive in their community; subsequent adaptation may be essential to permit them to persist and coexist. Speciation through the development of Bateson-Dobzhansky-Muller incompatibilities, and through many forms of sexual selection and mate recognition differentiation, will produce new species that have small or no ecological differences from their progenitors. In other words, speciation can produce new species that fill the entire gamut of species types in communities, from coexisting to walking-dead to neutral species.

Because I am not concerned with the mechanics of establishing reproductive isolation, I have not considered formal models of the process here (see Gavrilets 2004 for an excellent introduction to those models). For present purposes, the critical issues are how the traits that influence ecological performance have changed in the various mechanisms creating new species relative to their progenitors, and in what communities are new species produced relative to the communities occupied by their progenitors.

Change in ecologically important traits will determine the starting point for any adaptation going forward for the new species. In the framework I established in chapter 3, this could mean two things. First, the new species may have changed in the trait that is being modeled, which is the trait that influences its performance in interactions with other species and influences the other fitness components. Alternatively, the parameters of the models may have changed to some degree as well. One can think of this latter case as the situation where the phenotypic background for the trait being modeled has changed, since change in these parameter values implies change in other phenotypic traits not explicitly being modeled. So, two comparable species with different parameter values in their fitness function are phenotypically dissimilar from one another in traits not being modeled, and larger parameter value differences imply larger phenotypic differences.

New species introduced by speciation also may not have changed in these other traits, and thus the underlying parameters for their various fitness components are the same. If the trait being modeled did change, any of its subsequent evolution would occur in the same phenotypic background that the progenitor possesses. However, if the trait being modeled also did not change during the speciation process (e.g., Bateson-Dobzhansky-Muller incompatibilities developed or there was a change in reproductive traits that only influenced mating success), the new species would be ecologically identical to its progenitor. Subsequent evolution may cause these ecologically identical species to diverge. Subsequent evolution may also make ecologically dissimilar species converge. I explore these issues in the next chapter.

Moreover, speciation does not necessarily introduce any new species into communities directly. Only habitat/host shifts and sympatric speciation (including sympatric ecological speciation, hybridization, and polyploidy) will immediately introduce new species into a local community. Allopatric and parapatric speciation are both simply the modification of species that are already members of different local communities across the landscape or seascape. If allopatric and parapatric speciation are the most common geographical configurations for speciation (Coyne and Orr 2004), the logical conclusion must be that most speciation events increase the regional or biogeographic species pool, but do not affect local species richness at all.

CLIMATIC AND GEOLOGICAL INSTABILITY, AND COMMUNITY DISAGGREGATION AND RECOMBINATION

How then have local communities come to be so diverse if allopatric and parapatric speciation do not introduce new species into local communities? This implies that new species produced in one area routinely migrate out of that area and invade communities found in other areas. However, if they are created because dispersal or ecological boundaries exist in the first place, how do these new species breach these boundaries? Moreover, once breached, the new species must then contend with the invasibility criteria of the intact communities they are invading. Nothing in the ecological communities in which they arose has necessarily made them particularly adapted to the communities they are invading. Alternatively, if the invaders are simply neutral species with respect to some residents, they start at a huge frequency disadvantage relative to their resident neutral partners and will most likely go locally extinct by simple community drift.

Thinking about these subjects frequently leads one to imagine a fairly static or at least slowly changing landscape of local communities. Temporal variability that is driven either by intrinsic dynamics or slow extrinsic forcing is frequently considered as the only temporal ecological variability, but how can significant mixing of species across large biogeographic areas ever happen? Also, in the extreme, the consequences of wholesale obliteration and reorganization of communities are rarely if ever even postulated.

Yet, many disciplines outside of community ecology have conclusive data showing that the assemblages we study today did not exist only a few thousand years ago, even though many of the species have existed for millions of years. This is because of periodic climate upheavals caused by Milankovitch cycles in Earth's orbit (Hays et al. 1976, Zachos et al. 2001, Lisiecki and Raymo 2005, Palike et al. 2006, Cheng et al. 2013) and other major ecological perturbations to large geographic areas caused by geological activity (e.g., the closing of the Isthmus of Panama to the Caribbean marine realm [Bartoli et al. 2005]). For example, over much of the Quaternary period (i.e., the last ~2.6 million years), the northern reaches of the northern hemisphere were continuously glaciated, but these glaciers retreated and then almost immediately advanced again, at periodicities initially of about 41,000 years and then, more recently, at about 100,000 years (Petit et al. 1999, Lisiecki and Raymo 2005); in other words, the norm for the last few million years has been globally cool periods punctuated by spikes in global temperature of 4–8°C for a few thousand years that then plunged back into the next cool period (fig. 4.5). Glacial dynamics at the northern pole also had major effects on climate in the tropics because of major changes in ocean currents and continental rainfall patterns, so no area of the globe was immune from the consequences of these major perturbations (Wang et al. 2004, Trauth et al. 2005, Punyasena et al. 2008, Cheng et al. 2013). Given that molecular phylogenetic data show that more than 80% of extant species are more than a million years old (McPeek 2007) and that fossil data support this conclusion (Prothero 2014), almost all extant species today have lived through multiple major climatic events.

The signatures of response to rapid climatic change on the organization of Earth's biota show up almost everywhere one cares to look. For example, as the northern hemisphere glaciers retreated beginning approximately 18,000 years ago, the distribution of vegetation in North America and Europe changed continuously for about 10,000 years as species recolonized deglaciated areas and global temperatures increased on average by 8°C (Davis 1983, Huntley and Webb III 1989, Delcourt and Delcourt 1991, Overpeck et al. 1992, Jackson and Overpeck 2000, Jackson and Blois 2015, Maguire et al. 2015). Changes in local species assemblages consistent with these massive range movements are also seen in data covering the last 65,000 years in tropical America (Correa-Metrio et al. 2012, Correa-Metrio et al. 2014). Along with the movement of trees, ranges of mammal species showed huge shifts across large geographic areas (Van Devender 1986, Heaton 1990, Graham et al. 1996, Badgley et al. 2014). Beetles also show such huge range shifts (Coope 1979, Schwert and Ashworth 1988, Coope 1994, Ponel et al. 2003). Even ostracods at the bottom of the Atlantic Ocean show the same responses; fourfold spikes in local species richness of ostracods were associated with interglacial global temperature spikes over the past 500,000 years in the deep ocean, and these richness changes are due almost completely to range shifts (Cronin and Raymo 1997, Cronin et al. 1999, Yasuhara et al. 2009). In addition, the genetic structure across the ranges of most species today clearly show the phylogeographic signatures of large and taxonomically idiosyncratic range shifts in response to the retreating northern hemisphere glacier (Hewitt 1999, 2000).

An unmistakable feature of these massive shifts in species distributions is that assemblages and communities of species did not move en masse; species displayed very individualistic migration responses to climatic upheaval that completely reorganized assemblages and communities across the landscape. For

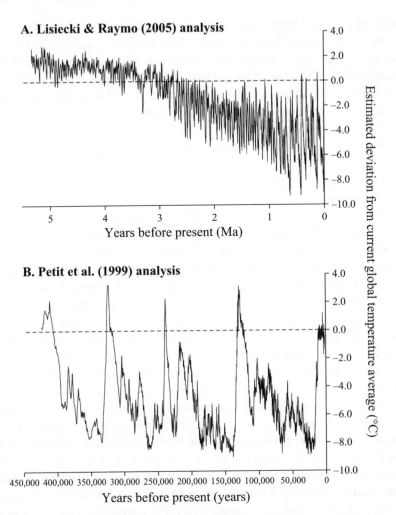

FIGURE 4.5. Estimated global average temperature profiles over the past (*A*) 5 million and (*B*) 400,000 years. Panel *A* shows data from Lisiecki and Raymo (2005) for the last 5 million years, and panel *B* shows data from Petit et al. (1999) for the last 450,000 years. The *dashed lines* at 0 in both panels show the current average global temperature as a reference.

example, Coope (1994) provided numerous examples of beetles that co-occurred in the British Isles 10,000 years ago and are now extinct there, but today have allopatric ranges in disparate parts of Eurasia (e.g., arctic Siberia, the Tibetan plateau, and the Caucasus Mountains). Graham et al. (1996) assembled a continent-wide database of fossil mammal distributions over the past 50,000 years or so and showed that many North American mammal assemblages from only a few thousand years ago have no modern analogs anywhere, even though most of the species are still extant: "Species dispersed diachronically, in divergent directions, and at variable rates. Mammal communities are continually, and unpredictably, emergent." (Graham et al. 1996, p. 1605). The same is true for vegetation: "Species assemblages have been disaggregated and recombined, forming a changing array of vegetation patterns on the landscape" (Jackson and Overpeck 2000 p. 194). First, these data prove that Gleason (1926) and Ricklefs (2008) were right about communities not being integrated units. More importantly, the *disaggregation and recombination of communities*, to borrow Jackson and Overpeck's (2000) phrase, on huge geographic scales has been a significant component of the history of today's biota.

The fossil record also offers examples of major environmental disruptions having significant impacts on community membership. For example, O'Dea and Jackson (2009) provide an excellent example of walking-dead species in the fossil record. The closing of the Isthmus of Panama four to five million years ago caused a series of environmental changes in the southwestern Caribbean basin that drove macroevolutionary changes in the cupuladriid bryozoans. Immediately after the closing, a round of speciation was sparked to raise species richness from 12 to 25 species. Then, for the next two million years, 5 of the species present before the closing and 7 of the brand new species slowly decreased in relative abundance until all 12 went extinct. All the species that went extinct were clonal, and the ultimate survivors were aclonal. Clonal species require high resource productivity to survive, and so their demise was attributed to the new oligotrophic conditions that prevailed in the basin after the closing (O'Dea et al. 2007, O'Dea and Jackson 2009). The important points here are that their extinction was driven by ecological changes in the community, and their extinction took two million years to finally occur, with their abundances declining all the while. The slow decline of many species to extinction seems to be a general feature found in the fossil record (Foote et al. 2007, Liow and Stenseth 2007). The walking dead may be among us for a very long time.

Pulses of extinction immediately followed by speciation in response to major environmental disruptions is a long-term signal in the fossil record (Foote 2000, 2005; Alroy 2008; Jablonski 2008). However, there is little evidence that the climatic turbulence over the last two million years significantly altered speciation and extinction rates overall (e.g., Faith and Behrensmeyer 2013).

Given that the experiences of species "in the last few million years" (Jackson and Johnson 2000) were to radically shift their ranges in most environments on Earth, we must begin to incorporate these facts into how we view the processes that have structured extant communities (Dynesius and Jansson 2000). Every 100,000 years or so for the past few million years, northern hemisphere glaciers retreated and then quickly advanced again (fig. 4.5), and local communities across the globe (particularly in North America and Eurasia, but all these various data also suggest that the perturbations had worldwide effects in all ecological realms) were disaggregated and recombined as species idiosyncratically moved in response to the rapidly changing climate. Each time, species ranges would shift great distances as their population dynamics tracked favorable environmental conditions. Those species that could not keep up would go extinct. Because species responded to these changes idiosyncratically, the local community that formed at any point on the globe at the end of an interglacial period (i.e., the end of a temperature spike in fig. 4.5) would represent new combinations of species being thrown together to interact in the local environmental conditions.

Once a new glacial period began, local community dynamics would again define species as coexisting, neutral, sink and walking-dead constituents, depending on whether they could satisfy their local invasibility criteria. However, any new local community probably contained large numbers of species that did not occupy that ground only a few thousand years before and that had little evolutionary history with one another. Consequently, a fair amount of adaptive evolution may have occurred as the species adjusted to one another in their new community settings.

Think about the new collection of species that were thrown together; some will adapt and coexist, some may represent neutral guilds of species occupying one functional group, some may ultimately be driven extinct, and some are sink species maintained by dispersal from nearby locations where they can maintain source populations. However, none of the species in each new local assemblage had to invade an intact community, as the invasibility criterion prescribes. Invasibility is a theoretical construct for evaluating whether a species coexists in a community that has settled to its long-term dynamical regime. *Invasibility is not a statement about how most real communities we study today were assembled in the past.* If paleontology, paleoecology, and phylogeography are any guide (which I believe they are), today's communities in most areas were largely thrown together over a few thousand years as glaciers retreated in the northern hemisphere starting 18,000 years ago. This is certainly true for communities that now occupy locations that would have been wholly under the glacier, but it is also true for other areas given the massive swings in global average temperature and the attendant shifts in rainfall patterns on land and in current circulation patterns in the oceans.

These repeated cycles of disaggregation and recombination also imply that present species ranges alone may provide few if any clues about the evolutionary history of a species, an assemblage or a clade and may in fact be downright deceptive. Remember Coope's (1994) beetles, Graham et al.'s (1996) mammals, Yasuhara et al.'s (2009) ostracods, and Jackson and Overpeck's (2000) trees. It also means that the implications of allopatric speciation for not increasing local species richness are largely moot; their mixing would only be delayed until the scrambling of geographic structures during the next interglacial period, when allopatrically produced sister taxa would be intermingled in the same communities as their ranges shifted in response to the changing climate.

Importantly, disaggregation/recombination would be a primary way of intermingling species that are effectively ecologically equivalent into the same community. One strong criticism of the possibility of neutral species is that it would be nearly impossible for one species to successfully invade a community that already contained an ecologically equivalent species at equilibrium abundance. However, if multiple ecologically identical species were produced in allopatry (e.g., Bateson-Dobzhansky-Muller incompatibilities or differentiation of the mate recognition systems) and subsequently colonized the same areas at the end of an interglacial period, none would start at a disadvantage, and all could potentially increase to substantial abundances before the community came to its abundance equilibrium. They may differentiate ecologically, but they may not (this will be addressed in the next chapter). If they do not, the expected time to extinction via community drift for one of them in this system might be orders of magnitude longer than the time to the next interglacial, and so both might persist together as a collection of neutral species in local communities across huge areas (McPeek and Gomulkiewicz 2005). This is in fact our hypothesis for the mechanism that may have generated and maintained the large number of neutral *Enallagma* species that have co-occurred for up to eight to nine million years in fish lakes across North America (McPeek and Brown 2000, Turgeon et al. 2005, McPeek et al. 2008, Siepielski et al. 2010, Callahan and McPeek 2016).

The brief interglacial periods in which species are colonizing new areas and moving across the landscape also provide opportunities for speciation not afforded during the glacial periods when communities would have approached or achieved ecological and evolutionary equilibria. For example, unique combinations of selection pressures and ecological conditions may prevail during these interglacials when the communities are scrambled, which may make habitat or host shifts more readily accomplished. When a community is at equilibrium, a particular herbivorous insect may be unable to persist on a particular host plant; the insect is somewhat maladapted to feeding on the plant (which may not supply it enough nutrition) and thus cannot maintain a population there in the face of its natural enemies (remember the criterion for coexistence in chapter 2). However, during an interglacial period, the herbivore and host plant may establish in an area where the former's natural enemies are either depressed in abundance or absent long enough for it to adapt to feeding on the plant. Once adapted, it can now maintain a population on the plant when its natural enemies are present. Thus, the problems of being a walking-dead species that also starts at low abundance may not be issues at all (as discussed in Gomulkiewicz and Holt 1995) if habitat shifts have occurred primarily during interglacial periods or other types of brief periods when communities are being reassembled. Also, remember that many of our best examples of habitat shifts are associated with these periods of climatic upheaval (e.g., Schluter and McPhail 1993, Rundle et al. 2000, Turgeon et al. 2005).

Modeling results also suggest that reproductive character displacement speciation is most likely to occur as the differentiating species (e.g., R_2 in fig. 4.4) is colonizing a community where the resident, reproductively isolated species from which it will locally differentiate (i.e., R_1 in fig. 4.4) is already at high abundance (McPeek and Gavrilets 2006). An interglacial period in which species are colonizing new regions is an ideal time to have such sequential colonization of a local area occur. Moreover, imagine that species R_2 colonizes areas already occupied by different species (e.g., R_5, R_6, R_7), and R_2 differentiates from each in unique ways to make every variant of R_2 a new species that is will not mate with the original R_2 or any of the other new R_2 variants (McPeek and Gavrilets 2006). In this way, a radiation of the mate recognition system to produce many species simultaneously may occur during one interglacial period. Moreover, since all these newly originated and ecologically equivalent species are colonizing new areas as they are differentiating, community invasibility is again not an issue for having multiple neutral species co-occurring in communities over large areas.

INVASIVE SPECIES AND HABITAT ALTERATION TODAY

The great expansion of human movement and trade over the past few centuries has accelerated the movement of species across what were previously dispersal barriers to invade new biogeographic regions. The consequences of invasive species on local communities is now a worldwide concern (Davis 2009). Many species that are dispersed into a new area never establish, some establish and have little effect, and a few cause major local extinctions and fundamentally change community structure (Davis 2009).

Obviously, these species being introduced today are shaped by the same ecological and evolutionary processes that influenced "native" species for their entire histories. Invasive species may adapt to their new ecological environs, and they must satisfy the invasibility criteria of the local community to establish and persist. Thus, the entire topic of this work is directly applicable to understanding the dynamics of invasive species today.

Like community assembly at the end of interglacial periods, invasive species today may also not be pure tests of invasibility that ecological theory envisions.

For many, they are entering communities that are being perturbed by human activity and so may not be close to their long-term ecological or evolutionary equilibria. Rather than this being an anomalous intervention into the operation of communities, human-assisted dispersal of species around the globe may simply be an extension of the processes that have punctuated the history of Earth for millions of years. Along with habitat alteration, this is propelling the world's biota into a period of heightened extinction and probably a mass extinction. If the paleontological record is any guide (which again, I think it is), a period of heighted speciation will follow, but it may take thousands to millions of years to begin. And *Homo sapiens* may not be one of the species that survives to see this burst of speciation.

NOW THAT WE HAVE ALL THESE SPECIES

In this chapter, I have focused specifically on the processes that create new species and the ways those new species may be introduced into communities. I have always been struck by how insular the discipline of community ecology has been about the sources and types of new species that can be introduced into communities and the ways these species are actually introduced. Even a cursory examination of the speciation literature would show that many more mechanisms than just ecological speciation have been involved in creating the species we study today. In addition, any perusal of the paleoecology, paleontology, or phylogeography literatures shows that local communities across the globe have been fundamentally changed over the last several thousand years, even before the influences of humans began. Thus, the science of community ecology must begin to embrace a much broader vision of how communities have been assembled and the sources of different types of species.

As I've considered the consequences of adaptation and speciation to the structure of the community in which those species reside, I have also always been struck by how insular the study of those topics has become. The myriad processes creating new species and introducing different types of species into communities also indicate that we must pay much greater theoretical attention to the ecological and evolutionary dynamics following speciation and introduction. For example, new species introduced via habitat/host shifts or via hybridization and polyploidy can for all intents and purposes be considered new species with random phenotypes relative to established community members and relative to the ecological opportunities that are available. Whether they immediately satisfy the invasibility criteria and their evolutionary dynamics that result upon their establishment will determine whether they persist for the long term or become extinct.

   Or imagine a new species arising because of reproductive character displacement with no ecological differentiation occurring in the speciation process. Once
this new species is formed, it may disperse back into areas inhabited by its progenitor, or the two may be thrown together because of their dispersal during a cycle
of community disaggregation and recombination. Once together, what will these
two ecologically identical species do? Will selection operate to differentiate them
from one another so that they exploit different ecological opportunities, or will
selection operate to keep them identical (i.e., two neutral species exploiting the
same ecological opportunity in a community)? In addition, differentiated species
may in fact converge to be ecologically more similar when inhabiting the same
community. The ecological and evolutionary dynamics immediately following the
various speciation mechanisms and immediately following community disaggregation and recombination (or species invasion of a largely intact community) are
critical determinants of ultimate community structure that we too often ignore.
The conditions before the ultimate steady state is reached are where our focus
often needs to be, particularly when evolutionary issues are considered. I will
take up these issues in the next chapter to see what ecological and evolutionary
conditions may shape the evolution of community structure in these brief periods
of change.

CHAPTER FIVE

Differentiating in the Community

As you hike though nature, you will find many closely related species living together in the communities you traverse. In the lakes, many centrarchid fish species utilize the diverse resources available to them; bluegills (*Lepomis macrochirus*) feed on zooplankton in the open water, pumpkinseeds (*L. gibbosus*) specialize on snails that crawl over plants growing around the margins, and green sunfish (*L. cyanellus*) eat larger prey in vegetation in shallow waters (Werner 1977; Werner et al. 1977; Mittelbach 1981, 1984; Osenberg and Mittelbach 1989). Likewise, many *Daphnia* species utilize different resources in the lake, but also are differentially sensitive to the presence of predators (Leibold 1991, Leibold and Tessier 1991, Tessier and Leibold 1997). *D. pulicaria* inhabits the cold waters in the middle of the lake where bluegills are absent and phytoplankton is not very plentiful. In contrast, *D. galeata* spends the daytime in the cold deeper waters of lakes, but at night when the fish are asleep it ventures into the warm upper waters where algae is more plentiful (Leibold 1991, Leibold and Tessier 1991, Tessier and Leibold 1997). All these species clearly occupy different functional groups within the lake community.

Not all closely related species found living together show such clear ecological segregation. For example, while different genera of damselflies in these same lakes as the *Lepomis* and *Daphnia* show great ecological differences (McPeek 1998, Stoks and McPeek 2006, Siepielski et al. 2011b), each genus contains multiple species living together that show no discernable differences in ecological performance (Siepielski et al. 2010). Along the trail through the woods, multiple lacewing species (*Chrysoperla*) might be found in the same bush even though they are indistinguishable except for their mating calls (Wells and Henry 1992). These ecologically undifferentiated neutral species are occupying the same functional position in their respective communities. In fact, the advent of molecular methods in systematics have forced biologists to recognize that multiple cryptic species are common (Bickford et al. 2007, Pfenninger and Schwenk 2007). Many of these cryptic species may also be neutral species embedded in the interaction networks of their communities, while many others may have differentiated in subtle ways (Hebert et al. 2004). However, neutral species need not be closely related to one another (Shinen and Navarrete 2014), which suggests that selection may actually drive ecological convergence in some situations (Abrams 1986, 1987, 1990).

Whether closely related species are ecologically differentiated from one another in a local community depends on their evolutionary histories and the selection pressures they have experienced since they diverged from a common ancestor. In chapter 4, I explored the degree to which various speciation modes generate ecological differentiation among species at the time of speciation. Because speciation occurs as a by-product, sympatric ecological speciation creates new species that are ecologically differentiated from one another as well as coexisting with other members of the local community. In this case, the dynamics of ecological differentiation are in large measure the dynamics of speciation (plus the dynamics of associated magic traits or other traits that confer reproductive mating advantages in parallel with ecological differentiation—for example, see van Doorn et al. 2009.

Speciation via hybridization and polyploidy produces new species that are ecologically differentiated to some degree from their progenitors (because they are typically phenotypically different from them), but are not necessarily well adapted to their community, and thus are not necessarily coexisting species when they arise. Finally, speciation via Bateson-Dobzhansky-Mueller incompatibilities, sexual selection, or reproductive character displacement may produce species that have little or no ecological differentiation among them. Even if new species were produced allopatrically, the disaggregation and recombination of communities during recent interglacial periods—including our current one—can potentially throw a heterogeneous mix of species produced by myriad different mechanisms together in local communities.

Speciation can, therefore, introduce coexisting, walking-dead, or neutral species from the outset. Whether species remain extant and in their current state will depend critically on how they may subsequently adapt to the other species and the local environment. Coexisting species may adapt somewhat to increase average fitness, but they will persist since they already meet the invasibility criteria of the local community. In contrast, walking-dead species must adapt to persist at all, and they will follow analogous dynamics outlined in chapter 4 for habitat/host shifts (fig. 4.3). If they cannot adequately adapt in time to satisfy the invasibility criteria of the local community, extinction will occur.

Perhaps the most interesting situation occurs when neutral species are put together into a local community. The common intuition of many ecologists is that natural selection should always act to differentiate species that are at the same functional position in a community in order to reduce competition among them; that is, character displacement should occur (Brown and Wilson 1956, Roughgarden 1976, Slatkin 1980, Taper and Case 1985, 1992, Price and Kirkpatrick 2009). Even if species have only recently come back into sympatry with one another, natural selection can act in only a few generations to change species in response to local selection pressures (Hendry and Kinnison 1999). Such differentiating species will be directed by natural selection to different equilibria on the fitness topography, and the result will be a set of coexisting species in the community. In this chapter I explore the joint ecological and evolutionary dynamics of this process.

However, differentiation of similar or identical species is by no means universally guaranteed to happen. Natural selection generated by interactions among species will cause those at the same functional position in the community to differentiate only when the interactions generate different fitness relationships that will drive divergence. When natural selection favors more than one functional type of species at a given trophic level, these species will often be evolving in response to a potentially multipeaked fitness surface. In this case, differentiation implies that the species ultimately end up on different adaptive peaks. If species begin very far from one another in phenotypic space so that each begins within the domain of attraction of different peaks (e.g., as with a habitat shift or hybridization/polyploid speciation), natural selection will easily drive each directly to a different peak.

The more interesting question is whether species beginning very close to one another so that they are initially under the influence of the same adaptive peak can differentiate to end up on different adaptive peaks. The first person to consider such an evolutionary problem was Sewall Wright, who developed the shifting balance theory to explain how this might occur via random genetic drift, which would cause a population to shift from one adaptive peak to another (Wright 1931, 1932). In Wright's formulation of the problem, the fitness surface is unchanging (Whitlock et al. 1995). Likewise, while he recognized the dynamical nature of a "selection landscape" and that the fitness landscape could change, Simpson (1944) described his "quantum evolution" as accomplishing a peak shift in very similar terms to Wright's shifting balance theory. Others have shown how stochasticity in various genetic, phenotypic, and fitness attributes of the population can facilitate a peak shift, but again these have all assumed that fitnesses associated with different genotypes are invariant (Kirkpatrick 1982; Barton and Rouhani 1987; Rouhani and Barton 1987a, 1987b; Whitlock 1995, 1997).

In this chapter, I take a decidedly different approach. As we saw in chapter 3, the fitness landscape experienced by a species has an ecological dynamic. I will explore how the dynamics of the fitness surfaces experienced by these species with similar phenotypes can drive their differentiation to accomplish peak shifts within a community context, and what ecological and evolutionary conditions hinder or enhance the likelihood of this differentiation occurring. This work follows a path blazed by previous research on evolution in multispecies interactions with models following the framework used in chapter 3 (e.g., Roughgarden 1976;

Slatkin 1980; Milligan 1985; Taper and Case 1985; Abrams 1986; Milligan 1986; Taper and Case 1992; Abrams et al. 1993; Abrams 2000; Abrams and Chen 2002; Křivan 2003; Křivan and Eisner 2003; Křivan 2007; Price and Kirkpatrick 2009; Abrams and Fung 2010a, 2010b; Křivan 2013), but models assuming other evolutionary frameworks are relevant here as well (e.g., Eshel 1981a, 1981b; Brown and Vincent 1992; Dieckmann et al. 1995; Dieckmann and Law 1996; Doebeli 2011). As we will see, the dynamics of the fitness surfaces in the component species in the community drive ecological differentiation—fitness surfaces are not static entities. Ecological differentiation is possible for any of the types of traits introduced in chapter 3 that may influence species interactions, but differentiation is more likely over broad areas of parameter space for some but not other trait types. The likelihood of differentiation also depends on the species' functional positions within the community. In addition, the differentiation of species at one trophic position (e.g., prey) may in turn drive the differentiation of species at other trophic positions (e.g., their predators) in the community. A fundamental ecological result to emerge from this work is that, when differentiation does occur, the ecological structure of the resulting community depends critically on the type of traits that underlie species interactions.

When ecological differentiation does not occur, natural selection will direct these similar species to the same adaptive fitness equilibrium. In these situations, the result will be a collection of neutral species occupying the same functional position in the community.

The dynamics explored in this chapter are also interpretable in a number of contexts. For example, they offer an evolutionarily explicit description of the microevolutionary dynamics that underlie the assumptions of branching events in adaptive dynamics models of sympatric ecological speciation (Geritz et al. 1997, Geritz et al. 1998, Doebeli and Dieckmann 2000, de Mazancourt and Dieckmann 2004, Doebeli 2011). The dynamics considered here can also be interpreted as the evolutionary responses of ecological traits following a sympatric speciation event involving differentiation of the mate recognition system. They can also be interpreted as the dynamics occurring after the establishment of a community containing neutral species at one functional position following a disaggregation/recombination cycle. In all these cases, the question is whether two or more ecologically identical species will differentiate from one another, and conversely whether natural selection will cause ecologically disparate species to converge in their ecological capabilities. This is why I cautioned the reader to not simply equate differentiation with the process of speciation. The two are conceptually completely separable and each should be considered in its own right, as I will do here. Species enter the ecological stage for many different reasons, but once there they must act in the evolutionary play in which they find themselves.

A GENERAL MODEL OF EVOLUTIONARY AND ECOLOGICAL DYNAMICS IN A COMMUNITY

To begin our exploration of adaptive differentiation at multiple trophic levels, we must first define the modeling framework. The full model is simply the models presented in chapters 2 and 3 to include multiple species at three different trophic levels and to permit the top level to feed on the basal resources (intraguild predation). In order to prevent the trait dynamics equations from becoming unwieldy, I make the simplifying assumptions that each consumer and predator has the same conversion efficiency and the same handling time for each prey that it consumes on a specific trophic level. Therefore, a top predator has the same handling time and conversion efficiency for every consumer it eats, and the same handling time and conversion efficiency for every resource it eats, but these parameters can be different between its prey at the two trophic levels. Because the attack coefficients depend on the trait values of both predator and prey, the attack coefficients can differ between every species pair. Also, as specified in chapter 3, each resource still has a density-dependent intrinsic birth rate that depends on the resource's trait value, and each consumer and predator has a density-dependent intrinsic death rate that depends on its trait value. The same functional forms used in chapter 3 are used for these here.

With multiple species at each trophic level and allowing intraguild predation under the assumptions stated in the last paragraph, the full ecological and evolutionary model is

$$\frac{dP_k}{dt} = P_k \left[\frac{n_k \left(\sum_{j=1}^{q} m_{jk}(\bar{z}_{N_j}, \bar{z}_{P_k})N_j \right) + w_k \left(\sum_{i=1}^{p} v_{ik}(\bar{z}_{R_i}, \bar{z}_{P_k})R_i \right)}{1 + \sum_{j=1}^{q} m_{jk}(\bar{z}_{N_j}, \bar{z}_{P_k})l_k N_j + \sum_{i=1}^{p} v_{ik}(\bar{z}_{R_i}, \bar{z}_{P_k})u_k R_i} - x_k(\bar{z}_{P_k}) - y_k P_k \right]$$

$$\frac{dN_j}{dt} = N_j \left[\frac{b_j \left(\sum_{i=1}^{p} a_{ij}(\bar{z}_{R_i}, \bar{z}_{N_j})R_i \right)}{1 + \sum_{i=1}^{p} a_{ij}(\bar{z}_{R_i}, \bar{z}_{N_j})h_j N_j} - \sum_{k=1}^{s} \frac{m_{jk}(\bar{z}_{N_j}, \bar{z}_{P_k})P_k}{1 + \sum_{j=1}^{q} m_{jk}(\bar{z}_{N_j}, \bar{z}_{P_k})l_k N_j + \sum_{i=1}^{p} v_{ik}(\bar{z}_{R_i}, \bar{z}_{P_k})u_k R_i} - f_j(\bar{z}_{N_j}) - g_j N_j \right]$$

$$\frac{dR_i}{dt} = R_i \left[c(\bar{z}_{R_i}) - d_i R_i - \sum_{j=1}^{q} \frac{a_{ij}(\bar{z}_{R_i}, \bar{z}_{N_j})N_j}{1 + \sum_{i=1}^{p} a_{ij}(\bar{z}_{R_i}, \bar{z}_{N_j})h_j N_j} - \sum_{k=1}^{s} \frac{v_{ik}(\bar{z}_{R_i}, \bar{z}_{P_k})P_k}{1 + \sum_{i=1}^{p} v_{ik}(\bar{z}_{R_i}, \bar{z}_{P_k})u_k R_i + \sum_{i=1}^{p} v_{ik}(\bar{z}_{R_i}, \bar{z}_{P_k})u_k R_i} \right]$$

$$\frac{d\bar{z}_{P_k}}{dt} = V_{P_k} \left[\frac{n_k \left(\sum_{j=1}^{q} N_j \frac{\partial m_{jk}(\bar{z}_{N_j}, \bar{z}_{P_k})}{\partial z_{P_k}} \right) + w_k \left(\sum_{i=1}^{p} R_i \frac{\partial v_{ik}(\bar{z}_{R_i}, \bar{z}_{P_k})}{\partial z_{P_k}} \right) + (n_k u_k - w_k l_k) \sum_{j=1}^{q} \sum_{i=1}^{p} R_i N_j \left(v_{ik}(\bar{z}_{R_i}, z_{P_k}) \frac{\partial m_{jk}(\bar{z}_{N_j}, \bar{z}_{P_k})}{\partial z_{P_k}} - m_{jk}(\bar{z}_{N_j}, z_{P_k}) \frac{\partial v_{ik}(\bar{z}_{R_i}, \bar{z}_{P_k})}{\partial z_{P_k}} \right)}{\left(1 + \sum_{j=1}^{q} m_{jk}(\bar{z}_{N_j}, z_{P_k})l_k N_j + \sum_{i=1}^{p} v_{ik}(\bar{z}_{R_i}, z_{P_k})u_k R_i \right)^2} \right.$$

$$\left. - \frac{\partial x_k(z_{P_k})}{\partial z_{P_k}} \right|_{z_{P_k} = \bar{z}_{P_k}}$$

$$\frac{d\bar{z}_{N_j}}{dt} = V_{N_j} \left[\frac{b_j \left(\sum_{i=1}^{p} R_i \frac{\partial a_{ij}(\bar{z}_{R_i}, z_{N_j})}{\partial z_{N_j}} \right)}{\left(1 + \sum_{i=1}^{p} a_{ij}(\bar{z}_{R_i}, z_{N_j})h_j N_j \right)^2} \right|_{z_{N_j} = \bar{z}_{N_j}} - \sum_{k=1}^{s} \frac{P_k \frac{\partial m_{jk}(z_{N_j}, \bar{z}_{P_k})}{\partial z_{N_j}}}{1 + \sum_{j=1}^{q} m_{jk}(\bar{z}_{N_j}, \bar{z}_{P_k})l_k N_j + \sum_{i=1}^{p} v_{ik}(\bar{z}_{R_i}, \bar{z}_{P_k})u_k R_i} \right|_{z_{N_j} = \bar{z}_{N_j}} - \frac{\partial f_j(z_{N_j})}{\partial z_{N_j}} \right|_{z_{N_j} = \bar{z}_{N_j}}$$

$$\frac{d\bar{z}_{R_i}}{dt} = V_{R_i} \left[\frac{\partial c(z_{R_i})}{\partial z_{R_i}} \right|_{z_{R_i} = \bar{z}_{R_i}} - \sum_{j=1}^{q} \frac{N_j \frac{\partial a_{ij}(z_{R_i}, \bar{z}_{N_j})}{\partial z_{R_i}}}{1 + \sum_{i=1}^{p} a_{ij}(\bar{z}_{R_i}, \bar{z}_{N_j})h_j N_j} \right|_{z_{R_i} = \bar{z}_{R_i}} - \sum_{k=1}^{s} \frac{P_k \frac{\partial v_{ik}(z_{R_i}, \bar{z}_{P_k})}{\partial z_{R_i}}}{1 + \sum_{i=1}^{p} v_{ik}(\bar{z}_{R_i}, \bar{z}_{P_k})u_k R_i + \sum_{i=1}^{p} v_{ik}(\bar{z}_{R_i}, \bar{z}_{P_k})u_k R_i} \right|_{z_{R_i} = \bar{z}_{R_i}}$$

$$(5.1)$$

All parameters are as described in chapter 3, but here are subscripted to identify species. This set of equations appears quite unwieldy, but all the underlying principles considered in chapters 2 and 3 still apply. Per capita fitness for each species is given by the terms in parentheses for the abundance dynamics equations in equations (5.1), and each term is a fitness component due to a species interaction. Fitness maxima and minima on the fitness topography for a species correspond to trait values that make the terms in parentheses for the trait dynamics in equations (5.1) sum to zero, given the combination of abundances and trait values for all other species in the community. Moreover, each term in parentheses of a trait dynamics equation is the selection gradient generated by a species interaction, and their total sum determines whether the trait value will increase, decrease, or remain unchanged in the next time interval. The relationships between trait values and model parameters must also be specified. I maintain the same functional forms as used in chapter 3.

Obviously, an exhaustive exploration of parameter space is impossible. In what follows, I will explore specific relationships among parameters to consider how various combinations of ecological and evolutionary conditions shape the structure of a local community.

DIFFERENTIATION IN A TWO-TROPHIC-LEVEL COMMUNITY

To refine our thinking on the subject, let us begin this exploration by considering a community with only resources (R_i) and consumers (N_j). With only two trophic levels, differentiation will result in community modules involving resource and apparent competition only (see chapter 2). After thoroughly examining the factors that promote and inhibit species differentiation of both resources and consumers, I will add the predators (P_k)—the third trophic level—in order to explore differentiation in the full set of possible community modules.

First, consider the evolutionary dynamics of differentiation in a specific example where two resources with identical underlying parameters and nearly identical initial trait values are fed upon by one consumer (see also Abrams 2000, Schreiber et al. 2011) (fig. 5.1, see the animation of the fitness component surfaces at http://press.princeton.edu/titles/11175.html). Imagine one species of Darwin's finch feeding on the seeds of two plant species. In this example, the attack coefficients are determined by bidirectional-dependent traits, but the general features of the process of differentiation are the same regardless of the trait types involved. In the beginning, both resources have trait values above that of the consumer; natural selection from predation on their death fitness components favors increasing their trait values, but selection on their birth fitness components favors decreasing their trait values, because both are above their intrinsic birth optima. In the first few iterations, the balance of these selection pressures favors increasing trait values in both resources, and selection favors increasing the trait value in the consumer (see fitness surfaces for iteration 5 in fig. 5.1D).

As the consumer evolves toward the resources in the first few iterations of this example, the selection gradients on the resources' death fitness components become weaker, and the selection gradients on their birth fitness components become stronger as their trait values move away from $\bar{z}_{R_i}^c$ (see analogous results for the dynamics of trait cycling in chapter 3). Eventually, the selection gradients on their death fitness components become less than the selection gradient on their birth fitness components, and both resources begin to evolve toward the consumer. At this point, the fitness peak at the higher trait value has disappeared for both resource species, and natural selection begins pushing both resources toward the lower fitness peak (e.g., iteration 450 in fig. 5.1E). Remember that when the traits of R_1 and the consumer evolve to exactly match, the selection gradient on that resource's death fitness component is zero, and so selection only acts via its birth fitness component (at that point in the process).

Once the resource closest in trait value to the consumer (R_1) passes the trait value of the consumer (just after iteration 450), the consumer reverses course and begins to evolve lower trait values in order to balance the birth contributions from the two resources that now phenotypically straddle it. As a consequence, the fitness valley between the two peaks reforms for both resources, and selection now favors their differentiation because they are now on different sides of the adaptive valley (e.g., iteration 600 in fig. 5.1F). Selection proceeds until all three species reach trait values that cause both the fitness components and the selection gradients acting on those fitness components in each and every species to exactly balance. This results in the simultaneous demographic (fitness components balancing) and evolutionary (selection gradients balancing) equilibrium.

In this specific case, also note that the consumer occupies an overall stable fitness minimum between the trait values of the two resources (fig. 5.1G: see also Abrams et al. 1993). This fitness minimum for the consumer is a stable equilibrium, because perturbations of its trait mean in either direction will cause changes in the resources' abundances that will alter the selection gradients on the consumer's two birth fitness components to return it to this equilibrium (analogous to the what was discussed for the resource at a stable fitness minimum in figures 3.2–3.3).

The critical evolutionary insight into this process is that the peak shift accomplished by R_1 occurred purely by the dynamical action of natural selection. The shifting balance process envisioned by Wright (1932) was not involved, and stochasticity in various aspects of the genetics, phenotypes, or fitnesses in the population also did not have to aid the selection process to accomplish the peak shift (e.g., Barton and Rouhani 1987; Rouhani and Barton 1987a, 1987b; Whitlock

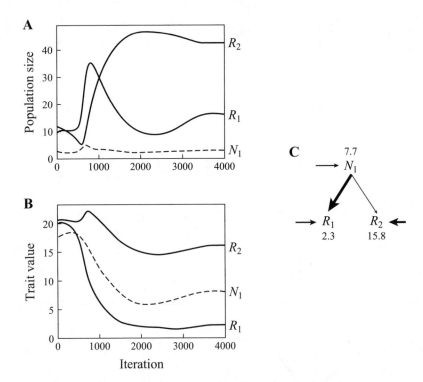

FIGURE 5.1. An example of a peak shift leading to the ecological differentiation of two resources in a simple, apparent competition module. The (A) population sizes and (B) trait values over the course of the simulation are shown. In panels A and B, the *solid lines* identify the resource values, and the *dashed lines* identify the consumer values. The final ecological structure of the module is given in panel C. The final phenotypic trait values for each species are next to the species labels, and the *arrows* illustrate the magnitudes of the resources' realized intrinsic birth rates, the consumer's realized intrinsic death rates, and the realized attack coefficients for the consumer feeding on each resource. Panels D–G show the overall fitness surfaces and the selective surfaces associated with the birth and death fitness components for each of the three species; these are shown at critical iterations to illustrate the dynamics of the fitness surfaces that cause the peak shift. Features in these panels are as described in figure 3.3. Panel H shows the isoclines for the three species at their demographic and evolutionary equilibrium; for simplicity of illustration the isocline shapes are illustrated for $h_1 = 0$, but all important features illustrated are as in the simulation with $h_1 = 0.1$. All symbols in panel H are as identified in figure 2.5. The parameters used for this scenario are $c_{0i} = 1.0$, $d_i = 0.02$, $a_{0ij} = 0.35$, $b_1 = 0.1$, $h_1 = 0.1$, $f_{01} = 0.2$, $g_1 = 0.0$, $\beta_{i1} = 5.0$, $\gamma_i = 0.005$, $\theta_1 = 0.005$, $\tilde{z}_{R_i}^c = 12.0$, $\tilde{z}_{N_i}^f = 5.0$, and $V_{z_{R_i}} = V_{z_{N_i}} = 0.2$. (This figure is redrawn from figure 7 of McPeek 2017, with permission.)

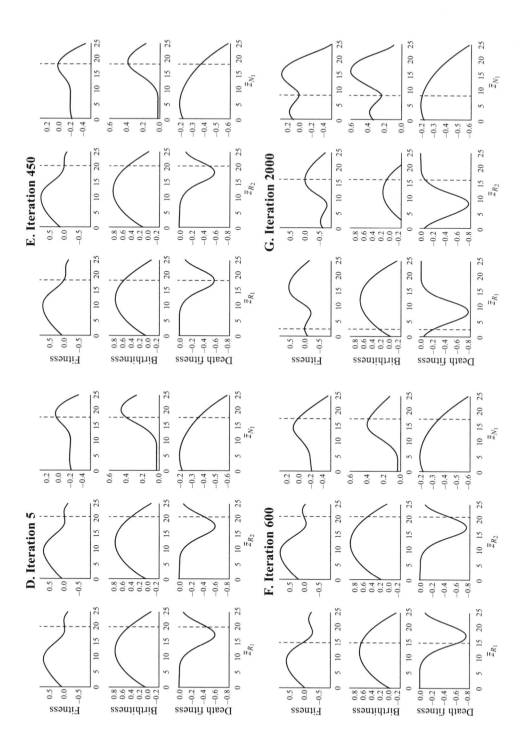

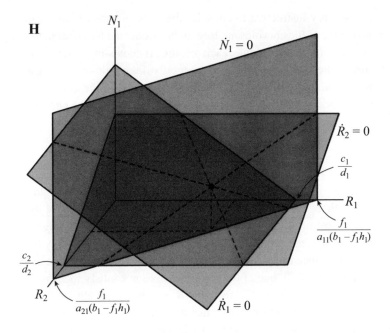

1995, 1997). The ecological dynamics of natural selection operating on R_1 completely accomplished the peak shift, and these dynamics resulted in the differentiation of R_1 and R_2, which have identical underlying parameters and therefore identical capacities for evolution, into functionally distinct, coexisting species. These dynamics are generated by changes in the magnitudes of both fitness components and the selection gradients on those fitness components caused by the joint demographic and coevolutionary responses of these species to one another. These dynamics also underlie previous models exploring the differentiation of resource competitors feeding on unimodal resource distributions (Roughgarden 1976, Slatkin 1980, Milligan 1985, Taper and Case 1985, Milligan 1986, Taper and Case 1992, Price and Kirkpatrick 2009).

Critical ecological insights are also apparent when considering this example. The result of this coevolution is an apparent competition community module of two resources coexisting with one consumer (fig. 5.1C). The resource R_1 has evolutionarily settled farther from its birth optimum than has R_2, and so R_1's realized intrinsic birth rate is lower. The consumer settles at a trait value that gives it a higher attack coefficient on R_1, because R_1 is closer to the consumer's intrinsic death optimum (fig. 5.1C). Consequently, R_1 equilibrates at a lower population abundance than R_2 (fig. 5.1A).

It is also very instructive to consider the coexistence criteria for this system over the course of the coevolution. Remember from chapter 2 that in this apparent competition community module, each resource is coexisting—in the strict ecological sense of the term—with the consumer only if the ratio of its realized intrinsic birth rate to its realized attack coefficient is greater than the consumer's equilibrium abundance, when only the other resource is present in the community:

$$\frac{c(\bar{z}_{R_i})}{a(\bar{z}_{R_i}, \bar{z}_{N_i})} > N^*_{1(j)}. \tag{5.2}$$

At the equilibrium, these ratios are $c(\bar{z}_{R_1})/a(\bar{z}_{R_1}, \bar{z}_{N_1}) = 0.51/0.10 = 5.1$ and $c(\bar{z}_{R_2})/a(\bar{z}_{R_2}, \bar{z}_{N_1}) = 0.93/0.03 = 35.0$ for the two resources, which explains the disparity in their equilibrium abundances; both satisfy their invasibility criterion. In fact, for most of the time covered by the iterations pictured in figure 5.1, both resources satisfy their criterion (note that all the values in inequality (5.2) change as the species' mean trait values change). However, for a brief period between iterations 480 and 650, one of the resources does not satisfy inequality (5.2) *and it is not the one you think*. The resource R_1 satisfies inequality (5.2) through the entire course of its evolution, even when its trait value exactly matches that of the consumer. In contrast, during the brief window immediately after R_1 evolves past the consumer's trait value, R_2 is not coexisting with the consumer because R_1 is inflating $N^*_{1(1)}$ to a level such that inequality (5.2) is not met by R_2. In this window, $c(\bar{z}_{R_1})/a(\bar{z}_{R_1}, \bar{z}_{N_1}) > c(\bar{z}_{R_2})/a(\bar{z}_{R_2}, \bar{z}_{N_1})$ as expected. As the consumer continues to evolve away from R_2, which decreases the attack coefficient between N_1 and R_2, the criterion is again met for both resources after iteration 650, and once at equilibrium, R_2 has a higher realized $c(\bar{z}_{R_i})/a(\bar{z}_{R_i}, \bar{z}_{N_j})$ ratio.

Another surprising ecological feature of coexistence in this example has to do with the invasibility criteria at equilibrium. At this point, both resources are satisfying inequality (5.2) because the equilibrium abundance of the consumer when feeding on either one alone (given their phenotypes at the equilibrium) is negative. This means that when considered in isolation, the consumer isocline does not intersect either of the resources isoclines when the other resource is at zero abundance. Mathematically, this means that the consumer isocline crosses each resource axis above the equilibrium abundance of each resource in the absence of the consumer (i.e., $f_i/(a_{i1}(b_i - f_i h_i)) > c_i/d_i$; fig. 5.1H) for both resources, given the trait values they have at this equilibrium. In other words, if either resource suddenly went extinct (imagine a disease suddenly wiped out one of the resources) and the two remaining species were prevented from evolving (imagine that neither had any additive genetic variation for their traits [i.e., $V_{R_i} = 0$]), the consumer would then go extinct, because it cannot support a population on the other resource alone. This community could not be assembled through a purely ecological process of either resource invading when rare and when the other two species are at their equilibrium—our usual conception of what coexistence of two resources means in an apparent competition module (Holt 1977). The only assembly route imaginable if only ecology were considered is the consumer invading a system where both resources are already present. By explicitly considering the coevolutionary dynamics, a much broader set of processes for community assembly has been provided. Moreover, the final community structure by itself provides little insight into the process of how this community was assembled.

The two resources considered in this example are identical in their underlying parameters, and yet they evolve to differentiate substantially from one another in their susceptibility to the predator. These two resources with identical parameters and identical trait values cannot be considered neutral species since they both will increase when rare, and when the consumer is absent, regardless of the presence of the other resource. As we will see in the next section, differentiation of identical species at a trophic level is fostered under particular ecological conditions, but differentiation of such species is not inevitable. Change some features of the environment that alter their fitness returns, or alter the underlying selection gradients on the various fitness components, and the species will not differentiate. If the ecological conditions (i.e., parameters here) were altered so that the peak shift of R_1 does not occur, and so these two species do not differentiate, both resources would be directed by natural selection to the same trait value under the same fitness peak.

DIFFERENTIATION OF SPECIES WITH
IDENTICAL UNDERLYING PARAMETERS

To explore the evolutionary capabilities for differentiation more generally, consider a simple community composed of two resources and two consumers in which species on each trophic level are identical in their underlying parameters and so have the same ecological and evolutionary capabilities. At the start, the consumers are true neutral species occupying the same functional position, but because each resource can increase when rare (i.e., logistic growth), the resources are ecologically identical but not truly neutral species. For all three trait types defining the attack coefficients between consumer and resource, consistent areas of parameter space exist where species will differentiate from one another at each trophic level. However, the sizes of these areas differ between the trait types, and the structures of the resulting communities are different. I will address these issues in turn.

The parameters of these models reflect features of the environment in which the community is found—these parameters are positions on the abiotic axes of the conceptual framework described by figure 3.13. Thus, asking whether species will differentiate from one another in different areas of parameter space is querying whether different sets of environmental conditions foster or retard ecological differentiation of similar or identical species. Species differentiation occurs over the largest range of environmental conditions (i.e., over the largest area of parameter space) in communities where bidirectional-dependent traits define the attack coefficients among consumers and resources. I consider differentiation in communities with this trait type first.

The productivity of the ecosystem fueling the basal resources has a large influence on whether species at the two trophic levels will differentiate. Again, resources growing in more productive environments (e.g., locations with more water and nutrients and more favorable temperatures) should have higher intrinsic birth rates (c_{0i}). Higher values of productivity (i.e., greater c_{0i}) result in greater abundances of species at both trophic levels when comparing across communities (fig. 5.2B), and differentiation at both trophic levels critically depends on the abundances of the species. In communities with the lowest productivities that will support both consumers and resources ($0.05 < c_{0i} < 0.14$ in fig. 5.2A–B) species at neither trophic level differentiate from one another; the consumers' abundances are too low to generate two peaks in the fitness surfaces of the resources, because the selection gradients on the resources' death fitness components are too shallow (fig. 5.3A). In communities with somewhat higher productivities, the resources differentiate, but the consumers do not ($0.14 \leq c_{0i} < 0.34$ in fig. 5.2A–B). For communities in this range, the consumer abundances are sufficient to create strong enough selection gradients associated with the resources' death fitness components to generate two adaptive peaks in the resources' fitness surfaces. However, the resource phenotypes are not differentiated enough to generate two adaptive peaks in the consumers' fitness surfaces (Fig. 5.3B). Species differentiate at both trophic levels only in communities with high productivities ($0.34 \leq c_{0i}$ in fig. 5.2A–B). In these communities, the realized selection gradients on the resources' death fitness components drive their adaptive peaks far enough apart, to in turn generate two peaks in the consumers' birth fitness component surfaces (fig. 5.3C).

Other types of environmental conditions may cause stress in the consumer or reflect differences in other necessary resources or mortality sources that are not being modeled. These environmental features would influence the value of the consumers' minimum intrinsic death rates (f_{0j}). These environmental conditions also influence whether differentiation occurs for exactly the same reasons as ecosystem productivity. Higher values of f_{0j} cause the consumers to equilibrate at lower abundances, which reduce the strength of selection gradients on the resources' death fitness components (fig. 5.2D). Consequently, species at neither trophic level differentiate at high values of f_{0j} (fig. 5.2C). Differentiation at both

Bidirectional dependent traits

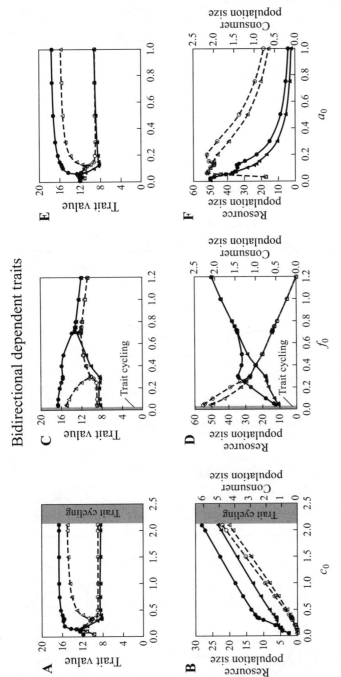

FIGURE 5.2. Differentiation of species with identical underlying parameters for two resources and two consumers in which the attack coefficients are defined by bidirectional-dependent traits in communities that differ in (A–B) the resources' maximum intrinsic birth rates (c_0), (C–D) the consumers' minimum intrinsic death rates (f_0), and (E–F) the maximum attack coefficients (a_0). Relationships for the resources are identified with *solid lines* and for consumers with *dashed lines*. The two different resources are identified with *closed, solid symbols* (● and ▲), and the two different consumers with *open symbols* (○ and △) connected by *dashed lines*. Parameter areas where trait cycling occurs are identified. The parameters used are $c_{0i} = 1.0$, $d_i = 0.02$, $a_{0ij} = 0.25$, $b_{ij} = 0.1$, $h_{ij} = 0.0$, $f_{0j} = 0.1$, $g_i = 0.1$, $\beta_{ij} = 5.0$, $\gamma_i = 0.01$, $\theta_j = 0.01$, $\bar{z}_{R_i}^c = 12.0$, $\bar{z}_{N_j}^f = 1.0$, and $V_{z_{-R_i}} = V_{z_{-N_j}} = 0.2$. (Panels A, C, and E are redrawn from figure 7 of McPeek 2017, with permission).

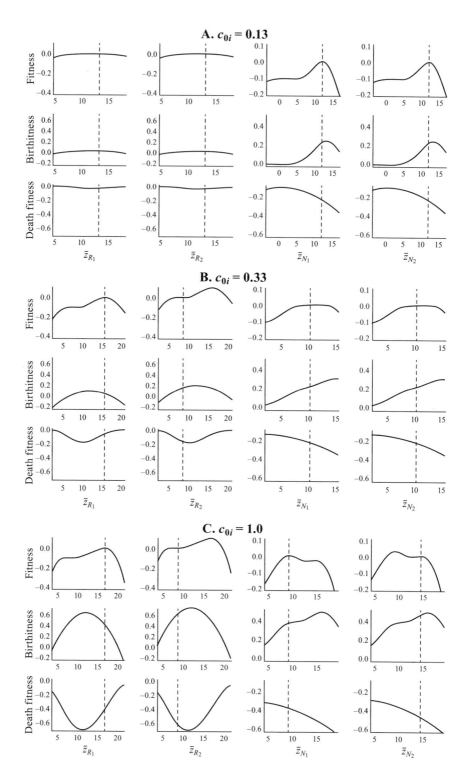

trophic levels occurs in communities in which the consumers have low values of f_{0j} where their high abundances impart strong selection gradients on the resources' death fitness components (fig. 5.2C). And again, only the resources differentiate at intermediate levels of f_{0j} because they are not differentiated enough to generate two peaks in the consumers' birth fitness topography. Again, this pattern of differentiation is generated by the differences in abundances and how these generate the strengths of selection gradients at the two trophic levels across communities. Only when the consumers are at high abundances can they generate adaptive peaks that are far enough apart for the resources that drive their own differentiation.

Still other structural features of the environment (e.g., water turbidity in aquatic and marine ecosystems, habitat structural complexity in all ecosystems) will influence the value of the maximum attack coefficient (a_{0ij}), and a similar pattern of differentiation is generated when these environmental conditions also create differences in the potential strengths of species interactions across communities as a result (fig. 5.2E–F). In this case, higher maximum attack coefficients drive differentiation not because of higher abundances, but rather because higher maximum attack coefficients directly generate steeper selection gradients associated with the resources' death fitness components and with the consumers' birth fitness components.

These results show that character displacement is not inevitable, but rather should occur with greater frequency under specific environmental conditions— namely, those supporting greater abundances of species. In chapter 3, we identified environmental conditions that generate steeper selection gradients on the fitness components of species throughout the food web, because the magnitudes of selection gradients increase with higher species abundances. Here, we see that these steeper selection gradients on fitness components associated with species interactions drive species differentiation when ecologically identical species are present. Thus, ecological differentiation is predicted to be more likely to occur among taxa in environments with higher productivity, in environmental conditions that are more benign to species at higher trophic levels, and in environments where the fundamental strengths of species interactions are higher.

These results also suggest that character displacement cascades up a food web along gradients of these environmental factors. In communities with low

---

FIGURE 5.3. The overall fitness surfaces and the underlying surfaces for the birth and death fitness components once the community reaches equilibrium for two consumers and two resources interacting in the community module studied in figure 5.2. The three sets of panels show the final fitness surfaces when the resources both have maximum intrinsic birth rates of (A) $c_{0i} = 0.13$, (B) $c_{0i} = 0.33$, and (C) $c_{0i} = 1.0$. Features in these panels are as described in figure 3.3. All other parameters are as in figure 5.2.

productivities, harsh conditions, or weak interactions, species at neither trophic level differentiate. In communities with intermediate values for these conditions, only the resources at the base of the food web differentiate. Greater levels of productivity, more benign conditions, and strong interactions are needed to drive differentiation at higher trophic levels. This analysis also shows that differentiation at lower trophic levels drives differentiation at higher trophic levels in the food web.

In addition to the environmental background in which the interactions occur, the steepness of the underlying relationships between fitness components and traits (i.e., the intrinsic fitness gradients for traits) also influence whether identical species will differentiate. For example, consider two different sets of consumers and resources interacting in communities with the same environmental conditions. The difference between them is that the consumers in one set are effective at feeding on a broader phenotypic range of resources (i.e., they have larger β_{ij} values; fig. 3.1E), but the consumers in the second set are efficient at feeding on only a very narrow range of resources (i.e., they have small β_{ij} values; fig. 3.1E). Is one set of consumers and resources more likely than the other to differentiate in a given environment, and why?

This is a question of how the parameters β_{ij}, γ_i, and θ_j influence species differentiation. Because the realized selection gradients associated with the various fitness components are influenced by both the abundances of the species and the steepnesses of these underlying relationships, the relative values of these parameters, not surprisingly, strongly influence whether identical resources and identical consumers will differentiate.

First, consider the effect of the steepness of the fitness relationship for the species interactions here—namely β_{ij}, which controls how rapidly the attack coefficient changes with a change in the difference in trait values between consumer and resource (fig. 3.1E). Large β_{ij} values mean that consumers feed efficiently on a broad phenotypic range of resources, and small β_{ij} values mean that consumers are efficient at feeding on only a narrow phenotypic range of resources. In a community having taxa in which the attack coefficient changes little with trait values (i.e., consumers are efficient on feeding on a broad phenotypic range of resources; $\beta_{ij} \geq 8.4$ in fig. 5.4E–F), neither trophic level differentiates because their other fitness components are more important in determining the shape of the overall fitness topography (i.e., a single fitness peak for each). In such a community, the selection gradient on the resources' death fitness components are so shallow that the overall shape of their fitness surfaces are dominated by the shape of the birth fitness component (i.e., a single peak), but the directional selection gradient for higher trait values from predation on their death fitness component causes the overall peak to be offset from $\bar{z}^c_{R_i}$ to a higher value (e.g., fig. 5.5A).

Bidirectional dependent traits

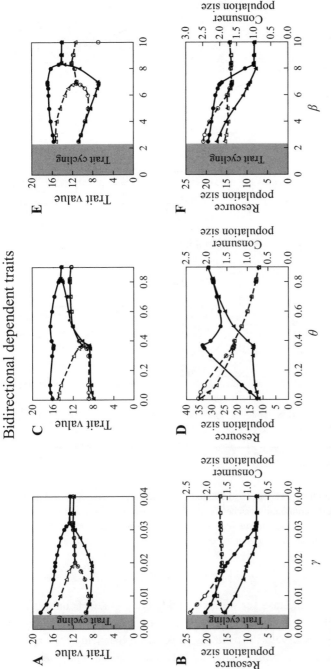

FIGURE 5.4. Differentiation of species with identical underlying parameters for two resources and two consumers, in which the attack coefficients are defined by bidirectional-dependent traits in communities that differ in the strength of the underlying relationship between (*A–B*) the resources' maximum intrinsic birth rate and their trait values (γ), (*C–D*) the consumers' minimum intrinsic death rate and their trait values (θ), and (*E–F*) the steepness of the change in attack coefficient with a change in the difference between the consumers' and resources' trait values (β). The selection gradients on the resources' birth fitness components are steeper for larger values of γ (fig. 3.1*A*), and the selection gradients on the consumers' death fitness components are steeper for larger values of θ (fig. 3.1*B*). The selection gradients generated by predation are steeper with smaller values of β (fig. 3.1*E*). All symbols and parameters are as described for figure 5.2. (Panels *A*, *C*, and *E* are redrawn from figure 8 of McPeek 2017, with permission).

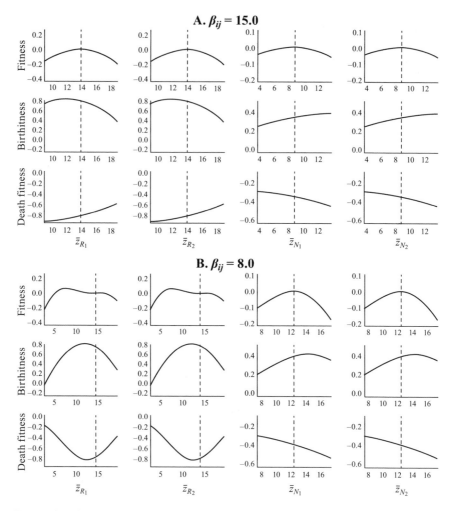

FIGURE 5.5. The overall fitness surfaces and the underlying surfaces for the birth and death fitness components once the community reaches equilibrium for two consumers and two resources interacting in the community module studied in figure 5.4. The three sets of panels show the final fitness surfaces when the steepness of the change in attack coefficient with a change in the

In a community having taxa for which the underlying gradients on the attack coefficients are slightly stronger (e.g., $6.9 \leq \beta_{ij} < 8.4$ in fig. 5.4E–F), the resources' overall fitness surfaces develop two fitness peaks from balancing the opposing selection gradients on their birth and death fitness components (fig. 5.5B). However, the consumers' overall fitness surfaces remain as a single peak (Fig. 5.5B).

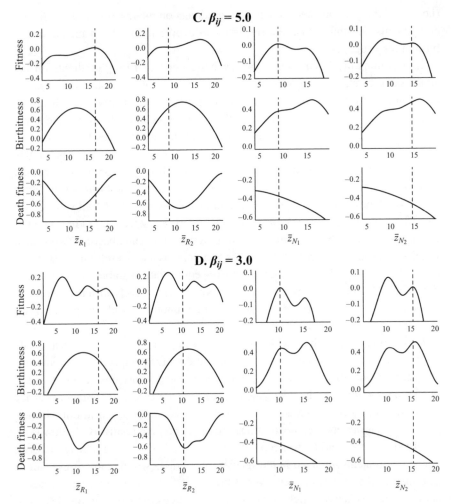

difference between the consumers' and resources' trait values are set to (A) β_{ij} = 15.0, (B) β_{ij} = 8.0, (C) β_{ij} = 5.0, and (D) β_{ij} = 3.0. The selection gradients generated by predation are steeper with smaller values of β (fig. 3.1E). Features in these panels are as described in figure 3.3. All other parameters are as in figure 5.2.

Consequently, the resources differentiate so that they have trait values that straddle their common intrinsic birth optima, but the consumers remain as neutral species with identical trait values; this results in an apparent competition community module with two neutral consumers and two differentiated resources. In this range of β_{ij}, communities or taxa with steeper underlying attack coefficient gradients (i.e., smaller β_{ij}) create steeper selection gradients on the resources' death fitness components and thus cause the resources to evolve farther from their intrinsic birth optima (fig. 5.4E–F).

In a community having taxa with even steeper underlying gradients on the attack coefficients (e.g., $\beta_{ij} < 6.8$ in fig. 5.4E–F), species at both trophic levels differentiate. At higher values of β_{ij} in this range, the consumers' birth fitness components do not have two peaks, but they do develop a saddle around the lower resource's trait value which causes their overall fitness surfaces to include two peaks (fig. 5.5C). A counterintuitive result in this range is that in communities with steeper attack gradients on the resources' death fitness components caused by predation, the fitness optima move closer to their intrinsic birth optima at $\tilde{z}^c_{R_i}$ (fig. 5.4E). The consumers also evolve to more closely match the trait value of the nearest resource (fig. 5.4E). With very high underlying selection gradients on the attack coefficients, the consumers' death fitness components become two peaks, which causes the resources' death fitness components and overall fitness surfaces to develop three peaks. Here, both resources equilibrate at stable fitness minima (fig. 5.5D). In this range with species at both trophic levels differentiated, the abundances of all four species are higher in communities with steeper intrinsic selection gradients (fig. 5.4F).

Comparing communities with taxa that vary in γ_i or θ_j, but have the same values for β_{ij}, shows the same results from the alternative perspective (i.e., altering the steepness of selection gradients for fitness components not involved in the species interaction). If the underlying relationships for the fitness components not involved in the species interaction of either the resources (i.e., birth fitness component) or consumers (i.e., death fitness components) are relatively shallow (i.e., small values for γ_i and θ_j, respectively), the realized selection gradients on their fitness components involved in the species interactions will be more important in determining the shapes of their overall fitness surfaces, and therefore species at both trophic levels will differentiate (i.e., small values for γ_i or θ_j in fig. 5.4A–D). However, if either of these underlying relationships are steep relative to the underlying relationship of the attack coefficient, species at neither trophic level will differentiate (i.e., large values for γ_i or θ_j in fig. 5.4A–D), and again intermediate values permit only the resources to differentiate (i.e., intermediate values for γ_i or θ_j in fig. 5.4A–D).

Thus, whether ecologically identical, and therefore neutral, species will differentiate depends not only on the environmental conditions in which the interactions take place, but also on the fundamental relationships defining how various fitness components change as trait values change in the interacting taxa. Here, the critical issue is which fitness components are more influential in determining the shape of the overall fitness surface (McPeek 1996a). If the fitness components associated with the species interaction have steep fitness gradients relative to those not involved in the species interactions, differentiation is fostered, because the fitness consequences of species interactions have a greater influence in determining overall fitness. In contrast, if the selection gradients generated by species interactions are weak relative to those not involved, identical species will not differentiate. In this latter case, because identical species have the same underlying topographies for these fitness components, they all evolve in response to the same adaptive peak.

The same patterns of differentiation occur among species with identical underlying parameters when unidirectional-dependent traits define the attack coefficients (e.g., traits like swimming speed or gape limitation define predator-prey interactions). For example, species at neither trophic level differentiate in communities with very low values for the maximum attack coefficients (a_{0ij}), only the resource species differentiate in communities with all a_{0ij} at intermediate levels, and both resource and consumer species differentiate in communities with all a_{0ij} having high values (fig. 5.6A–B). Likewise, increasing the underlying gradients on the attack coefficients defined by unidirectional-dependent traits (i.e., communities with higher values of α_{ij} [fig. 3.1D]) causes similar transitions as one compares communities. Namely, species at neither trophic level differentiate in communities with weak underlying gradients on the attack coefficients (i.e., low values of α_{ij}), only resource species differentiate in communities having moderate values of α_{ij}, and species at both trophic levels differentiate when the underlying gradients on the attack coefficients are strong (i.e., large values of α_{ij}); see figure 5.6C–D.

In contrast, differentiation of only the resources appears to be possible in a community when unidirectional-independent traits define the attack coefficients (e.g., traits like activity define predator-prey interactions; fig. 5.7) and resource differentiation occurs only in a small area of parameter space. When the resources do differentiate, one resource evolves a trait value that makes it invulnerable to the consumers (i.e., $a_{ij}(\bar{z}_{R_i}, \bar{z}_{N_j}) = 0$), while the other is vulnerable (fig. 5.7). This result is reminiscent of the many taxa that have alternative habitat use characteristics where one species utilizes a habitat free from predation but with low foraging returns, whereas another species utilizes another habitat where the risk of predation is higher but foraging returns are also higher (e.g., Sih 1980; Mittelbach 1981; Werner and Gilliam 1984; Gilliam and Fraser 1987; Brown 1988, 1989; Leibold 1990; Lima and Dill 1990).

Regardless of the types of traits defining the attack coefficients, the areas of parameter space where differentiation occurs have the same qualitative characteristics: the maximum intrinsic birth rates (c_{0i}) of the resources are high, the minimum intrinsic death rates (f_{0j}) of the consumers are low, the intrinsic phenotypic optima ($\tilde{z}^c_{R_i}$ and $\tilde{z}^f_{N_j}$) of consumers and resources are not too far apart, the

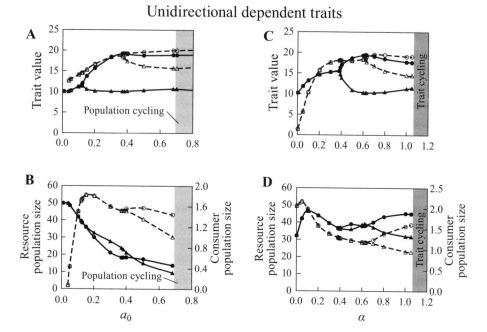

FIGURE 5.6. Differentiation of species with identical underlying parameters for two resources and two consumers in which the attack coefficients are defined by unidirectional-dependent traits in communities that differ in (A–B) the maximum attack coefficients (a_0) and (C–D) the underlying gradient strengths on the attack coefficients (α). The selection gradients generated by predation are steeper with larger values of α (fig. 3.1D). Symbols and lines identify the resources and consumers as given in figure 5.2. Parameter areas where trait cycling and population cycling occur are identified. The parameters used are $c_{0i} = 1.0$, $d_i = 0.02$, $a_{0ij} = 0.5$, $b_{ij} = 0.1$, $h_{ij} = 0.1$, $f_{0j} = 0.15$, $g_j = 0.1$, $\alpha_{ij} = 0.75$, $\gamma_i = 0.005$, $\theta_j = 0.005$, $\tilde{z}_{R_i}^c = 10.0$, $\tilde{z}_{N_j}^f = 1.0$, and $V_{z_{R_i}} = V_{z_{N_j}} = 0.2$. (Panel A is redrawn from figure 9 of McPeek 2017, with permission)

maximum attack coefficients (a_{0ij}) are high, the underlying strengths of selection on resources' intrinsic birth rates (γ_i) and the consumers' intrinsic death rates (θ_j) are weak, and the underlying selection strengths on the attack coefficients (α_{ij} for the unidirectional-independent, β_{ij} for the bidirectional-dependent, and ε_{R_i} and ε_{N_j} for unidirectional-independent traits) are strong. Specifically, differentiation requires that the underlying selection gradients on the resources' intrinsic birth rates and the consumers' intrinsic death rates are weak relative to the selection gradients on the functional responses driving their coevolutionary interactions.

The other striking result from this analysis of differentiation of species with identical underlying parameters is the ecological structures of the resulting

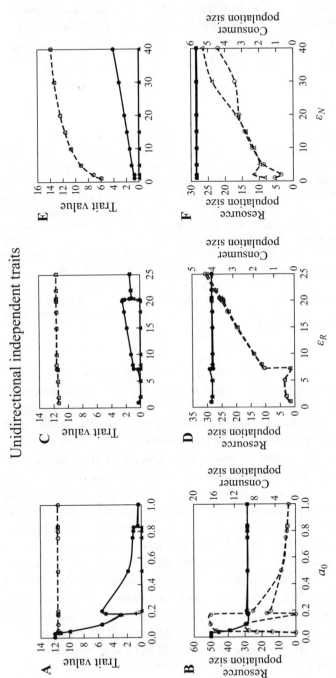

FIGURE 5.7. Differentiation of species with identical underlying parameters for two resources and two consumers in which the attack coefficients are defined by unidirectional-independent traits over gradients of (A–B) the maximum attack coefficients (a_0) and (C–F) the underlying strengths on the attack coefficients (ε_{R_i} and ε_{N_i}, respectively). The selection gradients generated by predation are steeper with smaller values of ε_{R_i} and ε_{N_i} (fig. 3.1C). Symbols and lines identify the resources and consumers as given in figure 5.2. The parameters used are $c_{0i} = 1.0$, $d_i = 0.02$, $a_{0ij} = 0.5$, $b_{ij} = 0.1$, $h_{ij} = 0.3$, $f_{0j} = 0.05$, $g_j = 0.0$, $\varepsilon_{R_i} = 15.0$, $\varepsilon_{N_j} = 15.0$, $\gamma_i = 0.003$, $\theta_j = 0.003$, $\tilde{z}_{R_i}^c = 12.0$, $\tilde{z}_{N_j}^f = 1.0$, and $V_{z_{R_i}} = V_{z_{N_j}} = 0.2$. (Panel A is redrawn from figure 9 of McPeek 2017, with permission.)

A. Unidirectional
independent traits

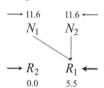

B. Unidirectional
dependent traits

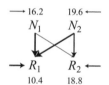

C. Bidirectional
dependent traits

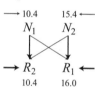

FIGURE 5.8. Examples of the ecological structure resulting from the differentiation of species with identical underlying parameters for two resources and two consumers in which the attack coefficients are defined by (*A*) unidirectional-independent traits, (*B*) unidirectional-dependent traits, and (*C*) bidirectional-dependent traits. The final phenotypic trait values for each species are next to the species labels, and the *arrows* illustrate the magnitudes of the resources' realized intrinsic birth rates, the consumers' realized intrinsic death rates, and the realized attack coefficients for each consumer feeding on each resource. The parameters used for this example are: (*A*) unidirectional-independent $c_{0i} = 1.0$, $d_i = 0.02$, $a_{0ij} = 0.2$, $b_{ij} = 0.1$, $h_{ij} = 0.1$, $f_{0j} = 0.05$, $g_j = 0.0$, $\varepsilon_{R_i} = 15.0$, $\varepsilon_{N_j} = 15.0$, $\gamma_i = 0.003$, $\theta_j = 0.003$, $\bar{z}_{R_i}^c = 12.0$, $\bar{z}_{N_j}^f = 2.0$, $V_{z_{R_i}} = V_{z_{N_j}} = 0.2$; (*B*) unidirectional-dependent $c_{0i} = 1.0$, $d_i = 0.02$, $a_{0ij} = 0.5$, $b_{ij} = 0.1$, $h_{ij} = 0.1$, $f_{0j} = 0.15$, $g_j = 0.1$, $\alpha_{ij} = 0.75$, $\gamma_i = 0.005$, $\theta_j = 0.005$, $\bar{z}_{R_i}^c = 10.0$, $\bar{z}_{N_j}^f = 1.0$, $V_{z_{R_i}} = V_{z_{N_j}} = 0.2$; and (*C*) bidirectional-dependent $c_{0i} = 1.0$, $d_i = 0.02$, $a_{0ij} = 0.25$, $b_{ij} = 0.1$, $h_{ij} = 0.0$, $f_{0j} = 0.1$, $g_j = 0.1$, $\beta_{ij} = 3.0$, $\gamma_i = 0.01$, $\theta_j = 0.01$, $\bar{z}_{R_i}^c = 12.0$, $\bar{z}_{N_j}^f = 1.0$, $V_{z_{R_i}} = V_{z_{N_j}} = 0.2$. (Panels *B* and *C* are redrawn from figure 10 of McPeek 2017, with permission.)

community modules when different trait types define the attack coefficients—namely, the pattern of interaction strengths among the species. For all three types of traits defining the attack coefficients, differentiation is caused by analogous changes in the shapes of the overall fitness surfaces (e.g., fig. 5.5). However, the resulting communities contain species with very different demographic structures when different trait types define the attack coefficients (fig. 5.8). Remember that in the bidirectional case, the differentiated resources straddle their intrinsic birth optima (e.g., figs. 5.2–5.5), which results in resources that have relatively similar realized birth rates at equilibrium (fig. 5.5C). Moreover, because of the bidirectionality of the attack coefficients, each consumer inflicts substantially greater mortality on the resource with the trait value that more closely matches its own (fig. 5.8C). Thus, bidirectional-dependent traits defining the attack coefficients permit the two consumers to differentiate so that each feeds more heavily on a different resource (fig. 5.8C). In contrast, with unidirectional traits (both independent and dependent) one resource experiences greater mortality inflicted by both consumers, but this resource also has a higher realized birth rate (figs. 5.8A–B). As a result, both consumers show the same diet hierarchy for feeding on the resources, and the resources display a trade-off between birth and death rates when compared.

Thus, when considering the trait types that define the attack coefficients, bidirectional-dependent traits favor consumer specialization onto different resources, whereas unidirectional traits (both dependent and independent) cause the same hierarchy of resource utilization among consumers. Also, considering this from a phenotypic perspective, bidirectional traits foster trade-offs among prey in their susceptibility to different coexisting predators, whereas unidirectional traits foster trade-offs between birth and death fitness components within the prey species. These are the recurrent patterns of phenotypic relationships and community structures that emerge from all these analyses, as we will see below.

DIFFERENTIATION OF SPECIES WITH DIFFERENT UNDERLYING PARAMETERS

In the last section, I considered the differentiation of species with identical underlying ecological capabilities (i.e., all the same parameters). When species differ in these underlying ecological capabilities, differentiation occurs over a much broader area of parameter space, because species have inherent underlying differences to push them apart. Moreover, many of these intrinsic differences that will foster differentiation may not be readily apparent, and this analysis should therefore also identify new avenues for the study of ecological differentiation. However, such inherent differences do not ensure differentiation. Species with large inherent differences may converge in trait values to make them ecological more similar and possibly even completely neutral with respect to one another.

In this section, I systematically explore whether interspecific differences in each parameter of the model can cause differentiation in a community module with three resources and three consumers. I do this by first setting the parameters for all species to be identical and in an area of parameter space that results in three identical consumers and three identical resources (i.e., outside the parameter space where differentiation occurs among identical species). Then for each parameter in turn, I set the values to be different for the species while maintaining all other parameters identical among the respective species. For each, I characterize when these differences in an underlying parameter results in species evolving to be appreciably differentiated over gradients of the other parameters, and whether the ecologically identical species (because they all have the identical parameters) at the other trophic level differentiate as a consequence. I repeat this process across all parameters for each of the trait types defining the attack coefficients.

When bidirectional-dependent traits define the attack coefficients, differences in some parameters for one species can cause the species at both trophic levels to differentiate. For example, consider the situations presented in figure 5.9, in

Bidirectional dependent traits

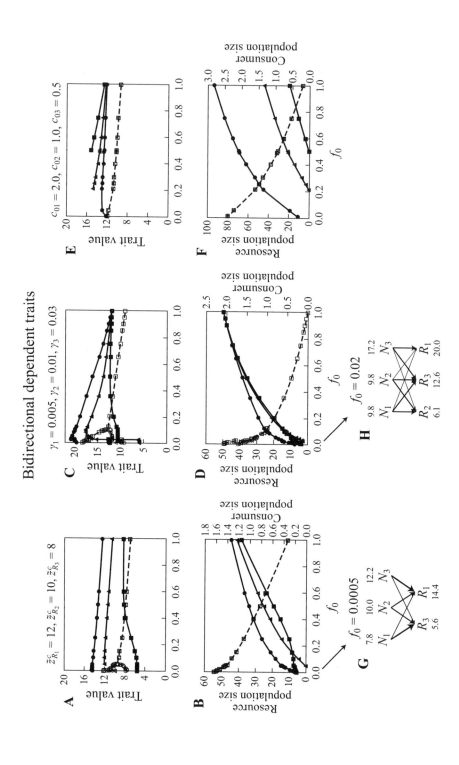

$\gamma_1 = 0.005, \gamma_2 = 0.01, \gamma_3 = 0.03$

$\tilde{z}^c_{R_1} = 12, \tilde{z}^c_{R_2} = 10, \tilde{z}^c_{R_3} = 8$

$c_{01} = 2.0, c_{02} = 1.0, c_{03} = 0.5$

A Trait value

B Resource population size / Consumer population size

C Trait value

D Resource population size / Consumer population size

E Trait value

F Resource population size / Consumer population size

G $f_0 = 0.0005$

N_1 7.8 N_2 10.0 N_3 12.2
R_1 14.4 R_3 5.6

H $f_0 = 0.02$

N_1 9.8 N_2 9.8 N_3 17.2
R_2 6.1 R_3 12.6 R_1 20.0

which one parameter differs among the resources, but all other parameters are the same for species at each trophic level. When the resources have different intrinsic birth optima $\left(\tilde{z}^c_{R_i}\right)$ and the consumers can reach high abundances because they have low minimum intrinsic death rates (f_{0j}), both the resources and consumers differentiate (fig. 5.9A–B). If the consumers have extremely low values of f_{0j}, the ecological structure of the resulting community module has three differentiated consumers and two differentiated resources; the resource with the intermediate $\tilde{z}^c_{R_i}$ cannot coexist in the resulting community (fig. 5.9G). The consumers with the most extreme trait values have a much higher attack coefficient on the resource nearest to them in trait value, while the consumer with the intermediate trait value feeds relatively similarly on both consumers.

In a community in which the consumers' minimum intrinsic death rate is somewhat higher (in this case, $f_{0j} > 0.049$), the consumers do not differentiate, but the resources do (fig. 5.9A-B); here, three neutral consumers with identical trait values feed on the three differentiated consumers in an apparent competition community module. The resource with the intermediate $\tilde{z}^c_{R_i}$ can only coexist in the system when the consumers do not differentiate. For a community with f_{0j} values above this level, the consumers' abundances and trait values decrease as their death fitness components become more prominent in defining both their demography and fitness surfaces, which allows the three resources to evolve to trait values closer to their respective intrinsic birth optima. In a community with high values of f_{0j} that do not permit the consumers to support populations, the three resources evolve to their respective $\tilde{z}^c_{R_i}$.

One fascinating result is that differences among the resources only in the underlying selection gradients on their intrinsic birth rates (γ_i) can also result in

---

FIGURE 5.9. Differentiation of species that differ in one parameter but have the remaining parameters identical for three resources and three consumers, in which the attack coefficients are defined by bidirectional-dependent traits in communities that differ in the consumers' intrinsic minimum death rate (f_{0j}). Panels A and B show when the resources differ in their intrinsic birth optima, panels C and D when they differ in the selection strengths on their intrinsic birth rates, and panels E and F when they differ in their intrinsic maximum birth rates. The specific parameter differences are identified above each column of panels. Relationships for the resources are identified with *solid lines* and for consumers with *dashed lines*. The three different resources are identified with *closed, solid symbols* (●, ▲, and ■) connected by *solid lines*, and the three different consumers with *open symbols* (○, △, and □) connected by *dashed lines*. Panels G and H illustrate two examples of the resulting ecological structure when the resources and consumers differentiate, as in figure 5.8. Only the relative magnitudes of the attack coefficients are illustrated. Unless otherwise specified, the parameters used are $c_{0i} = 1.0$, $d_i = 0.02$, $a_{0ij} = 0.25$, $b_{ij} = 0.1$, $h_{ij} = 0.0$, f_{0j} = variable, $g_j = 0.1$, $\beta_{ij} = 6.0$, $\gamma_i = 0.03$, $\theta_j = 0.03$, $\tilde{z}^c_{R_i} = 12.0$, $\tilde{z}^f_{N_j} = 1.0$, and $V_{z_{R_i}} = V_{z_{N_j}} = 0.2$.

differentiation at both trophic levels (fig. 5.9C–D). In other words, differentiation does not require that different trait values be favored for any particular fitness component; instead, merely having species that differ in the steepness of selection gradients on a fitness component will cause them to differentiate. In this case, the resources all have $\tilde{z}_{R_i}^c = 12.0$, but they differ in the underlying selection strengths around this optimum (γ_i), and this will drive their differentiation by causing different balances in the selection gradients to be struck among the various fitness components within each species. The resulting community module also has the consumers largely segregated onto different resources, based on their trait values (fig. 5.9H). Again, in communities in which the consumers have higher minimum intrinsic death rates, the resources evolve to trait values that are closer to $\tilde{z}_{R_i}^c$, which in this case means that they converge toward the same trait value (fig. 5.9C–D).

Differences among the resources in only their maximum intrinsic birth rates can also cause them to differentiate from one another (fig. 5.9E–F). Remember that in an apparent competition module, conditions analogous to inequality (5.2) determine whether a resource can coexist in the system (in this case with three neutral consumers). Thus, resources with lower maximum intrinsic birth rates can only invade, adapt, and coexist when the consumers' abundances and trait values permit; for example, resources with lower c_i values can only adapt and coexist when the three neutral consumers have higher minimum intrinsic death rates and thus lower abundances and lower attack coefficients because of lower trait values (Fig. 5.9E–F). I could find no combinations of parameters for this case that caused the consumers to differentiate in response to the resources differentiating, and so I conclude that consumer differentiation in response to variation in c_{0i} among the resources is, at best, very rare.

Interspecific differences in a number of other parameters also cause either the resources or consumers to differentiate when bidirectional-dependent traits define the attack coefficients. Differences in the maximum attack coefficients (a_{0ij}) can cause species on both trophic levels to differentiate. Because I specified that conversion efficiencies (b_j) and handling times (h_j) were properties solely of the consumers, variation in these only caused the consumers to differentiate. Also, parameters associated with the consumers' intrinsic death rates (f_{0j}, θ_j, or $\tilde{z}_{N_j}^f$) only caused the consumers to differentiate. Finally, no differentiation was caused by interspecific variation in either the strengths of density dependence on resource birth rates (d_i) or consumer death rates (g_j).

With unidirectional-dependent traits defining the attack coefficients, patterns of differentiation were generally the same as I described for bidirectional-dependent traits in the previous few paragraphs (fig. 5.10). The only difference seemed to be that interspecific variation among the resources in $\tilde{z}_{R_i}^c$ would drive differentiation of the resources but not cause differentiation among the consumers (fig. 5.10A–B).

Again, differentiated consumers in the resulting community have the same hierarchy of utilization across differentiated resources when unidirectional-dependent traits define the attack coefficients (fig. 5.10G).

When unidirectional-independent traits define the attack coefficients, interspecific differences at a single parameter only generate differentiation among the species that differ in the parameter values and do not cause differentiation of the species with identical parameters at the other trophic level. In addition, interspecific variation in the maximum attack coefficients (a_{0ij}) or the resources' underlying selection strengths on the attack coefficients (ε_{R_i}) generate only trivial differences among the resources. However, larger differences among the consumers could be generated by interspecific differences in the consumers' underlying selection strengths on the attack coefficients (ε_{N_j}). When multiple parameters are varied among species to generate differentiation at both trophic levels (e.g., intraspecific differences in both $\tilde{z}_{R_i}^c$ and $\tilde{z}_{N_j}^f$), while unidirectional-independent traits define the attack coefficients, consumers also have similar hierarchies of resource utilization.

DIFFERENTIATION IN A THREE-TROPHIC-LEVEL COMMUNITY

Now add a trophic level of predators (P_k) to the community and see how this might change the dynamics of differentiation and resulting community structure. The predators will certainly change both the abundances of the consumers and the selection pressures that impinge on the consumers' traits. The abundance and mean trait responses of the consumers to these changes will in turn influence the selection pressures on the resources at the bottom of the food web. Clearly, an exhaustive exploration of parameter space is impossible, so I will simply present results from specific examples to highlight the major features of differentiation here and how they differ from those when only two trophic levels are present.

DIFFERENTIATION OF SPECIES WITH
IDENTICAL UNDERLYING PARAMETERS

The general patterns of differentiation across parameter space with three trophic levels are similar in broad features to community modules with only two trophic levels. Differentiation of species with identical underlying parameters also occurs in similar areas of parameter space as when species are at two trophic levels (see above).

In addition to parameter differences, which species will differentiate depends on the abundances of species that result in that area of parameter space. For

Unidirectional dependent traits

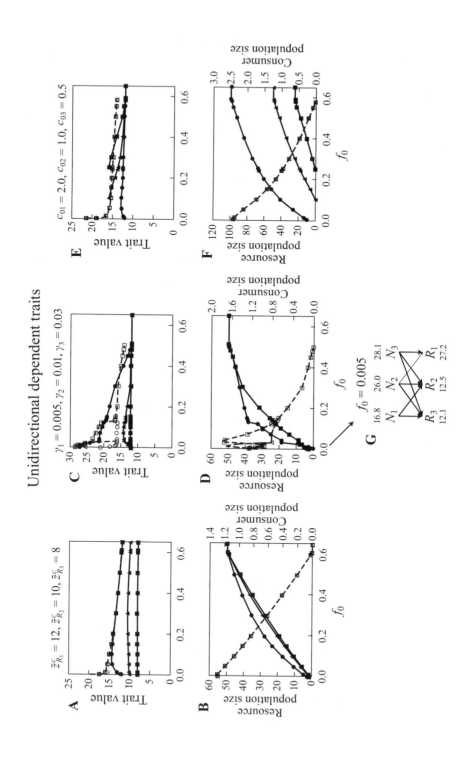

example, figure 5.11 illustrates the differentiation that occurs when three species with identical underlying parameters are at each of the three trophic levels
and bidirectional-dependent traits define all attack coefficients for communities
that differ in the minimum intrinsic death rates of the predators (x_{0k}). (Remember
that in any one community, all the top predators have the same minimum intrinsic death rates.) Trait cycles for this specific combination of parameters occur
when x_{0k} is very small (fig. 5.11C), and all three predators differentiate from one
another, but the consumers and resources do not. At the upper boundary where
trait cycling stops, the predators and consumers each differentiate into two functional groups (i.e., one of the functional groups has two neutral species with identical trait values and the other functional group is represented by a single species
with a different trait value), but the resources do not differentiate. For communities in this range (x_{0k} = 0.007–0.010), the consumers first differentiate so that
their trait values straddle that of the resources; this occurs to both increase consumption of one resource and simultaneously reduce mortality from the predator
(fig. 5.11A). After the consumers differentiate, the predators then differentiate to
specialize on each functional consumer group. At equilibrium in this example, the
predator functional group with the higher trait value (and two neutral species) is
at a fitness maximum, and the predator functional group with the lower trait value
(one species) is at a fitness minimum. The resulting module contains two predator
functional groups that feed differentially on each of the two consumer functional
groups, and the resources with identical trait values balance mortality from the
two consumer functional groups.

For a community with a slightly higher value of x_k (x_{0k} = 0.011–0.016), the
predators do not differentiate (i.e., three neutral species with identical trait values),
but the consumers do (fig. 5.11A). The realized death rates of the predators are
now high enough that only one fitness peak exists in their' overall fitness surfaces.

FIGURE 5.10. Differentiation of species that differ in one parameter but have the remaining
parameters identical for three resources and three consumers in which the attack coefficients are
defined by unidirectional-dependent traits over gradients of the consumers' intrinsic minimum
death rate (f_{0j}). Panels A and B show when the resources differ in their intrinsic birth optima,
panels C and D when they differ in the selection strengths on their intrinsic birth rates, and
panels E and F when they differ in their intrinsic maximum birth rates. The specific parameter
differences are identified above each column of panels. Species values are identified as in figure
5.9. Panel G illustrates an example of the resulting ecological structure when the resources and
consumers differentiate, as in figure 5.5. Only the relative magnitudes of the attack coefficients
are illustrated. Unless otherwise specified, the parameters used are c_{0i} = 1.0, d_i = 0.02, a_{0ij} =
0.25, b_{ij} = 0.1, h_{ij} = 0.0, f_{0j} = variable, g_j = 0.1, α_{ij} = 0.75, γ_i = 0.03, θ_j = 0.03, $\tilde{z}_{R_i}^c$ = 12.0,
$\tilde{z}_{N_j}^f$ = 1.0, and $V_{z_{R_i}} = V_{z_{N_j}}$ = 0.2.

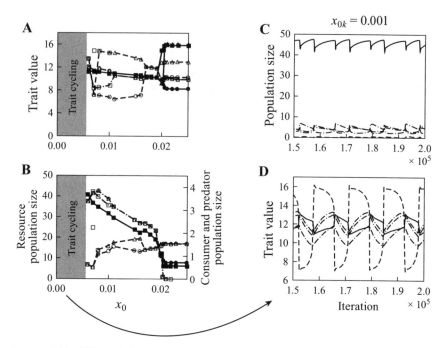

FIGURE 5.11. Differentiation of species with identical underlying parameters for three resources, three consumers, and three predators, in which the attack coefficients are defined by bidirectional-dependent traits over gradients of intrinsic minimum death rates for the predators (x_{0k}). Relationships for the resources are identified with *solid lines*, the consumers with *dashed lines*, and the predators with *dot-dashed lines*. The three different resources and consumers are identified as in figure 5.9, and the three different predators are identified with *open symbols* ($\circ$, $\triangle$, and $\square$). Parameter areas where trait cycling occurs are identified. Panels C and D illustrate the differentiation of the three predators in the area where trait cycling occurs. The predator values are *dot-dashed lines*; the consumers are *dashed lines*; and the resources are *solid lines*. Parameters used are $c_{0i} = 1.0$, $d_i = 0.02$, $a_{0ij} = 0.5$, $b_{ij} = 0.1$, $h_{ij} = 0.05$, $m_{0jk} = 0.05$, $n_{jk} = 0.1$, $l_{jk} = 0.05$, $f_{0j} = 0.05$, $g_j = 0.1$, x_{0k} = variable, $y_k = 0.001$, $\beta_{ij} = 5.0$, $\phi_{jk} = 5.0$, $\gamma_i = 0.02$, $\theta_j = 0.02$, $\delta_k = 0.02$, $\bar{z}^c_{R_i} = 12.0$, $\bar{z}^f_{N_j} = 5.0$, $\bar{z}^x_{P_k} = 4.0$, and $V_{z_{R_i}} = V_{z_{N_j}} = V_{z_{P_k}} = 0.2$.

In this range, the community structure collapses to take the functional form of a diamond community module in which the consumer with the lower trait value has a slightly lower attack coefficient on the resources, but experiences a slightly higher attack coefficient from the predators.

For a community with predators in the range of $x_{0k} = 0.017$–0.019, the consumers' death fitness components are reduced even further because of the decrease in the predators' abundances (fig. 5.11B), which removes one of the fitness peaks in the consumers' fitness landscapes. Consequently, the consumers also do not differentiate in this range, resulting in a single functional food chain with three neutral species at each of the consumer and predator trophic levels, and three ecologically identical resources (fig. 5.11A). The predators cannot support a population at $x_{0k} > 0.019$, and therefore the consumers and resources each differentiate into two functional groups in the absence of the predators (fig. 5.11A).

In general, adding the predator trophic level reduces the range of parameter space over which identical consumers and identical resources differentiate. In particular, resources differentiate over far less parameter space for all the trait types underlying attack coefficients, because the predators reduce the abundances of the consumers and thus reduce the impact of the consumers on resource fitnesses. Also, for species with identical underlying parameters, differentiation of species at more than one trophic level was more common for bidirectional-dependent traits than for unidirectional-dependent traits underlying the attack coefficients. For example, figure 5.12 shows one example area of parameter space over a gradient of minimum intrinsic death rates for the consumers (f_{0j}). With bidirectional-dependent traits determining the attack coefficients, both the consumers and predators differentiate into two functional groups at low consumer intrinsic mortality rates (fig. 5.12C–D), but only the consumers differentiate when unidirectional-dependent traits determine the attack coefficients (fig. 5.12A–B).

The range of parameter space in which species with identical underlying parameters differentiate is also reduced when unidirectional-independent traits define the attack coefficients. As with two trophic levels, differentiation typically is restricted to the resource trophic level with one species evolving to be essentially invulnerable and the others being more vulnerable to the consumers.

DIFFERENTIATION OF SPECIES WITH DIFFERENT UNDERLYING PARAMETERS

When species at the same trophic level differ in one of their underlying parameters, species in three-trophic-level community modules differentiate easily just as in two-trophic-level community modules. In addition, patterns of differentiation caused by interspecific parameter variation were generally similar to the two-trophic-level results discussed above. Not surprisingly, the number of species that could be supported in the system depended specifically on whether the resulting species satisfied the ecological coexistence criteria for the resulting community module; adaptation and differentiation cannot break the ecological laws of invasibility.

In general, if unidirectional-independent traits define the attack coefficients of all species, interspecific parameter differences causing differentiation at one trophic level do not foster differentiation of species with identical parameters at other

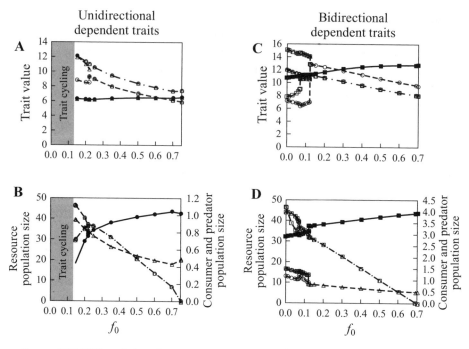

FIGURE 5.12. Differentiation of species with identical underlying parameters for two resources, two consumers, and two predators, in which the attack coefficients are defined by (A–B) unidirectional-dependent traits and (C–D) bidirectional-dependent traits over gradients of intrinsic minimum death rates for the consumers (f_{0j}). Relationships for the resources are identified with *solid lines*, those for consumers with *dashed lines*, and those for predators with *dot-dashed lines*. The different resources and consumers are identified as in figure 5.2, and the two different predators are identified with *open symbols* (○ and △). Parameter areas where trait cycling occurs are identified. Parameters used are for unidirectional-dependent traits $c_{0i} = 1.0$, $d_i = 0.02$, $a_{0ij} = 0.75$, $b_{ij} = 0.1$, $h_{ij} = 0.05$, $m_{0jk} = 0.5$, $n_{jk} = 0.1$, $l_{jk} = 0.05$, f_{0j} = variable, $g_j = 0.05$, $x_{0k} = 0.01$, $y_k = 0.001$, $\alpha_{ij} = 0.75$, $\rho_{jk} = 0.75$, $\gamma_i = 0.05$, $\theta_j = 0.05$, $\delta_k = 0.05$, $\bar{z}^c_{R_i} = 6.0$, $\bar{z}^f_{N_j} = 3.0$, $\bar{z}^x_{P_k} = 1.0$, $V_{z_{R_i}} = V_{z_{N_j}} = V_{z_{P_k}} = 0.2$; and for bidirectional traits $c_{0i} = 1.0$, $d_i = 0.02$, $a_{0ij} = 0.25$, $b_{ij} = 0.1$, $h_{ij} = 0.05$, $m_{0jk} = 0.1$, $n_{jk} = 0.1$, $l_{jk} = 0.05$, f_{0j} = variable, $g_j = 0.1$, $x_{0k} = 0.005$, $y_k = 0.0005$, $\beta_{ij} = 5.0$, $\phi_{jk} = 5.0$, $\gamma_i = 0.02$, $\theta_j = 0.02$, $\delta_k = 0.02$, $\bar{z}^c_{R_i} = 12.0$, $\bar{z}^f_{N_j} = 5.0$, $\bar{z}^x_{P_k} = 4.0$, $V_{z_{R_i}} = V_{z_{N_j}} = V_{z_{P_k}} = 0.2$.

trophic levels. However, differentiation caused by interspecific parameter variation at one trophic level could foster differentiation of species with identical parameters at higher trophic levels when both bidirectional-dependent and unidirectional-dependent traits define the attack coefficients, and this was much more prevalent when these traits were bidirectional. For example, if attack coefficients were determined by bidirectional-dependent traits, differences in parameters among the resources that caused them to differentiate could drive differentiation of both consumers and predators that were each identical in their underlying parameters.

A. Unidirectional independent traits

17.3   19.9   20.6
P_3 P_2 P_1

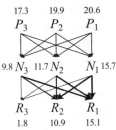

9.8 N_3 11.7 N_2 N_1 15.7

R_3 R_2 R_1
1.8   10.9   15.1

B. Unidirectional dependent traits

17.4   20.8   21.3
P_3 P_2 P_1

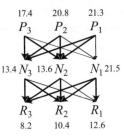

13.4 N_3 13.6 N_2 N_1 21.5

R_3 R_2 R_1
8.2   10.4   12.6

C. Bidirectional dependent traits

4.5   5.7   11.2
P_3 P_2 P_1

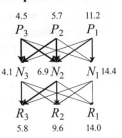

4.1 N_3 6.9 N_2 N_1 14.4

R_3 R_2 R_1
5.8   9.6   14.0

FIGURE 5.13. Examples of the ecological structure resulting from the differentiation of species with different underlying parameters for three resources, three consumers, and three predators, in which the attack coefficients are defined by (A) unidirectional-independent traits, (B) unidirectional-dependent traits, and (C) bidirectional-dependent traits. The final phenotypic trait values for each species are next to the species labels, and the *arrows* illustrate the magnitudes of the realized attack coefficients among species. Parameters used are for unidirectional-independent traits $c_{0i} = 1.0$, $d_i = 0.02$, $a_{0ij} = 0.5$, $b_{ij} = 0.1$, $h_{ij} = 0.05$, $m_{0jk} = 0.1$, $n_{jk} = 0.1$, $l_{jk} = 0.05$, $f_{0j} = 0.01$, $g_j = 0.01$, $x_{0k} = 0.001$, $y_k = 0.01$, $\varepsilon_{R_i} = 20.0$, $\varepsilon_{N_j} = 15.0$, $\eta_{N_j} = 20.0$, $\eta_{P_k} = 20.0$, $\gamma_i = 0.02$, $V_{z_{R_i}} = V_{z_{N_j}} = V_{z_{P_k}} = 0.2 = 0.02$, $\delta_k = 0.02$, $\tilde{z}_{R_1}^c = 16.0$, $\tilde{z}_{R_2}^c = 12.0$, $\tilde{z}_{N_1}^f = 4.0$, $\tilde{z}_{N_2}^f = 8.0$, $\tilde{z}_{N_3}^f = 5.0$, $\tilde{z}_{N_3}^f = 2.0$, $\tilde{z}_{P_1}^x = 7.0$, $\tilde{z}_{P_2}^x = 4.0$, $\tilde{z}_{P_3}^x = 1.0$, $V_{z_{R_i}} = V_{z_{N_j}} = V_{z_{P_k}} = 0.2$; for unidirectional-dependent traits $c_{0i} = 1.0$, $d_i = 0.02$, $a_{0ij} = 0.5$, $b_{ij} = 0.1$, $h_{ij} = 0.05$, $m_{0jk} = 0.5$, $n_{jk} = 0.1$, $l_{jk} = 0.05$, $f_{0j} = 0.15$, $g_j = 0.05$, $x_{0k} = 0.01$, $y_k = 0.01$, $\alpha_{ij} = 0.5$, $\rho_{jk} = 0.5$, $\gamma_i = 0.02$, $V_{z_{R_i}} = V_{z_{N_j}} = V_{z_{P_k}} = 0.2 = 0.02$, $\delta_k = 0.02$, $\tilde{z}_{R_1}^c = 12.0$, $\tilde{z}_{R_2}^c = 10.0$, $\tilde{z}_{R_3}^c = 8.0$, $\tilde{z}_{N_1}^f = 4.5$, $\tilde{z}_{N_2}^f = 4.0$, $\tilde{z}_{N_3}^f = 3.5$, $\tilde{z}_{P_1}^x = 3.5$, $\tilde{z}_{P_2}^x = 3.0$, $\tilde{z}_{P_2}^x = 2.5$, $V_{z_{R_i}} = V_{z_{N_j}} = V_{z_{P_k}} = 0.2$; and for bidirectional-dependent traits $c_{0i} = 1.0$, $d_i = 0.02$, $a_{0ij} = 0.25$, $b_{ij} = 0.1$, $h_{ij} = 0.05$, $m_{0jk} = 0.1$, $n_{jk} = 0.1$, $l_{jk} = 0.05$, $f_{0j} = 0.1$, $g_j = 0.1$, $x_{0k} = 0.01$, $y_k = 0.001$, $\beta_{ij} = 5.0$, $\phi_{jk} = 5.0$, $\gamma_i = 0.02$, $\theta_j = 0.02$, $\delta_k = 0.02$, $\tilde{z}_{R_1}^c = 14.0$, $\tilde{z}_{R_2}^c = 10.0$, $\tilde{z}_{R_3}^c = 6.0$, $\tilde{z}_{N_1}^f = 8.0$, $\tilde{z}_{N_2}^f = 5.0$, $\tilde{z}_{N_3}^f = 2.0$, $\tilde{z}_{P_1}^x = 7.0$, $\tilde{z}_{P_2}^x = 4.0$, $\tilde{z}_{P_3}^x = 1.0$, $V_{z_{R_i}} = V_{z_{N_j}} = V_{z_{P_k}} = 0.2$.

In contrast, I could only find parameter areas where differences among resources drive the differentiation at the consumer level for unidirectional-dependent attack coefficients when consumers and predators each had identical parameters. Also, interspecific differences at one trophic level only caused differentiation of species with identical parameters at higher trophic levels; for example, consumer differences could cause identically parameterized predators to differentiate, but could not drive resource differentiation.

Again, the structure of the resulting communities is different when different trait types define the attack coefficients (fig. 5.13). When unidirectional traits— both independent and dependent—define the attack coefficients, all consumers and predators have the same hierarchical impacts on their prey. For unidirectional-independent traits, prey with higher trait values are more exposed to their predators; each consumer and predator then has the largest realized attack coefficient on the prey with the largest trait value and the smallest realized attack coefficient on the prey with the smallest trait value (fig. 5.13A). Because larger trait values of prey reduce the attack coefficient determined by unidirectional-dependent traits, the opposite hierarchical relationships exist for consumer and predator impacts on their prey (fig. 5.13B). Thus, the same species experiences the most predation from all species at the next higher trophic level when unidirectional traits determine the attack coefficients.

In contrast, different prey species experience the greatest mortality inflicted by different consumers and predators when bidirectional-dependent traits define the attack coefficients (fig. 5.13C). In the example community illustrated in figure 5.13C, consumer N_1 has its highest realized attack coefficient on R_1, and its lowest on R_3; whereas N_2 and N_3 both have their highest attack coefficients on R_3, and their lowest on R_1 (fig. 5.13C). Comparable differences exist among the predators feeding on the consumers. Bidirectional-dependent traits permit predators and consumers to segregate onto different prey, and in so doing allow some degree of compartmentalization in community structure (Pimm and Lawton 1980, Krause et al. 2003).

The abilities of species to respond evolutionarily to one another can also alter how ecological properties of the community change for different parameter combinations. For example, with three trophic levels, increasing the intrinsic death rates of the consumers should generally decrease the abundances of both the consumers (i.e., their own death rates are higher) and predators (i.e., fewer consumers are available to be eaten) while increasing the abundances of the resources (i.e., fewer consumers to eat them); see chapter 2. This is the general pattern of response in the community illustrated in figure 5.14C–D in which bidirectional-dependent traits determine the attack coefficients. The three predators in this example can support populations over consumer intrinsic death rates of $f_{0j} < 0.30$, and their abundances decrease over the entire range.

However, the example with unidirectional-dependent traits determining attack coefficients that is illustrated in figure 5.14A–B shows how evolutionary responses can alter the predictions made on purely ecological grounds. In the latter example, when the intrinsic consumer death rates are very low ($f_{0j} < 0.059$), the three consumers are able to evolve trait values that prevent their predators from supporting populations (fig. 5.14A–B). As f_{0j} is increased above this level the predators in the example can sequentially invade and their abundances increase over the range $0.06 < f_{0j} < 0.20$, because the consumers evolve to have lower trait values, which increases the attack coefficients of the predators on all of the resource species. The system only responds above this level of consumer death rates ($f_{0j} > 0.020$) as the purely ecological considerations would predict, with declining predator and consumer abundances and increasing resource abundances (fig. 5.14A–B).

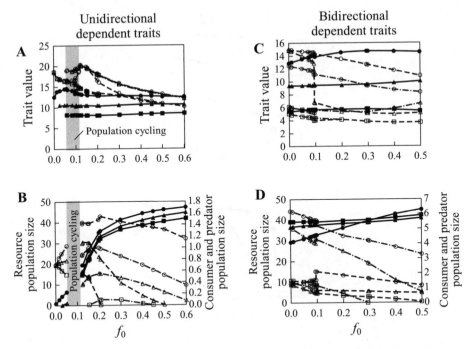

FIGURE 5.14. Differentiation of species that differ in one parameter but have the remaining parameters identical for three resources, three consumers, and three predators, in which the attack coefficients are defined by (A–B) unidirectional-dependent traits or (C–D) bidirectional-dependent traits over gradients of intrinsic minimum death rates for the consumers (f_{0j}). Parameter areas where population cycling occurs is identified. Symbols identifying species are as in figure 5.8. Parameters used are for unidirectional-dependent traits $c_{0i} = 1.0$, $d_i = 0.02$, $a_{0ij} = 0.5$, $b_{ij} = 0.1$, $h_{ij} = 0.05$, $m_{0jk} = 0.1$, $n_{jk} = 0.1$, $l_{jk} = 0.05$, $f_{0j} =$ variable, $g_j = 0.05$, $x_{0k} = 0.01$, $y_k = 0.01$, $\alpha_{ij} = 0.5$, $\rho_{jk} = 0.5$, $\gamma_i = 0.02$, $\theta_j = 0.02$, $\delta_k = 0.02$, $\tilde{z}^c_{R_1} = 12.0$, $\tilde{z}^c_{R_2} = 10.0$, $\tilde{z}^c_{R_3} = 8.0$, $\tilde{z}^f_{N_1} = 4.5$, $\tilde{z}^f_{N_2} = 4.0$, $\tilde{z}^f_{N_3} = 3.5$, $\tilde{z}^x_{P_1} = 3.5$, $\tilde{z}^x_{P_2} = 3.0$, $\tilde{z}^x_{P_3} = 2.5$, $V_{z_{R_i}} = V_{z_{N_i}} = V_{z_{P_i}} = 0.2$; and for bidirectional-dependent traits $c_{0i} = 1.0$, $d_i = 0.02$, $a_{0ij} = 0.25$, $b_{ij} = 0.1$, $h_{ij} = 0.05$, $m_{0jk} = 0.1$, $n_{jk} = 0.1$, $l_{jk} = 0.05$, $f_{0j} =$ variable, $g_j = 0.1$, $x_{0k} = 0.01$, $y_k = 0.001$, $\beta_{ij} = 5.0$, $\phi_{jk} = 5.0$, $\gamma_i = 0.02$, $\theta_j = 0.02$, $\delta_k = 0.02$, $\tilde{z}^c_{R_1} = 14.0$, $\tilde{z}^c_{R_2} = 10.0$, $\tilde{z}^c_{R_3} = 6.0$, $\tilde{z}^f_{N_1} = 8.0$, $\tilde{z}^f_{N_2} = 5.0$, $\tilde{z}^f_{N_3} = 2.0$, $\tilde{z}^x_{P_1} = 7.0$, $\tilde{z}^x_{P_2} = 4.0$, $\tilde{z}^x_{P_3} = 1.0$, $V_{z_{R_i}} = V_{z_{N_i}} = V_{z_{P_i}} = 0.2$.

INTRAGUILD PREDATION

No qualitative differences are caused by the predators being capable of feeding on both trophic levels below them (i.e., intraguild predation, with $v_{0ik} > 0$ in equation (5.1)). Obviously, the degree to which the predators feed on the resources will shape diversification on all trophic levels because of changes both in the selection pressures imposed directly by the predators on both the resources and consumers

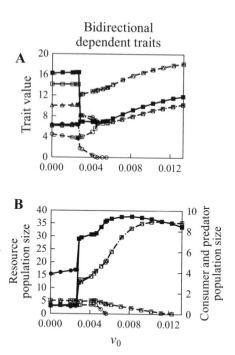

FIGURE 5.15. Differentiation of species with identical underlying parameters for three resources, three consumers, and three intraguild predators in which the attack coefficients are defined by bidirectional-dependent traits over gradients of intrinsic minimum death rates for the predators feeding on the resources (v_{0ik}). Symbols identifying species are as in figure 5.8. Parameters used are $c_{0i} = 1.0$, $d_i = 0.02$, $a_{0ij} = 0.5$, $b_{ij} = 0.1$, $h_{ij} = 0.05$, $m_{0jk} = 0.2$, $n_{jk} = 0.1$, $l_{jk} = 0.05$, $v_{0ik} =$ variable, $w_{ik} = 0.05$, $u_{ik} = 0.05$, $f_{0j} = 0.05$, $g_j = 0.05$, $x_{0k} = 0.02$, $y_k = 0.001$, $\beta_{ij} = 5.0$, $\phi_{jk} = 5.0$, $\psi_{jk} = 5.0$, $\gamma_i = 0.01$, $\theta_j = 0.02$, $\delta_k = 0.02$, $\tilde{z}^c_{R_i} = 10.0$, $\tilde{z}^f_{N_j} = 3.0$, $\tilde{z}^x_{P_k} = 2.0$, $V_{z_{R_i}} = V_{z_{N_i}} = V_{z_{P_i}} = 0.2$.

and indirectly on the consumers via changes in the resources' trait values and in the abundances of the interacting species.

For example, figure. 5.15 presents the communities resulting from three identically parameterized species at each trophic level differentiating in response to one another along a gradient of the maximum attack coefficient of the predators on the resources (v_{0ik}) when attack coefficients are determined by bidirectional-dependent traits. With no intraguild predation (all $v_{0ik} = 0.0$), the resources differentiate into two groups with different trait values, the three functionally different consumers differentiate, and only one functional predator (i.e., three neutral species) is present in the community at equilibrium. At very low levels of intraguild predation ($0.0 < v_{0ik} \leq 0.0028$) this structure is maintained. At $v_{0ik} = 0.0029$, predation on the resources from the predator is high enough to cause the three resources to converge to the same trait value, which causes the three consumers to collapse to two functional groups, and the three predators then differentiate into three functional groups (fig. 5.15A). As v_{0ik} is increased further above this range, the three predator groups converge in trait value until they coalesce into one functional group at $v_{0ik} = 0.00043$. Increases above this maximum attack coefficient for the predators on the resources decreases the consumer functional group abundances until they are driven extinct (at $v_{0ik} = 0.012$ in fig. 5.15) and only the predators and resources remain in the community. This is a two-trophic-level community with one functional group at each level (fig. 5.15).

EVOLVING ECOLOGICAL OPPORTUNITIES

These modeling results suggest that we should explore different routes by which species come to occupy different functional positions within biological communities. Ultimately, the dynamics of selection must favor differentiation of species for them to actually separate. However, species do not differentiate against fixed and unchanging fitness landscapes. In many circumstances, the dynamics of the fitness surfaces are actually what fosters species differentiation.

The first set of conditions that fosters species differentiation in a community is unsurprising. If species differ in the phenotypic traits that are favored by the underlying selection gradients, these species inherently experience different fitness surfaces almost without regard to the abundances of interacting species, and differentiation is consequently easy in this case. These differences may exist because different species have stabilizing selection gradients on a fitness component that have different optima (e.g., predators have different $\bar{z}_P^x$). Inherently, these species must then strike different balances among their fitness components and selection gradients, and species with different traits result.

Species may also strike different balances among selection gradients associated with various fitness components because of inherent differences in the magnitudes of fitness components. Species may differ in their abilities to convert ingested food into offspring (e.g., interspecific differences in c_i or b_j), because of differences in their physiological capacities for digestion or metabolic costs for other physiological processes they must fuel. Such differences will also necessarily result in different contributions of the various selection gradients to overall fitness among species, which again means that each will be evolving against inherently different fitness surfaces.

To me, one of the novel insights that these analyses reveal is that interspecific differences in the underlying gradients of one or more fitness components can also foster species differentiation. Once one considers the decomposition of an overall fitness surface into its component parts, this result is intuitively obvious (Arnold and Wade 1984a, 1984b; McPeek 1996a). If one species experiences substantially stronger selection on one fitness component, that fitness component will dominate the shape of its overall fitness surface, and thus limit the influence of selection on other fitness components (McPeek 1996a). Here again, what fosters species differentiation is the inherent differences among species in the fitness balances they strike.

For ecological differentiation to happen, fitness components associated with intrinsic capacities of the species must be relatively favorable to their overall fitness profile (e.g., relatively high intrinsic birth or relatively low intrinsic death rates) and experience relatively weak selection. At the same time, fitness components involved in the species interactions make high contributions to overall fitness (e.g., high intrinsic values of attack coefficients) and experience relatively strong selection. The abundances of the interacting species also influence whether differentiation will occur through their effects on determining the magnitude of fitness components; thus, resources will not differentiate when the intrinsic death rate of the consumer makes the consumer's equilibrium abundance too small (e.g., fig. 5.2C–D or 5.12).

This analysis also reveals that neutral species will differentiate from one another under the right ecological circumstances. This is why I have separated the analysis of ecological differentiation from speciation. Surely, the process of ecological speciation results from just such evolutionary dynamics as described in this chapter (Schluter 2000, Nosil 2012). However, as discussed in chapter 4, ecological speciation is by no means the only mode of speciation, and many modes actually result in species that are ecologically nearly identical to one another. Moreover, species may develop genetic incompatibilities in allopatry that do not affect their ecological performances, but subsequently find themselves in sympatry (e.g., community disaggregation/recombination cycles). In these latter cases, multiple neutral or nearly neutral species begin together in a community, and we still must consider whether they will differentiate or not, just as we still must consider whether they will coexist or not. For example, it is important to note that the two coexisting stickleback species in some Vancouver Island lakes were the result of two independent invasions of freshwater lakes by a marine ancestor (i.e., two separate speciation events) creating two species in the same lake that subsequently differentiated ecologically from one another to feed in the limnetic and benthic zones of the lake, and not in situ sympatric speciation in those lakes (Schluter and McPhail 1993).

When the differentiating species are identical in their underlying capacities, the dynamics of their fitness surfaces are key to their differentiation (e.g., fig. 5.1). How species accomplish a shift from one fitness peak to another has been an enduring question since Sewall Wright first proposed the idea of an adaptive landscape (Wright 1932, Charlesworth et al. 1982, Whitlock et al. 1995). The question has always been posed regarding how a species could evolve against a fitness gradient (i.e., downhill on the fitness surface) to traverse a fitness valley. For example, Whitlock (1995) showed how the shape of the average fitness surface varies in small populations due to fluctuations in phenotypic variance so that in some generations the fitness valley may disappear and selection will briefly push the population toward the alternative peak. What the present analysis shows is that the ecological dynamics of the system can create situations in which natural selection is the only driver needed to accomplish a peak shift. However, to understand this process we must begin to think of the fitness surfaces experienced by species as explicitly dynamical entities that change as the traits and abundances of all species in the community change (i.e., fig. 3.13).

We also tend to fixate on the conditions that favor the differentiation of species (Milligan 1985, Taper and Case 1985, Milligan 1986, Brown and Vincent 1992, Taper and Case 1992, Abrams 2000, Abrams and Chen 2002). Much of this focus is predicated on the assumption that species found together should be doing different things ecologically—in a word, coexisting. This certainly is the motivation for ideas about ecological character displacement (Brown and Wilson 1956; Schluter and Grant 1984;, Taper and Case 1985; Abrams 1986, 1987; Taper and Case 1992; Kawecki and Abrams 1999; Price and Kirkpatrick 2009). However, ecologists are now explicitly evaluating experimentally whether co-occurring neutral species can be found in communities, and are finding they are there (e.g., Siepielski et al. 2010, Shinen and Navarrete 2014). Circumstantial evidence for neutral species also continues to accumulate from other disciplines. For example, molecular systematic studies have identified innumerable examples of co-occurring cryptic species (Wells and Henry 1992, Bickford et al. 2007, Pfenninger and Schwenk 2007). Thus, the conditions that retard or prevent differentiation after neutral species are introduced into a community are just as important to our understanding of biological communities as those conditions that foster differentiation.

This exploration has also identified the ecological and evolutionary conditions in which an undifferentiated set of neutral species are most likely to occur in a community. Because differentiation at lower trophic levels occurs over a larger area of parameter space, the first prediction is that functional positions supporting neutral species should be more common at higher trophic positions in the community. Because higher productivity (i.e., higher values of c_{0i}) favors species differentiation, functional positions with neutral species are predicted to be more common in communities occupying lower productivity environments. Likewise, functional positions with neutral species should be more common when the interactions strengths (i.e., the realized values of attack coefficients) among species are relatively weak, which will result in relatively weaker selection gradients on fitness components influenced by species interactions. Finally, because ecological differentiation is least likely with unidirectional-independent traits defining the interactions among species, these may be the traits that undergird species interactions in many neutral species.

The converses of all these predictions suggest the ecological and evolutionary conditions that favor differentiation of ecologically very similar or identical species after being thrown together by the disaggregation and recombination processes considered in chapter 4, as well as the ecological and evolutionary conditions that would favor differentiation of a single species via sympatric ecological speciation directly. In other words, ecological character displacement/sympatric ecological speciation should be more favored to occur in species found in communities occupying more productive and more benign environments, when the strengths of species interactions are strong, when bidirectional and unidirectional-dependent traits underlie species interactions, and when selection gradients resulting from species interactions are also strong. Obviously, the genetic structures of species also will influence whether differentiation will occur. Sympatric ecological speciation will also require additional considerations as to how reproductive isolation is established along with ecological differentiation (Gavrilets 2004).

One obvious empirical relationship that all this suggests is the latitudinal diversity gradient in which species richness increases as one moves closer to Earth's equator. Recently, Schemske et al. (2009) proposed that greater diversity may result from the fact that species interactions are stronger in the tropics than in temperate and polar regions. The results presented here provide a mechanistic basis for how the strength of species interactions may actually drive diversification. Tropical regions may possess all the environmental hallmarks of those areas of parameter space that foster diversification; these include higher ecosystem productivity and more benign conditions—conditions that foster larger population sizes. One test of this would be whether selection gradients on fitness components associated with species interactions are steeper with decreasing latitude.

Trait types also differ in the likelihood of species differentiation. Again, differentiation of identical species occurs over a broader range of parameter space when bidirectional-dependent traits define species interactions, and is more likely to result at multiple trophic levels simultaneously. Likewise, when comparing unidirectional traits, those in which the fitness consequences in one species depend on the trait values of interacting species are more likely to foster differentiation than those in which the traits of each species have independent fitness consequences for the interactors. One immediate and general prediction that flows from this argument is that the diversity of functional groups in communities should depend on the types of traits involved in species interactions; functional diversity should be greatest when bidirectional-dependent traits determine the mechanisms of species interactions, intermediate with unidirectional-dependent traits, and least when independent traits define species interaction mechanisms. I hope this preliminary probe into the mechanisms of traits in coevolution will spark a more robust and mechanistic analysis of trait-centered community ecology as well.

Studying natural selection as an ecologically dynamical process that integrates the community context in which species are embedded will make this a truly predictive endeavor. The possibilities for predictions that result from this way of seeing species embedded in communities are endless, but let me identify a few possible foci. From the considerations outlined above, a first natural question is whether a set of closely related and coexisting species (e.g., the clade of Darwin's finches) have differentiated because they have inherent differences in the underlying parameters shaping their fitness components (e.g., as in the situations modeled in fig. 5.9), or despite them starting with identical underlying parameters (e.g., as in the situations modeled in fig. 5.2). Conversely, for co-occurring neutral species (e.g., *Enallagma* damselflies), do they also have properties of their underlying fitness components that would hinder the differentiation of identical species, or is it the ecological properties of the communities they occupy that retard their differentiation? In particular, many interesting questions will involve comparing the dynamics of natural selection in co-occurring and coexisting species, and explaining how the interactions in which they are involved shapes the selection pressures they experience. One of my main goals in writing this book is to spark a much richer empirical analysis in real biological communities about the operation of natural selection than simply tallying whether selection is operating and, if so, what qualitative form it takes.

Intriguingly, many of the best examples of character displacement and ecological differentiation that foster coexistence among taxa involve bidirectional-dependent traits in foraging interactions with other species. Darwin's finches feeding on seeds of various sizes is an obvious example (Grant and Grant 1982, Schluter and Grant 1984, Grant and Grant 2006). Likewise, multiple desert rodent species show similar patterns of differentiation for feeding on seeds of different sizes (Brown and Lieberman 1973, Bowers and Brown 1982). Limb and body size matching to various tree branch diameters that augment foraging capabilities in Caribbean *Anolis* lizards also represents the result of ecological diversification based on bidirectional-dependent traits (Losos 1990a, 1990b). Another example is two stickleback species diverging in body size and gill raker size, both bidirectional-dependent traits, to feed on small zooplankton in the limnetic zone or large invertebrates in the benthic zone of some Vancouver Island lakes (Schluter and McPhail 1993). Interestingly, two differentiated stickleback species are found in only six Vancouver Island lakes, whereas most lakes on the island have only one stickleback species (Schluter and McPhail 1993). Might those be the only six lakes on Vancouver Island with environmental conditions that would foster ecological differentiation (e.g., those that are more productive or more benign for sticklebacks to increase their population sizes, or less turbid waters that would make species interactions stronger)?

In stark contrast to these exemplars of coexistence and ecological differentiation stands my own work on the 38 *Enallagma* damselfly species of eastern North America. As I described in chapter 4, although the last common ancestor of the clade dates to eight to nine million years ago (Turgeon et al. 2005, Callahan and McPeek 2016), the vast majority of diversification in this clade was accomplished during the last 500,000 years by reproductive character displacement speciation (McPeek et al. 2008, McPeek et al. 2009, McPeek et al. 2011) that produced a large set of phenotypically nearly identical species in lakes with sunfish (McPeek 2000, 2004; Stoks and McPeek 2006). Up to 12 co-occurring *Enallagma* species can be found in every lake with sunfish as the top predators from the Gulf of Mexico to southern Canada (Johnson and Crowley 1980; McPeek 1990b, 1998), and field experiments have shown that these are in fact neutral species occupying a single functional position within the lakes' food webs (Siepielski et al. 2010).

These neutral *Enallagma* species occupy one intermediate-consumer functional position in a diamond community module where they suffer low mortality from fish predators but are poor at utilizing resource species (McPeek 1998, 2004). These ecological performance capabilities permit them to coexist as a group with a set of two to three neutral *Ischnura* species at the other intermediate-consumer functional position in this module (e.g., fig. 2.10) (McPeek 1998, 2004). *Enallagma* larvae of all species avoid fish predators by being very inactive when they perceive no predators and nearly cease activity when a predator is very near them; these are unidirectional-independent traits for survival in the face of fish predation (McPeek 1990a, Stoks et al. 2003, McPeek 2004). Their poor utilization of resources results from the strong physiological stress responses to the presence of fish that all of them express (McPeek et al. 2001a, McPeek 2004), which is also a unidirectional-independent trait that affects their conversion efficiency for feeding on resources (i.e., b_j). Thus, the major species interactions in which fish-lake *Enallagma* engage are both defined by unidirectional-independent traits—the trait type that is least likely to foster ecological diversification, based on the models presented here.

Understanding the properties of species and communities that foster or retard differentiation also provides compelling insights into community structure. To me, the most obvious feature of communities that these considerations inform is the prevalence of neutral versus coexisting species. Differentiation of species necessarily implies that the ecological conditions they experience are creating selection pressures to make them different, and as this occurs, one or both are moving to occupy different ecological opportunities in the community. Most ecologists assume that such differentiation is driven by resource competition (e.g., MacArthur and Levins 1967; Roughgarden 1976; Pacala and Roughgarden 1982; Milligan 1985; Abrams 1986, 1987; Taper and Case 1992). However, as I hope has been made abundantly clear in this chapter, the same kinds of coevolutionary dynamics can be driven in any position in any type of community module. Whether neutral or coexisting species will inhabit a functional position in a community depends on both the properties of the species (e.g., trait types and underlying selection gradients), properties of the community (e.g., species abundances and trait values), and properties of the ecosystem (e.g., productivity). "Where did these species come from?" and "how did they acquire the phenotypic properties they have now?" are fundamental questions for understanding community structure.

As I also hope was abundantly clear from this chapter, the properties of species that are relevant to understanding community structure do not simply involve fitness components and fitness gradients. A fundamental aspect of understanding community structure is understanding both the nature of species' traits that define their abilities to engage in interactions with other species, and the different types of mechanisms of species interactions that these traits engender. Community ecology has become fascinated with trait-based approaches in the last 20 years. However, these analyses can be just as superficial as tallying the frequency of directional versus stabilizing selection among species. Moreover, the traits that are studied in trait-based analyses of community structure are often simply the traits that are easily measured (e.g., body size or leaf shape), and not those that have been shown to have the greatest influence on the fitness of individuals in the species.

Differentiation of species where dissimilar trait types underlie interactions results in very different community structures. For example, specialization of coexisting consumers to feed on different resource species is only possible if those interactions are defined by bidirectional-dependent traits (e.g., figs. 5.8C and 5.13C). This is the type of differentiation that has been imagined and considered in community ecology in the context of niche differentiation, species packing, and limiting similarity approaches that focus on resource competition (Hutchinson 1958, MacArthur and Levins 1967, Roughgarden 1974). Darwin's finches can specialize on seeds of different sizes because their bills determine optimal sizes of seeds for each, and seeds that are both larger and smaller than that optimum are more difficult for a bird to consume (Schluter and Grant 1984). *Anolis* lizards can specialize on different areas of a tree because their body and leg dimensions define optimal proportions to utilize different diameters of perches (Losos 1990a, 1990b).

In contrast, unidirectional traits (both independent and dependent) underlying species interactions produce the same hierarchies of interaction strengths among all consumers (figs. 5.8A–B and 5.13A–B). The time a prey species spends in activity will influence its attack coefficient with all its predators the same way. Likewise, if escape speed primarily defines the probability that prey will evade their predators, all predators will have the same hierarchy of feeding on the prey in their diets. If species have the same hierarchical ranking of interactions with other species, these would produce a nested pattern of species interactions, exactly like those seen in plant-animal mutualism interaction networks (Bascompte et al. 2003, Jordano et al. 2003, Vázquez and Simberloff 2003, Bascompte et al. 2006). This suggests the prediction that performance in these interactions (primarily plant-pollinator interactions) is underlain by species traits that influence their fitnesses unidirectionally. Thus, the traits of species define the patterns of interaction strengths that will emerge in communities. Other interaction network metrics are also probably influenced by the particular types of traits that shape species interactions (e.g., Blüthgen et al. 2008, Ings et al. 2009, Fortuna et al. 2010).

One final thought centers on the issue of whether ecological opportunities in a community exist independently of the evolutionary potentials of the species that fill them. In one sense they do. Imagine some community module with a configuration of species already present. Given the connections and strengths of interactions among them, one could conceivably describe the entire set of possible species that satisfy the invasibility criteria for this community. This would be sufficient if species could not coevolve. However, as soon as one allows the possibility of coevolution among the species, the limited area of parameter space permitting invasibility expands into a much broader area in which the resident species alter the "ecological opportunity" they are exploiting in response to the invader. Ecological opportunities emerge out of, and are reoriented by, this evolutionary process for both residents and invaders. Once complete, the structure of a community may provide a very false picture of how a species invaded and came to reside in the community. We cannot simply consider the present-day ecology of a community to understand how it was assembled in the past. We cannot really understand the structure of a community without understanding the role that past evolutionary interactions among the component species have played.

CHAPTER SIX

Moving among Communities

As you walk along the trail through the woods, you will immediately notice you are not the only creature moving across the landscape. Look up and you will see birds flying from tree to tree, and others soaring high above. Insects patrol the fields, and some are up above the trees with the birds on long-distance movements. If you look closely, you will also see seeds wafting on the wind and spiders ballooning across the countryside. If you could see very minute objects, you would glimpse the tiny eggs of fairy shrimp and water fleas blowing across the ground and stuck to the feet of those birds that passed overhead. All of these species are on the move.

Much of this movement occurs within a community. Animals and many microbes can move through their environment in search of resources (i.e., water, minerals, and prey) or in order to not become resources for others. Bees move from flower to flower in search of pollen. Dragonfly larvae stalk the stems of macrophytes in the pond in search of prey. Plants move by casting their offspring spores and seeds to the wind. These movements do not remove individuals from the breeding population in which they were born, nor do they remove individuals from the geographical extents of the populations of species with which they are interacting.

Other movements, however, clearly constitute organisms moving from one community to another. A water strider adult leaves one pond and flies to another pond 20 kilometers away. A coconut falls from a tree on one oceanic island, is washed into the water, and drifts to another island where it sprouts into a growing tree. A fairy shrimp egg is blown from the spot where one vernal pond typically forms during the rainy spring to another suitable spot kilometers away, while others are blown to spots where no vernal pond will ever form. In these cases, an individual has moved permanently from one breeding population to another; part of its life cycle is spent in the demographic and selective regime of one population, and the rest is spent in the new regime (Ronce 2007). This movement is typically characterized as *dispersal* between populations for this species. Some species disperse as embryos, some as juveniles, and some as reproductive adults, but the consequences of that movement are all qualitatively the same for populations.

Still other movements of organisms may be somewhere between these two extremes. For many species in a creek, each pool along the creek constitutes a nearly unique breeding population, whereas the kingfisher forages up and down the length of the creek across all these pools. A bear may wander over thousands of square kilometers to forage, and in so doing move among many local populations of plants, insects, and other prey species. Likewise, a meadow may constitute a relatively closed population for many of the plants found there, but individual bees may forage in multiple isolated meadows on any given day. These species that repeatedly move among geographic areas that constitute different populations for some of their interacting species *link* the population dynamics of these other species over space. The frequency of such movements may be rapid (e.g., daily feeding forays) or slow and periodic (e.g., seasonal migration). All along this continuum, the population dynamics of what seem to be isolated populations are indirectly coupled via interaction with a linking species.

Because populations of different species encompass very different spatial scales, the boundaries of communities are rarely clearly defined. Even for those where the boundary could be made definitively, communities in different geographic locations are still demographically and evolutionarily linked by dispersal of their component species. Thus, for some situations there may be little difference between dispersing and link species, because individuals entering (immigrating) or leaving (emigrating) a population will potentially affect the demographic rates and evolutionary trajectories of that population. Thus, we must explore how these various levels of movement between populations affect both the ecological and evolutionary dynamics of local communities, and how structure across these communities may in turn shape movement patterns.

A number of researchers are now championing the importance of movement among local communities, under the rubric of *metacommunities*, to understanding both local and regional structure (Holyoak et al. 2005). They identify four perspectives in which recognizing these linkages alters our perceptions about the processes influencing local community structure. The first is the patch dynamic perspective, which recognizes the potential importance of link species and the interplay of population dynamics within local communities and dispersal among communities. The second is the species sorting perspective, which recognizes that species will assort into different local communities based on their ecological capabilities. The third is the mass effects perspective, which emphasizes that dispersal among local communities can perturb local abundances away from what is favored by local regulatory mechanisms, including the existence and potential importance of sink species in local community dynamics (Mouquet and Loreau 2003). The last is the neutral perspective, which acknowledges the existence and potential ubiquity of neutral species in shaping patterns of species richness and other metrics of community structure across the landscape. Some authors are also beginning to consider how species might evolve in the metacommunity context, particularly in response to climate change (e.g., Urban et al. 2012).

In this chapter, I present a more refined analysis of the effects of dispersal on local communities for this metacommunity perspective. Movement among communities certainly does perturb local abundances away from what is favored by local interactions, but how and when? This movement also can create sink species in a community. The propensity to move among communities can itself evolve in response to the constellation of fitness values that a species may experience in different communities. Movement will also shape the evolution of those traits that determine fitness within each community as well. I explore all of these issues here.

The community module perspective (Holt 1997a) is also useful here because it places community dynamics into the same conceptual framework as the exploration of the ecological and evolutionary consequences of individual movements among interaction networks. The behavioral choices of link species moving among local areas and the evolutionary dynamics caused by dispersing individuals among different populations are both understood in the context of the demographic and fitness returns for either remaining in place or moving to another location. Likewise, the invasibility determination questions whether the fitness returns are great enough in a particular interaction network for a species to establish a population and identifies why or why not. This is also the first criterion to evaluate whether an individual should move to this location. Once the world is divided into those places with negative and positive fitness returns, the individual then must consider the subtler differences among the places with positive fitness returns. The evolutionary actors must decide on which ecological stages to perform.

AN EXAMPLE OF THE DEMOGRAPHIC
CONSEQUENCES OF DISPERSAL

Immigration and emigration generate population demographic rates comparable to birth and death rates, respectively, and these rates of movement can generate various forms of density dependence that interact with local mechanisms of population regulation to shape local abundances and population dynamics (Murdoch et al. 1992, Amarasekare 1998, Neubert et al. 2002, Abrams et al. 2007). Consequently, dispersal can perturb local abundance away from what is favored by local mechanisms of population regulation (Holt 1985).

To begin our exploration, consider the exchange of dispersers among two populations of one species. For now, I will ignore all the complications of species interactions that generate local fitness relationships and population regulation, but

I will return to those complications later. However, assume that all these species interactions I am ignoring generate negative density dependence such that the local populations are regulated to stable equilibria (e.g., Reynolds and Brassil 2013, Arditi et al. 2015). Initially, also assume that individuals have an innate propensity, $\omega_{1(j)}$ (in the subscript, the 1 identifies the species and j the population), to emigrate from their natal population, and all individuals have the same dispersal propensity in all populations (i.e., $\omega_{1(1)} = \omega_{1(2)} = \omega$, in this case). These simple assumptions imply a model for two populations linked by dispersal:

$$\frac{dR_{1(1)}}{dt} = (1 - \omega)\left(c_{(1)} R_{1(1)} - d_{(1)} R_{1(1)}^2\right) - \omega R_{1(1)} + \omega R_{1(2)}$$
$$\frac{dR_{1(2)}}{dt} = (1 - \omega)\left(c_{(2)} R_{1(2)} - d_{(2)} R_{1(2)}^2\right) - \omega R_{1(2)} + \omega R_{1(1)}$$
$$\text{(6.1)}$$

This model is a modification of Holt's (1985) analysis, but here ω is the propensity to disperse (i.e., $0 \leq \omega \leq 1$) instead of a rate. In the absence of dispersal (i.e., $\omega = 0$), each population is regulated to an equilibrium abundance of $R_{1(j)}^* = c_{(j)}/d_{(j)}$. If $c_{(1)}/d_{(1)} = c_{(2)}/d_{(2)}$, at equilibrium this form of dispersal will have no effect on local abundances because the same number of individuals (i.e., $\omega R_{1(j)}^*$) will be simultaneously entering and leaving each population. Table 6.1 summarizes the new parameters for this chapter.

In contrast, if $c_{(1)}/d_{(1)} \neq c_{(2)}/d_{(2)}$, dispersal will alter the abundances of both populations away from these equilibria. Holt (1985) provides an excellent general presentation of this issue, and I follow his analysis using this model to illustrate the main points here. Assuming that $c_{(1)}/d_{(1)} > c_{(2)}/d_{(2)}$ and $\omega > 0$, more individuals emigrate from population 1 than from population 2, making the former a net exporter of individuals and the latter a net importer. The system of two populations comes to a new set of equilibrium abundances when net dispersal balances the local population growth rate in each patch, such that

TABLE 6.1. Additional state variables and parameters in the models of species movement among localities. Variables and parameters that are common to multiple species types are shown only for resource species. All other variables and parameters are as listed in table 2.1.

State Variable	Description
$R_{1(j)}$	Population abundance of resource species 1 in population j
Parameter	Description
$\omega_{1(j)}$	Probability that an individual of species 1 will emigrate from population j and enter the pool of dispersers
ξ	Proportion of emigrants that survive to immigrate to another population

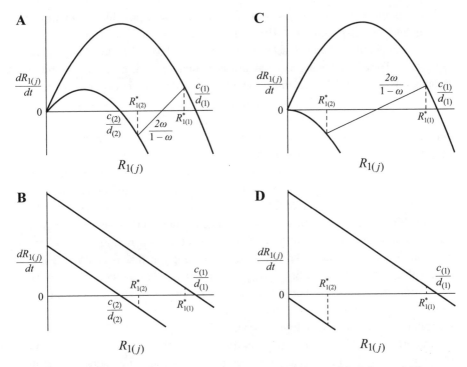

FIGURE 6.1. Graphical representation of a simple two population model of dispersal. The representation here follows Holt (1985). Panel A shows the relationships between total population growth rates and abundance for the two populations linked by passive dispersal (see equations (6.1). In the absence of dispersal, each population equilibrates at an abundance of $c_{(j)}/d_{(j)}$. Holt (1985) showed that passive dispersal causes new equilibrium abundances (i.e., $R^*_{1(1)}$ and $R^*_{1(2)}$), and given the model analyzed here, these equilibria are determined by a line with slope $2\omega/(1-\omega)$ (*narrow line* in panel A) that connects points of total growth rate in the two populations that are equal in magnitude but opposite in sign. Panel B represents the same relationships using per capita growth rate—average individual fitness—as the metric. At these new equilibria with dispersal, individuals in population 1 have positive fitnesses and individuals in population 2 have negative fitnesses. Panels C and D show the same relationships for the case in which population 2 is a sink.

$$\frac{c_{(2)}}{d_{(2)}} < R^*_{1(2)} < R^*_{1(1)} < \frac{c_{(1)}}{d_{(1)}} \tag{6.2}$$

(Holt 1985). Greater dispersal propensity (i.e., larger ω) will cause these new equilibria to be farther from the values favored by local population regulation, and thus closer to one another (fig. 6.1). As a result, individuals in population 1 perpetually have positive fitnesses at equilibrium to match the net export of individuals, and individuals in population 2 perpetually have negative fitnesses because of the net import of individuals (fig. 6.1).

This simple model illustrates a number of important possible demographic consequences of dispersal (Holt 1985). The most important is that dispersal can distort local population abundances away from those favored by local population regulation. These distortions make abundances more similar across populations. Populations with abundances above the average tend to be net exporters of individuals and populations below the average tend to be net importers of individuals. This will tend to cause individuals in populations with high abundances to perpetually have positive fitnesses and individuals in low abundance populations to perpetually have negative fitnesses (fig. 6.1B).

When taken to its extreme, such demographic effects of dispersal are the reason that sink species exist in communities (Pulliam 1988, Pulliam and Danielson 1991). Even if a species cannot support a population in a local community, continual immigration into that community can maintain a population at a stable equilibrium abundance. Figure 6.1C–D illustrates the same relationships for the populations in equations (6.1), but with the assumption that population 2 is a sink with $c_{(2)} < 0$. The equilibrium $R^*_{1(2)}$ remains stable despite the local negative population growth rate. The only requirement for this sink to exist is that the net immigration of individuals into this population must be greater than $|c_{(2)}|$, the magnitude of the sink population's growth rate at $R_{1(2)} = 0$.

The presence of this sink species in the local community would then distort the abundances of other species in the community through interactions with it. If it was a prey species, consumers' abundances may be inflated. Likewise, if it was a mutualist, it may inflate the abundances of mutualistic partners. If it was a predator, it may depress the abundances of other prey species.

EVOLUTION OF DISPERSAL IN THIS
TEMPORALLY CONSTANT ENVIRONMENT

Dispersal strategies may evolve, but the evolution of dispersal is a much more complex problem than the types of traits considered in chapters 3 and 5. Natural selection on dispersal rate is generated by the distribution of simultaneous fitnesses across the multiple populations connected by dispersal. Fitness variation through time also plays an important role in shaping natural selection on dispersal strategies. Because dispersal scatters individuals across multiple populations, the genetic influences on the rate of dispersal evolution, such as in equations used in chapters 3 and 5, are difficult to specify. Consequently, writing simple equations to describe the evolution of dispersal is difficult except in the most trivial cases.

Individuals of different species also employ myriad strategies to decide when and where to disperse (Amarasekare 2007, Ronce 2007). The first decision to be made is whether to emigrate from the population in which the individual resides. For some species, this may not be a decision at all; individuals may simply passively disperse because they have some innate propensity to emigrate from the population in which they live—as in the model just considered (e.g., Gadgil 1971, Roff 1975, Comins et al. 1980, Vance 1980, Kuno 1981, Hastings 1983, Metz et al. 1983, Levin et al. 1984, Holt 1985, McPeek and Holt 1992, Doebeli and Ruxton 1997, Metz and Gyllenberg 2001, McNamara and Dall 2011). For other species, individuals may respond to some estimate of the fitness returns that might be expected in this population to set their probability of emigration; they would then emigrate in response to some cue (e.g., conspecific or heterospecific density, resource availability, other potential indicators of fitness) (e.g., Asmussen 1983, Vance 1984, Ruxton and Rohani 1999, Amarasekare 2007, Ronce 2007). Individuals may also choose among populations in which to immigrate based on similar fitness criteria—that is, through habitat selection (e.g., Fretwell and Lucas 1969, Parker and Sutherland 1986, Moody et al. 1996, Ruxton and Humphries 1999, Holt and Barfield 2001, Cressman et al. 2004, Cressman and Křivan 2006, Cantrell et al. 2010, Cressman and Křivan 2010). The fitnesses of individuals may also be altered because of dispersing (e.g., dispersing individuals may die in transit or expend significant energy while moving). Although this seems to be a dizzying array of alternative possibilities, all lead to a few simple qualitative conclusions about the evolution of individual movements among populations and when individual movements among communities will distort abundances and trait evolution away from what the local ecological system favors.

Return to the model populations of one species embedded in two local communities like those specified by equations (6.1). To understand analytically how dispersal propensity will evolve in this case, assume that both of these populations have no density regulation (i.e., $d_{(1)} = d_{(2)} = 0$) and therefore grow exponentially (following Appendix II of Holt 1985). The combined populations of this species then grow at a rate that is the dominant eigenvalue of the matrix

$$\begin{bmatrix} c_{(1)} - \omega & \omega \\ \omega & c_{(2)} - \omega \end{bmatrix},$$

which is

$$r_1 = \frac{c_{(1)} + c_{(2)}}{2} - \omega + \sqrt{\omega^2 + \left(\frac{c_{(1)} - c_{(2)}}{2}\right)^2}. \tag{6.3}$$

Since $\partial r_1 / \partial \omega < 0$ if $c_{(1)} \neq c_{(2)}$, decreasing the propensity to disperse always increases fitness in this system (Holt 1985). The general result is that the average fitness of the species will be higher for a dispersal propensity that results in more individuals in the population with positive fitness and fewer individuals in the population with lower fitness. This can be seen visually in figure 6.1, since decreasing dispersal propensity (ω) will move the equilibrium abundances of both populations closer to what is favored by local population regulation. In this case, fitness will be maximized when no individuals disperse between the populations. Given that this is true generally for any fitness values substituted for $c_{(1)}$ and $c_{(2)}$, this result can be extended to any form of density regulation by simply updating the local fitness values as the abundances of the two populations change (Holt 1985). Note that no dispersal is favored under these circumstances even if dispersal is cost free.

Also, note that $\partial r_1 / \partial \omega = 0$ if $c_{(1)} = c_{(2)}$, meaning that dispersal propensity is a neutral trait if all populations have identical fitness determining relationships, and dispersers have no fitness costs. If all locations are identically regulated, individuals have no advantage or disadvantage from being in any particular population or moving among populations.

Hastings (1983) provided a general proof of these results for more than two populations, any form of population regulation, and any combination of patch-specific dispersal propensities (i.e., $c_{(1)} = c_{(2)}$). Balkau and Feldman (1973) analyzed an explicit population genetic formulation of this problem (see also Teague 1977, Asmussen 1983). In fact, Hastings (1983) and Holt (1985) both showed that no dispersal is an *evolutionarily stable strategy* when fitnesses vary in space but fitness relationships do not vary through time, meaning that no other dispersal strategy would have higher fitness in these environmental conditions (Maynard Smith 1983).

The general result of all these analyses is that if fitnesses vary in space but the relationships determining fitness do not vary through time (fitness may vary with local abundance but the relationships between abundance and fitness does not vary) at any location, natural selection favors no dispersal. However, this is not strictly true. What selection favors is a dispersal strategy that does not distort population abundances away from what local demographic processes favor. When all the local populations of a species with no dispersal reach demographic equilibrium, the fitness of all individuals in all population are the same. Any dispersal strategy that results in every individual having the same fitness in every population when they are at demographic equilibrium will be favored by natural selection.

Over a larger class of possible dispersal strategies, this same result can be derived using the logic first expounded by Fretwell and Lucas (1969) in the *ideal free distribution* to explain habitat selection in birds. They reasoned that the suitability of a habitat patch decreases as the number of individuals in that patch increases (fitness is negatively density dependent), that each individual selects the habitat patch "best suited to its survival and reproduction" (the ideal component), and that individuals can enter "any habitat on an equal basis with residents" (the free component). Their hypothesis predicts that individuals should apportion themselves among habitat patches so that all receive the same fitness (Fretwell and Lucas 1969). The ideal free distribution is also an evolutionarily stable strategy (Maynard Smith 1983), because no individual can increase its fitness by using another strategy to determine which habitat patch it should occupy.

In figure 6.1B, if we equate Fretwell and Lucas's (1969) "suitability" with local per capita fitness, individuals that select habitat occupancy based on the ideal free distribution should apportion themselves among the two populations so that $dR_{1(1)}/R_{1(1)} dt = dR_{1(2)}/R_{1(2)} dt$ always. This should remain true as population sizes increase or decrease via population regulation until the two populations equilibrate at $dR_{1(1)}/R_{1(1)} dt = dR_{1(2)}/R_{1(2)} dt = 0$. At this joint equilibrium, $R^*_{1(j)} = c_{(j)}/d_{(j)}$ in both populations, which are in fact the equilibrium abundances in the absence of dispersal.

Nowhere is this more apparent than with sink species in local communities (figs. 6.1C and D). If individuals can choose among the available habitats, an ideal free distribution would never include habitat sinks as long as some source areas (i.e., locations that can maintain viable populations in the absence of immigration) are available. Even if individuals are unable to identify and choose among possible habitats and only have a passive propensity to disperse, natural selection in this type of environment would favor no dispersal. Immigration would then not be available to maintain the abundances of sink species in a community, and sink species would be driven locally extinct by local population regulation. Sink species in such temporally constant environments are, therefore, prime evidence that evolution cannot or has not yet optimized dispersal rates of this species.

LOCAL ECOLOGICAL CONDITIONS ARE
NOT ALWAYS TEMPORALLY CONSTANT

Until now, I have considered only the case where local ecological conditions do not fundamentally vary through time within a local community; the abiotic environmental conditions do not vary, and the only variability in population abundances is due to transient dynamics as the community approaches its equilibrium. Under these conditions, natural selection should favor no (net) dispersal among populations, even if dispersal is cost free.

However, ecological conditions are not necessarily static in any local community. Temporal variability in ecological conditions will cause temporal variability in fitness for species in that community. Two different mechanisms can generate temporal variability in local fitness. The first is *exogenous* influences on population dynamics. The per capita birth and death rates of species are regulated by the interactions they experience with other species and with the abiotic environment. Fluctuations in the abiotic environment may directly impact these rates for a given species (e.g., by causing a change in the intrinsic birth rate of the resource) or indirectly by directly affecting another species and changing its abundance (e.g., the change in the intrinsic birth rate of the resource would change its abundance and thus alter the birth rate of its consumer). Changing these parameters changes the positions of the isoclines those parameters influence in the community, which changes the position of the equilibrium abundance of potentially every species in the community. Because the average fitness of the species is a function of the difference between its actual abundance and this equilibrium abundance, its fitness will then vary. Conceptually, variation in fitness is caused by how abiotic fluctuations alter the positions of species isoclines and thus the position of the local community abundance equilibrium (i.e., moving the slicing planes in fig. 3.13 along abiotic axes).

The second mechanism that can generate temporal variability in fitness is *endogenous* drivers of population dynamics—namely, limit cycles and chaotic population dynamics. In both limit cycles and chaos, population abundances orbit around an equilibrium. Again, because the average fitness of the species is defined by the difference between its actual abundance and this equilibrium, average fitness will vary temporally as the species' abundance orbits around this equilibrium. Here, the positions and shapes of abundance isoclines do not change, but rather the positions of these isoclines inherently generate changes in population sizes.

Such fitness variability is important to our present concerns because such temporal fitness variability is critical to positively favoring the movement of individuals among populations. I begin this exploration by again considering a passively dispersing species in which each individual in a population has a propensity to emigrate from its natal patch with probability $\omega_{1(j)}$ and enter the dispersal pool. Once in the pool, each individual has an equal probability of immigrating to all the other populations in the region (i.e., no habitat selection). Dispersal may also come with some cost; here I assume any cost is expressed as only the fraction of individuals, ξ, that enter the dispersal pool and who ultimately survive to immigrate to another patch.

Also, throughout this section I assume the species in question has nonoverlapping generations, and consequently a difference equation formulation of population dynamics is more appropriate. I make this assumption because the arguments based on fitness considerations are more intuitive to understand in this context, and also to maintain more similarity with much previous work on the evolution of dispersal in a temporally variable environment (e.g., Kuno 1981, Metz et al. 1983, McPeek and Holt 1992, Doebeli and Ruxton 1997). However, the results from continuous time analyses give comparable results (e.g., Cantrell et al. 2007, Cantrell et al. 2010, Schreiber 2012).

Temporal fitness variability alone does not favor the evolution of dispersal. Imagine a collection of populations in which the environmental factors driving changes in species' parameter values operated identically in all populations simultaneously (e.g., seasonal or yearly variation). In this case, fitnesses in populations would change through time but would be constant across space (i.e., temporal but no spatial fitness variation). This form of spatiotemporal fitness variation has the same consequence for dispersal evolution as if the environment is completely constant in space and time.

What is needed for some nonzero dispersal propensity to be favored is simultaneous spatial and temporal variation in fitness among populations (Gadgil 1971, Kuno 1981, Motro 1982, Metz et al. 1983, Levin et al. 1984, Cohen and Levin 1991, McPeek and Holt 1992, Doebeli and Ruxton 1997). In our two populations, now assume that the abiotic environmental conditions fluctuate over time such that both populations have the same long-term average equilibrium abundances (i.e., $\bar{R}^*_{1(1)} = \bar{R}^*_{1(2)}$), but the fitnesses vary through time in an uncorrelated fashion. In this case, four qualitative fitness relationships exist between the two populations. At some times, both populations are above their equilibrium abundances, and at other times they are both below. At these two types of points in time, dispersers would gain a small fitness advantage by moving from the population with lower to higher fitness. However, large fitness gains accrue to dispersers in the other two types of time points, when one population is above its equilibrium abundance and the other is below its equilibrium abundance. Because these four types of qualitative fitness relationships occur with equal frequency and the population that has higher fitness at any given point in time is also random, half of the time it would pay to move to the other patch; therefore, a dispersal propensity of $\omega_{1(1)} = \omega_{1(2)} = 0.5$ is favored by selection in the absence of a cost to dispersal (Gadgil 1971, Cohen and Levin 1991, McPeek and Holt 1992). Specifically, for dispersal between the two populations to be favored, individuals must be able to gain a fitness advantage from moving in one direction in some years and from moving in the other direction in other years. In general, with J populations, a dispersal propensity of $\omega_{1(j)} = (J-1)/J$ is favored in all populations under these conditions (Kuno 1981).

Extending this to multiple local populations is also instructive. Assume now that the region contains J local populations of this species. For each population, $(1 - \omega_{1(j)})R_{1(j)}$ individuals do not disperse, and the emigrants from population j are apportioned equally among the other populations so that each receives $\xi\omega_{1(j)}R_{1(j)}/(J-1)$ immigrants from population j (note that this assumes a dispersal cost of only a fraction of individuals surviving dispersal). Here, I will also assume that the species has discrete and nonoverlapping generations (as in Kuno 1981, McPeek and Holt 1992, and Doebeli and Ruxton 1997). Given these propositions, the average fitness of a type with dispersal propensity $\omega_{1(j)}$ across all the populations in this region for one generation is

$$
W_1 = \frac{\displaystyle\sum_{j=1}^{J} R_{1(j)} \left[W_{1(j)} + \omega_{1(j)} \left(\xi \frac{\displaystyle\sum_{k=1(k\neq j)}^{J} W_{1(k)}}{J-1} - W_{1(j)} \right) \right]}{\displaystyle\sum_{j=1}^{J} R_{1(j)}}. \tag{6.4}
$$

A type that does not disperse at all (i.e., all $\omega_{1(j)} = 0$) is favored in this generation if the term in the inner brackets of the numerator is negative. Note that the terms in square brackets in the numerator compare the fitness in the current population to the arithmetic average fitness across all other populations decreased by the cost of dispersal. No dispersal is favored overall if this term is negative on average across populations and across generations. Some nonzero dispersal propensity is favored if this term is on average positive.

In any given generation, some fraction of populations has lower fitness than the average fitness in the other populations decreased by the dispersal cost, making dispersal from these populations advantageous. In the other populations where local fitness is higher than this cost-decreased average fitness, not dispersing is advantageous. The balance between these two over many generations influences the exact dispersal propensity that is favored.

This arithmetic averaging of fitnesses across populations within a generation also has another key fitness advantage (Kuno 1981, Metz et al. 1983). The expected variation in fitness through time (e.g., across generations) for a given type is reduced by dispersal among the J populations. Kuno (1981) was the first to show that the underlying fitness variation in $\ln(W_{1(j)})$ is normally distributed with variance $\sigma^2_{1(j)}$ (i.e., the intergenerational fitness variation experienced by a nondispersing type in any one of the populations), which is approximately equal to $Var(W_{1(j)})/\bar{W}_1^2$, where $Var(W_{1(j)})$ is the variance in fitness within each population on the original scale, and $\bar{W}_1$ here is the long-term arithmetic average of fitness (i.e., the expected value) for each population. In contrast, the fitness variance (on the log scale) experienced by a type that disperses equally among the J populations each generation is $\sigma^2_{1(j)}/J = Var(W_{1(j)})/J\bar{W}_1^2$. This is also approximately true if fitness variation follows other statistical distributions (Metz et al. 1983). Thus, dispersal reduces the temporal variance in fitness because of this spatial averaging.

This reduction in temporal fitness variance by spatial arithmetic averaging within a generation has a tremendous benefit to the long-term growth rate of the population, and thus to the long-term average fitness of individuals. Because fitnesses are multiplicative across generations, the appropriate measure of the long-term fitness of a given type when fitness varies through time is the *geometric mean fitness* across generations (Lewontin and Cohen 1969). The long-run geometric mean fitness of a nondispersal type (i.e., all $\omega_{1(j)} = 0$) within one patch is

$$\overline{\ln(W_1)} \cong \ln(\bar{W}_1) - \frac{Var(W_{1(j)})}{2\bar{W}_1^2} \qquad (6.5)$$

(Lewontin and Cohen 1969, Turelli 1977, Lande et al. 2003). The type with the highest geometric mean fitness is favored in an environment with temporal fitness variation. Because the geometric mean is reduced to a much greater extent by low values, types that experience a greater variance in fitness but the same arithmetic average fitness through time have lower geometric mean fitness (Lewontin and Cohen 1969, Turelli 1977, Vance 1980, 1984, Schreiber 2012). Spatial fitness averaging in the context of stochastic fitness relationships across populations is what decreases $Var(W_{1(j)})$. (As an aside, note that the previous results for a spatially variable but temporally constant environment can be understood in this same context, since dispersal in such an environment decreases average fitness each generation [i.e., equation (6.3)] but this spatial averaging does not decrease the temporal variance in fitness, because it is already zero.)

In this environmental type with no dispersal cost and J populations, the expected fitness each generation for all dispersal propensities, including no dispersal, is $\bar{W}_1$. The dispersal type that reduces fitness variation to the greatest extent via spatial averaging has $\omega_{1(j)} = (J-1)/J$ for all populations (Kuno 1981, Metz et al. 1983). This disperser's geometric mean fitness is

$$\overline{\ln(W_1)} \cong \ln(\bar{W}_1) - \frac{Var(W_{1(j)})}{2J\bar{W}_1^2}, \qquad (6.6)$$

(cf. equations (6.5) and (6.6)). (Here again, these calculations assume that $\ln(W_{1(j)})$ is normally distributed through time, but other distributions give qualitatively similar results [Metz et al. 1983]. These calculations also assume all populations experience the same fitness variation, and there is no covariation in fitnesses across populations.)

This comparison shows that the geometric mean fitness advantage of this optimal dispersal type over a nondisperser increases with the number of populations in the region and with the magnitude of within-population fitness variation. More populations mean that the disperser has more populations over which to average within-generation fitness. This is analogous to the statistical property that the standard error of the mean decreases with increasing sample size, whereas the standard deviation does not change. The fitness advantage of this disperser over a nondispersal type is also larger with greater fitness variation, again because of the ameliorating effect of spatial averaging of fitnesses within each generation on long-term geometric mean fitness. Vance (1980) showed similar results for continuous-time formulations of local population dynamics with density regulation (see also Vance 1984, Schreiber 2012).)

Equation (6.6) assumes no dispersal cost. If some fraction of dispersers die before reaching a new population (i.e., $0 \leq \xi < 1$), the geometric fitness of a disperser that expresses the same propensity to disperse from every population (i.e., $\omega_{1(j)} = \omega$ for all j) is then

$$\overline{\ln(W_1)} \cong \ln(\bar{W}_1(1 - \omega(1 - \xi))) - \frac{\left((1-\omega)^2 + \frac{\xi^2\omega^2}{J-1}\right)Var(W_{1(j)})}{2\bar{W}_1^2(1 - \omega(1 - \xi))^2} \quad (6.7)$$

(note that this simplifies to equation (6.6) with $\omega = (J-1)/J$ and $\xi = 1$). The average fitness of a dispersal type is now decreased by the fitness cost payed by dispersers each generation (first term on the right-hand side of equation (6.7)). Moreover, dispersal types with $\omega_{1(j)} \neq (J-1)/J$ have higher fitness variances through time because of reduced spatial averaging each generation. Figure 6.2 illustrates the dispersal propensities favored by natural selection as a function of the number of populations linked by dispersal and the dispersal cost. Greater fitness variation favors a higher dispersal propensity in the face of increasing dispersal costs.

Equations (6.4) through (6.7) are general forms and do not consider how fitness variation is generated. For example, if per capita fitness is density dependent within each population, the above forms of fitness variation would correspond to all populations being regulated to the same equilibrium abundance on average (i.e., $\bar{R}_{1(j)}^* = \bar{R}_{1(k)}^*$ for all population combinations). Fitness variation could be generated by either variation in the underlying parameters to generate variation in equilibrium abundances through time, or by having populations that express endogenous population cycles or chaotic population dynamics. Simulation studies of such models using a number of different frameworks for competing dispersal types give comparable results (McPeek and Holt 1992, Holt and McPeek 1996, Doebeli and Ruxton 1997, Massol and Debarre 2015).

Fitness variation can take other spatiotemporal forms as well—namely, the equilibrium abundances may vary temporally within each local population but be regulated to different long-term average abundances. Returning to our two-population example, assume that the long-term average equilibrium abundance is higher in population 1 than in 2 (i.e., $\bar{R}_{1(1)}^* > \bar{R}_{1(2)}^*$), but both vary temporally in an uncorrelated fashion. With this form of spatiotemporal fitness variation, dispersal can still be favored because in every generation fitness will be higher in

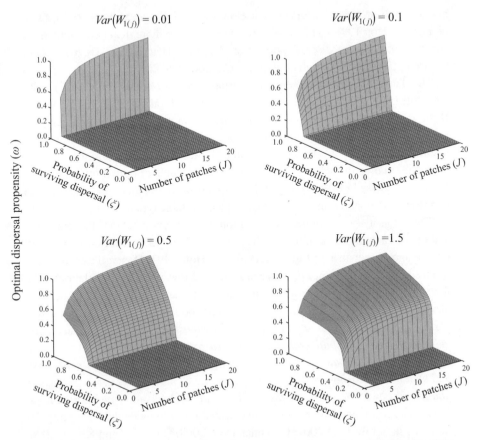

Figure 6.2. Surfaces showing the optimal dispersal propensity, ω, for various combinations of the number of patches linked by dispersal and the probability of surviving dispersal. Each panel shows a surface for a specified level of temporal variance in per capita fitness within every population. Surfaces are calculated using equation (6.7). All populations experience the same distribution of fitnesses through time, and all calculations presented here assume $\bar{W}_{1(j)} = 1$.

one population than in the other, but now the movement of individuals among populations can again distort local fitnesses substantially because of the net flux of dispersers. Under these circumstances, a dimorphism of a low and a high dispersal type is favored if individuals must express the same dispersal propensity in both populations. However, if dispersers can express different dispersal propensities in different populations, one type with conditional dispersal propensities that satisfy

$$\frac{\omega_{1(1)}}{\omega_{1(2)}} = \frac{\bar{R}^*_{1(2)}}{\bar{R}^*_{1(1)}} \tag{6.8}$$

has the highest fitness (McPeek and Holt 1992, Holt and McPeek 1996, Doebeli and Ruxton 1997). This result can be extended to more than two populations and to continuous or discrete population dynamics (Padron and Trevisan 2006, Cantrell et al. 2007, Cantrell et al. 2010, Cantrell et al. 2013).

This latter form of spatiotemporal variation also offers the only opportunity when utilizing a sink habitat may be favored by natural selection (Holt 1997b). If the range of fitness variation in populations that are on average sources (i.e., $\overline{\ln(W_{1(j)})} > 0$) overlaps the range of fitness variation in sink populations (i.e., populations with $\overline{\ln(W_{1(j)})} < 0$), at some points in time fitness returns in the sink will be greater than in the source. If this is true, some degree of spatial fitness averaging will decrease $Var(W_{1(j)})$, and if dispersal costs are not too high, dispersing to and from sink populations may be evolutionarily stable (Holt 1997b).

The type satisfying criteria like equation (6.8) also will balance dispersal into and out of each population; that is, the number of emigrants and immigrants will on average be equal over time (McPeek and Holt 1992, Doncaster et al. 1997, Cantrell et al. 2007). You may have also noticed that when all populations have the same long-term average abundance, the dispersal type favored by selection has the same dispersal propensities from every population. *Thus, the hallmarks of all these passive dispersal strategies that are most favored by natural selection under any regime of spatiotemporal fitness variation are those that (1) maximize long-term geometric mean fitness by spatially averaging fitness each generation and thereby decreasing the year-to-year variance in fitness, and (2) do not distort abundances away from what is favored by local mechanisms of population regulation.* The result is again an ideal free distribution of individuals among populations (Holt and Barfield 2001, Cressman et al. 2004, Cressman and Křivan 2006, Křivan et al. 2008, Cantrell et al. 2010). Even though individuals do not assess the fitness consequences of their movement in any of the strategies considered thus far, they end up apportioned among patches as though they do. From a population perspective, individuals are also apportioned among populations as though only local density-dependent processes regulated populations and no dispersal occurred at all (Cressman et al. 2004, Cressman and Křivan 2006, Armsworth and Roughgarden 2008, Křivan et al. 2008, Cressman and Křivan 2010).

As an aside, let me briefly comment on another selection pressure that supposedly favors dispersal—namely, competition among siblings for gaining a site occupied by a parent (Hamilton and May 1977). In this mechanism, competition among the siblings favors dispersal to then compete for and occupy sites elsewhere. However, this mechanism fosters the evolution of a nonzero dispersal rate by exactly the same logic as causes the evolution of dispersal in a variable environment, as described above. In their classic formulation of the problem, Hamilton and May (1977) considered the situation where each population consists of a few individuals, typically a breeding asexual parent (or two parents if sexual) and the offspring that the parent(s) produced. Offspring are essentially waiting to take over the breeding site occupied by the parent when the parent dies, and so each population is of extremely small size. In their model, the authors posited no special detriment for competing with siblings as opposed to competing with non-kin (they mention inbreeding, but it plays no role in their model). Sibling competition is not the driver of the evolution of dispersal in their model, but rather competition among siblings is reduced merely as a by-product of dispersal from the natal site. Clearly, these effects would be most important when local population sizes are very small.

By dispersing to another location, the offspring of a parent are competing for multiple sites that vary in their fitness potentials, instead of only one (i.e., their natal site occupied by their parent). Although the environment is characterized as "stable," the death of a parent opens a site for occupancy, which is a huge change in local population size and thus causes large spatial and temporal variation in local fitnesses. Thus, the scenario that Hamilton and May (1977) consider fits exactly into the logic of fitness characterized by equations (6.4) through (6.7). A type that does disperse is increasing its geometric mean fitness by averaging fitness across all populations.

MORE THAN SIMPLY PASSIVE DISPERSAL STRATEGIES

Up to now, I have considered only two types of dispersal strategies. These include (1) passive dispersal, in which individuals have some innate propensity to emigrate, and which may differ among populations; and (2) an omniscient fitness assessment strategy that results in the ideal free distribution. However, myriad other strategies can be imagined between these two extremes, and a tremendous diversity have been considered in models of population dynamics and dispersal evolution (see reviews in Abrams et al. 2007, Amarasekare 2007, Ronce 2007, Lowe and McPeek 2014). Reviewing all the consequences of the variations of dispersal strategies is beyond the scope of this presentation, but I do want to identify a number of important issues to consider when dealing with different strategies.

The first is that the effects of various dispersal strategies on local population abundances and dynamics are caused by the form of local density dependence they generate (Murdoch et al. 1992, Amarasekare 1998, Neubert et al. 2002, Amarasekare 2004, Abrams 2007). As illustrated in figure 6.1 for one specific dispersal strategy, net emigration or immigration may perturb local abundances away from the values favored by local dynamics alone. However, it is more important to realize that emigration and immigration are themselves demographic rates that generate components of density and frequency dependence in local population dynamics, just as birth and death do, and thereby either enhance or reduce local stability (Murdoch et al. 1992, Amarasekare 1998, Neubert et al. 2002, Amarasekare 2004, Abrams 2007).

The simplest illustration of these effects is to consider various forms of density-dependent immigration into a sink in which the population declines exponentially:

$$\frac{dR_{1(1)}}{dt} = -c_{(1)}R_{1(1)} + \omega(R_{1(1)})^{\vartheta}, \tag{6.9}$$

where ω is a scaling parameter for dispersal and ϑ controls the form of density dependence for immigration (Amarasekare 2004). In this model, note that local population decline is density independent (i.e., $dR_{1(1)}/R_{1(1)}\,dt = -c_{(1)}$) when $\omega = 0$, and so local per capita population growth rate is independent of local population abundance. With $0 \leq \vartheta < 1$, immigration is a negatively density-dependent demographic rate (i.e., the per capita rate of immigration decreases with increasing local population abundance), and this negative density dependence causes a stable equilibrium that balances the loss of individuals, due to local population dynamics, with the addition of immigrants; if $\omega > 0$ this stable equilibrium always exists. With $\vartheta = 1$, the immigration rate is density independent, and so the population either grows or declines exponentially depending on the sign of $\omega - c_{(1)}$. Finally, with $\vartheta > 1$, the immigration rate is positively density dependent, and either $R_{1(1)} = 0$ is a stable equilibrium with an unstable equilibrium at a higher value, or $R_{1(1)} = 0$ is the only equilibrium, is unstable, and the population grows exponentially without bounds. Mortality or other changes in fitness during dispersal can also affect local population dynamics (Amarasekare 1998).

Numerous studies have explored how various forms of dispersal can either dampen or stop population cycling, or create cycling in a previously stable population. Density-independent dispersal (e.g., passive dispersal in which each individual has some probability of emigrating from a population) can significantly perturb equilibrium abundances (see above) but typically does little to alter the stability properties of a system (Rohani et al. 1996, Ruxton 1996b). Positive density-dependent immigration—for example, aggregation of predators (e.g., Rohani et al. 1996) or emigration (e.g., Ruxton 1996a) can change the dynamics of a system from a global, stable equilibrium to limit cycles or chaos. In contrast, negative density dependence in immigration (e.g., Murdoch et al. 1992) or emigration (e.g., Neubert et al. 2002) can dampen populations to promote stability. Dispersal can also interact with other density-dependent mechanisms to affect stability. For example, prey aggregation creates positive density dependence if the predators have saturating functional responses (Abrams 2007).

One form of dispersal—namely, if individuals have a bias for moving from lower to higher fitness populations—typically ensures that population abundances are not perturbed away from the values that local regulation mechanisms favor. Such fitness-climbing dispersal strategies typically ensure that individuals of a species can achieve an ideal free distribution among populations (Abrams et al. 2007). Fitness-climbing dispersal strategies (in effect, various forms of habitat choice) will also outcompete passive dispersal strategies, even when both can achieve the same ideal free distribution. This is because habitat selection further reduces temporal fitness variation by biasing individual movements from low to high fitness populations, and in so doing creates an even better averaging of fitness across space (Vance 1980, 1984; Cantrell et al. 2013). In the extreme, an omniscient and perfect habitat selector could reduce temporal fitness variation to zero because it could instantly respond to fitness differences among populations to equilibrate fitnesses across space.

Moreover, by equalizing fitnesses across the landscape, fitness-climbing strategies do not perturb local abundances away from the equilibria that would result without dispersal (Cressman et al. 2004, Cressman and Křivan 2006, Armsworth and Roughgarden 2008, Křivan et al. 2008, Cressman and Křivan 2010). This turns out to be a general result for the evolution of dispersal under all environmental conditions; if an ideal free distribution can be achieved, this evolutionarily stable dispersal strategy will not perturb local abundances away from what is favored by local population dynamics in the absence of dispersal. If selection can achieve a dispersal strategy that results in an ideal free distribution for each species in every local community, each community can essentially be considered in isolation, even though individuals may move among them. However, if natural selection cannot or has not yet achieved the very best dispersal strategy for each species, local abundances may not be what local population regulation favors.

LINK SPECIES EFFECTS ON LOCAL ABUNDANCES VIA THEIR MOVEMENT STRATEGIES

Not all interacting species operate on the same spatial scales. Frequently, the rapid and continual movement of one species among nearby locations can link population dynamics of other species that are segregated into these different areas—in effect, linking these geographically separated populations of species to the same community. The issues that arise because of these link species are identical to those arising from dispersal among local communities. As with dispersal, the strategies used by individuals of these link species that influence how they move among different areas affect the fitnesses of other species and can cause local abundances and dynamics to be quite different from what local regulation mechanisms would favor. In fact, drawing a distinct demarcation between link species and dispersal is difficult. This has been a fertile area of inquiry, and like dispersal, all the possible permutations of the positions of link species in various geographic configurations of a community and various decision rules for link species moving among area are not possible to explore here. However, I will discuss two studies of link species in different community module contexts to illustrate some general features.

Holt (1984) studies a model of a simple apparent competition community module with two resources and one consumer that occupy two habitat patches, and the resources are segregated between the two patches. If the consumers move randomly between the two patches, which is equivalent to the passive dispersal scenario discussed above with a constant propensity to leave each patch, the population dynamics of the two spatially segregated resources are coupled with the less productive resource being underexploited (i.e., the resource equilibrates at an abundance below what it would have if it were only present with the consumer in this one patch) and the more productive resource being overexploited. Figure 6.1 is a representation of this scenario as well. In this case, the dynamics of the spatially segregated resources are linked, so that environmental changes in one patch will influence the abundance of the other resource, because of the effects this will have on the consumer's abundance as an example of spatial apparent competition module. If the resource in the more productive patch is abundant enough, it may inflate the consumer's abundance to great enough levels to cause the resource in the other patch to be driven extinct (i.e., as in fig. 2.6) (Holt 1984).

In contrast, if each consumer can choose to occupy the patch that gives it the highest yield of resource so that the population of consumers can achieve an ideal free distribution among the patches, the population dynamics of the resources in the two patches become decoupled from one another (Holt 1984). Even though consumers may move among the patches continually, the resource and consumer abundances equilibrate in the two patches as though they were separate communities. Consequently, if the consumers can achieve an ideal free distribution, environmental perturbations that alter equilibrium abundances in one patch do not affect the abundances favored in the other (Holt 1984).

Amarasekare (2007) explored a model in which resources are spatially segregated among three patches, and these resources are eaten by an intraguild prey and intraguild predator that can move among the patches. She assumed environmental differences among the patches, such that in the absence of movement among them only the intraguild prey could coexist with the resource in one patch, only the intraguild predator could coexist with the resource in the second patch, and all three species could coexist in the third patch (e.g., fig. 2.12). Furthermore, she explored a number of different movement rules for the intraguild prey and predator among the patches, including random movement (i.e., passive dispersal); habitat quality dependent movement, such that leaving is a decreasing function of resource productivity and entering is more likely with higher resource productivity; density-dependent movement where the patch leaving and entering rates depend on intraspecific densities; and fitness-dependent movement where these rates depend on the fitness differences among patches.

When the intraguild prey and predator move among patches either randomly or based on their densities among the patches, the system equilibrates at abundances that are quite different from those that result when no movement is permitted (Amarasekare 2007). In contrast, when they move among the patches based on either resource productivity or their overall fitness, the species equilibrate at abundances that are the same as when the three patches are isolated communities (Amarasekare 2007). When the intraguild prey and predator use different movement criteria, the intraguild predator's movement has a greater impact on the intraguild prey's distribution than vice versa. As a result, if the intraguild predator can achieve an ideal free distribution because of its movement rule, the intraguild prey will achieve a distribution among patches that closely approximates an ideal free distribution regardless of its movement rules, but the converse is not true (Amarasekare 2007).

If species can achieve ideal free distributions across space with both the movement of link species and dispersal among separated local communities, each location will have abundances of interacting species that are equivalent to those favored by local regulatory mechanisms. Consequently, local communities will act as independent ecological entities, even though individuals move freely among them. An ideal free distribution is almost guaranteed if individuals move among locations in a biased fashion up fitness gradients, but other movement strategies can achieve an ideal free distribution in many situations, even without being able to assess fitness differences among populations.

If ideal free distributions cannot be achieved, movement of link species and dispersal will displace community structures away from what is favored by local regulatory processes (Holyoak et al. 2005). Now local community structure must be understood as the interplay between local abundance regulation and connections to other communities in the region. Each local community is now embedded in a regional network that makes sink species more likely. Also, connections among local communities can prolong the persistence of walking-dead species in the entire system by reducing their rate of decline through spatial averaging of fitness and recolonization of areas where they have already been extirpated (McPeek and Gomulkiewicz 2005).

INTERPLAY OF LOCAL ADAPTATION AND DISPERSAL EVOLUTION

Individuals moving among populations not only influence the demographics of where they go, but also carry with them trait distributions that may be different from the populations they are entering or leaving. Dispersal can, therefore, perturb trait distributions away from what is favored by local natural selection. Consequently, in the context of this dispersal, species must simultaneously adapt to the local communities in which they can maintain populations, and also potentially adapt their dispersal strategies to move among those local communities. In the end, each species will strike a *migration-selection balance* in populations across its entire range. In this section, I explore some of the overt features of this migration-selection balance within populations and examine the reciprocity between local adaptation and dispersal evolution.

I also adapt a model of the effects of gene flow on the evolution of quantitative traits in multiple populations (Tufto 2000; Ronce and Kirkpatrick 2001; Tufto 2001, 2010). (Other models have been used to study how dispersal influences adaptation along ecological gradients [e.g., Pease et al. 1989, García-Ramos and Kirkpatrick 1997, Kirkpatrick and Barton 1997], but these do not lend themselves to investigating the joint evolution of local adaptation and dispersal evolution because they model dispersal as a diffusion process.)

I begin by assuming that equations (6.1) describe the negative density dependence that regulates the populations of a resource species embedded in two different local communities, and individuals of this species have some propensity, ω, to emigrate from their natal population to the other without cost. As in chapters 3 and 5, the species' trait influences the value of the intrinsic birth rate, $c_{(j)}(\bar{z}_{1(j)})$, according to equation (3.10). The populations can differ in the intrinsic birth rate, $c_{0(j)}$; the intrinsic birth optimal phenotype, $\bar{z}^c_{R_{1(j)}}$; and the strength of density dependence, $d_{(j)}$.

As with fitnesses among populations, dispersal causes an averaging of the phenotypes among populations. The trait mean in a population after dispersal is approximately the weighted average of the trait means of the immigrants and residents who did not emigrate:

$$\bar{z}'_{1(j)} = \frac{(1-\omega)R_{1(j)}\bar{z}_{1(j)} + \omega R_{1(k)}\bar{z}_{1(k)}}{(1-\omega)R_{1(j)} + \omega R_{1(k)}}. \tag{6.10}$$

Selection is then applied to this new average trait (Tufto 2001). These assumptions lead to the trait dynamics being described by

$$\frac{d\bar{z}_{1(j)}}{dt} = V_{z_{R_1}}\left[\frac{\omega R_{1(k)}(\bar{z}_{1(k)} - \bar{z}_{1j})}{(1-\omega)R_{1(j)} + \omega R_{1(k)}} - 2c_{0(j)}\gamma(\bar{z}'_{1(j)} - \bar{z}^c_{R_{1(j)}})\right], \tag{6.11}$$

where the first term in the square brackets is the change in the mean phenotype in population j caused by dispersal, and the second is the amount of change caused by phenotypic selection applied to this new average phenotype (Tufto 2000, 2001, 2010).

SPATIAL BUT NO TEMPORAL ENVIRONMENTAL VARIANCE

As we saw above, if the populations differ from one another but do not vary over time, passive dispersal causes the abundances to be displaced from the values favored by local regulation (fig. 6.1). Figure 6.3 shows the outcome of local adaptation for fixed values of the dispersal propensity when various features of the populations are different. If the same trait value yields the highest fitness in both populations, but the populations differ in the maximum intrinsic birth rate at this optimal value, the system acts as described in Holt (1985). Local adaptation moves both populations to match exactly the optimal phenotype favored by the local selection regime at all dispersal propensities (fig. 6.3A). Consequently, immigrants have the same phenotype as residents, and dispersal does not cause any phenotypic deviation from what local selection favors. However, higher dispersal causes the two populations to become more similar in abundances, with one below and the other above the abundances favored by the local regulation mechanisms (fig. 6.3B). As a consequence, local per capita fitnesses of individuals are positive in the high abundance population (population 1 in figs. 6.3A and B) and are negative in the low abundance population (see fig. 6.1).

Figures 6.3C and D show results when the populations differ in the optimal phenotype for the intrinsic birth rate, but the populations have the same maximum intrinsic birth rate at their respective phenotypic optima. With no dispersal ($\omega = 0$), each population evolves to the local intrinsic birth optimum (i.e., $\tilde{z}^c_{R_{1(j)}}$) and equilibrates at the expected abundance (i.e., $c_{0(j)}/d_{(j)}$). As dispersal propensity is increased, the average trait values in the two populations become more similar; immigrants continually displace each population away from what local selection favors, and the magnitudes of this displacement increase with higher dispersal propensities. Because the average traits are farther from their respective optima, both populations equilibrate at lower abundances. The difference in the abundance when the population adapts without and with dispersal is a measure of the *migration load* to adaptation that dispersal causes in each population (Tufto 2010). However, in this case, dispersal does not distort local abundances from what local population regulation mechanisms favor (i.e., no migration load), and so the per capita fitnesses are at replacement and are the same in the two populations.

If the populations differ in both the trait values giving highest fitness and the fitness maxima at those optimal trait values, higher dispersal propensities again cause both populations to have more similar average phenotypes (i.e., displaced farther from their phenotypic optima). However, both are biased toward the optimal phenotype of the population with the higher equilibrium abundance because of the disparity in the number of dispersers moving in each direction (fig. 6.3E). The low abundance population (2 in this case) is displaced farther from its local

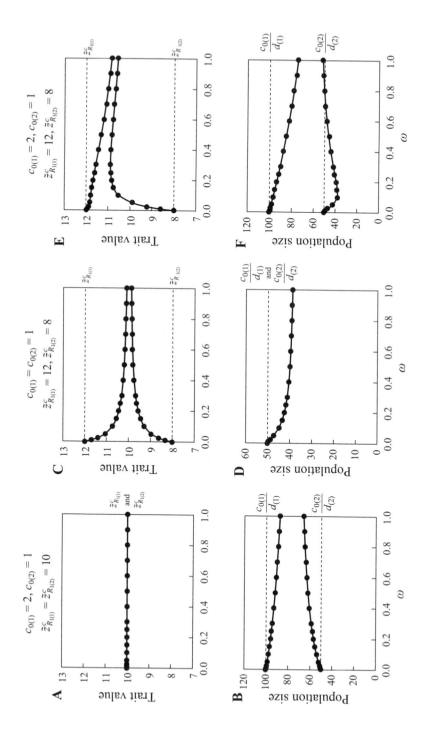

optimal phenotype, and so this population equilibrates at lower abundances than without dispersal (fig. 6.3F). As a result, the average per capita fitness is more negative in the low abundance population and more positive in the high abundance population, when compared to the situation where the same trait value is favored in both populations (cf. figs. 6.3A and B with figs. 6.3E and F).

These all involve local adaptation with dispersal among populations, but without dispersal propensity being able to evolve. As expected, if types with different dispersal propensities compete against one another and are simultaneously allowed to locally adapt to the conditions in the two populations, the type with $\omega = 0$ will drive all other dispersal propensities extinct, and this dispersal type will perfectly adapt to the conditions in each population (i.e., $\bar{z}_{1(j)} = \bar{z}^c_{R_{1(j)}}$). In all cases illustrated in figure 6.3, selection moves dispersal propensity to lower values until $\omega = 0$.

When both local adaptation and dispersal evolution can occur in a spatially variable but temporally constant environment, types with lower dispersal propensities have two advantages. The first is that they displace local abundances less and so have more equitable fitnesses across space. This is important when the populations differ in maximum fitnesses, and thus abundances (e.g., $c_{0(1)} \neq c_{0(2)}$). The second advantage is that they can evolve to have higher fitnesses in all populations because the displacements caused by the phenotypic averaging effect are less with less dispersal. This is important when different phenotypes are favored by the selection regimes in different populations (e.g., $\bar{z}^c_{R_{1(1)}} \neq \bar{z}^c_{R_{1(2)}}$).

SPATIAL AND TEMPORAL ENVIRONMENTAL VARIANCE

Simultaneous spatial and temporal variation in environmental conditions favors the evolution of dispersal among populations. However, such environmental variation implies that the selection regimes experienced by traits that determine a population's demographic performance will also vary both spatially and temporally; in other words, the shape of the fitness topography will vary.

FIGURE 6.3. Local adaptation of a resource in two populations that are linked by dispersal. Population abundances are regulated and dispersal occurs according to equations (6.1), where $c_{(j)} = c_{0(j)}\left(1 - \gamma\left(\bar{z}_{1(j)} - \bar{z}^c_{R_{1(j)}}\right)^2\right)$ and ω is the propensity to disperse to the other population. Trait dynamics in the two populations are governed by equations (6.11). Panels A and B show the equilibrium trait values and population sizes for the two populations over a gradient of dispersal propensity, when the maximum intrinsic birth rates for the two populations are $c_{0(1)} = 2$ and $c_{0(2)} = 1$, and both populations have $\bar{z}^c_{R_{1(j)}} = 10$. Panels C and D show the same relationships when the two populations both have $c_{0(j)} = 1$, but differ in having $\bar{z}^c_{R_{1(1)}} = 12$ and $\bar{z}^c_{R_{1(2)}} = 8$. Panels E and F show the same when the two populations differ for both the maximum intrinsic birth rate and intrinsic birth optimum. Both populations have $d_{(j)} = 0.02$, $\gamma = 0.05$, and $V_{R_1} = 0.2$.

Variation in the shape of the fitness surface implies that the optimal phenotype and the strength of selection on the phenotype will change through time in a population. In such an environment, a population will evolve in response to the fitness surface defined by the long-run geometric mean fitness, as in equation (6.5), of each trait value (Lande 2007, Lande et al. 2009). Local environmental variation may alter the geometric mean fitness surface in two ways. First, variation may simply change the heights of the local fitness peaks on the geometric mean topography if the same phenotypic values give locally maximum geometric mean fitness. Imagine the topography simply moving up and down through time. In this case, the rates that populations adapt to the local ecological milieu will be different, but the traits that are locally favored will not. Once a population evolves to the optimal phenotype, its temporally averaged fitness (e.g., $\ln(\bar{W}_1)$ in equation (6.6)) will be maximized, and the geometric mean fitness will be shaped primarily by the variance in fitness through time.

Environmental variation may also make the position of the optimal phenotype a random variable. Selection should push an isolated population to the local maximum in the geometric mean fitness topography. However, the population will only very rarely be at the optimum for the fitness topography that the population experiences at any point in time; in other words, the long-term geometric mean fitness peak does not necessarily correspond to the optimal trait value at any point in time, and so the population will always be on a hillside of the current fitness topology. Such fluctuations will thus decrease $\ln(\bar{W}_1)$ as compared to the value that could be obtained if the population could track the optimal phenotype perfectly.

When dispersal among populations is added, gene flow may move local populations away from the long-term geometric mean optima as well. Spatial trait averaging then decreases $\ln(\bar{W}_1)$ even further for dispersal types moving among populations with different long-term geometric mean optima, and greater dispersal causes greater displacements from those optima.

All of these insights taken together provide the intuition for understanding how dispersal and traits determining local fitness should jointly evolve. As an illustration, consider the system with two populations connected by dispersal defined by equations (6.1) and (6.11) but permit various parameters in this model to vary probabilistically according to normal distributions with specified parameters. Dispersal is also assumed to be cost free ($\xi = 1$) here to keep the presentation as simple as possible. First, consider the situation in which the maximum intrinsic birth rates ($c_{0(j)}$) or strengths of density dependence ($d_{(j)}$) vary through time in the two populations in an uncorrelated fashion with the same means and variances; but the same trait value gives the highest intrinsic birth rate in both populations ($\tilde{z}^c_{R_{1(1)}} = \tilde{z}^c_{R_{1(2)}}$) and does not vary through time. Both populations evolve to match the optimal trait value, and a dispersal propensity of $\omega = 0.5$ is favored regardless of the amount of variation. Here, fitness variation is caused exclusively by the fitness surface moving up and down but not changing fundamentally in position.

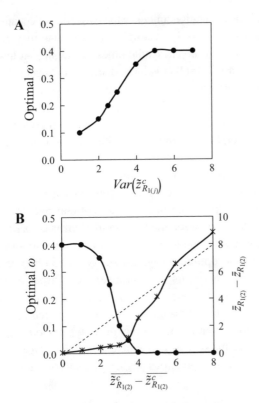

FIGURE 6.4. The joint evolution of dispersal propensity and local trait values in a two-population model governed by equations (6.1) and (6.11), in which specified parameters for both populations are allowed to vary temporally. In all circumstances considered here, both populations have $d_{(j)} = 0.02$, $\gamma = 0.05$, and $V_{R_1} = 0.2$. Panel A shows the optimal dispersal propensity that will drive all others extinct for different levels of variance in $\tilde{z}^c_{R_{1(j)}}$; here, each iteration of the simulation, $\tilde{z}^c_{R_{1(j)}}$, is drawn from a normal distribution with specified mean and variance. In all simulations in this panel, $c_{0(1)} = c_{0(2)} = 5$, the mean optimal trait values are the same at $\overline{\tilde{z}}^c_{R_{1(1)}} = \overline{\tilde{z}}^c_{R_{1(2)}} = 30$, and $Var(\tilde{z}^c_{R_{1(1)}}) = Var(\tilde{z}^c_{R_{1(2)}})$ as specified. In panel B, all parameters are the same except that both $c_{0(j)}$ and $\tilde{z}^c_{R_{1(j)}}$ vary according to normal distributions and the two populations have $\overline{\tilde{z}}^c_{R_{1(1)}} \neq \overline{\tilde{z}}^c_{R_{1(2)}}$ as specified. For the maximum intrinsic birth rate, $\overline{c_{0(1)}} = \overline{c_{0(2)}} = 5$ and $Var(c_{0(j)}) = 1$. Both the optimal dispersal propensity (*filled circles* and the left axis) and the difference in the trait values that evolve in the two populations (*x's* and right axis) are shown. The *dotted line* shows the 1:1 line where $z_{R_{1(2)}} - z_{R_{1(1)}} = \overline{\tilde{z}}^c_{R_{1(2)}} - \overline{\tilde{z}}^c_{R_{1(1)}}$ for comparison.

If instead, $c_{0(j)}$ and $d_{(j)}$ are constant but the $\tilde{z}^c_{R_{1(j)}}$'s both vary with the same mean and variance, the optimal dispersal propensity increases with increasing variance in the optimal trait values (fig. 6.4A). The optimal propensity eventually plateaus at higher variances, but the asymptote is only at $\omega = 0.4$. These differences in the optimal dispersal propensity with variance in the optimal phenotypes result from how dispersal and local adaptation influence the two terms of equation (6.6). Higher levels of dispersal decrease $\ln(\bar{W}_1)$ via trait averaging, but they also decrease the variance in fitness through time via spatial fitness averaging. At low levels of variance in $\tilde{z}^c_{R_{1(j)}}$, the decrease in $\ln(\bar{W}_1)$ is more detrimental to geometric mean fitness than the decrease in fitness variance is beneficial. As the variance in $\tilde{z}^c_{R_{1(j)}}$ is increased, this balance shifts, but the optimal dispersal propensity becomes an asymptote at a slightly lower value (as compared to situations in which fitness varies but the optimal phenotype does not) because of the effect on decreasing $\ln(\bar{W}_1)$.

Finally, consider situations in which the optimal trait on the geometric mean fitness topographies are different between the two populations (i.e., when the optimal phenotypes vary through time with different means in the two populations: $\overline{\tilde{z}^c_{R_{1(1)}}} \neq \overline{\tilde{z}^c_{R_{1(2)}}}$). When the difference in the mean optimal phenotypes are small, a high dispersal propensity between the two populations is favored, and as a result, the average traits in both populations are similar to one another and relatively far away from the local optimum (fig. 6.4B). As this difference is increased, a relatively abrupt transition occurs to favor a very low dispersal propensity (e.g., $0 < \omega < 0.01$ at values of $\overline{\tilde{z}^c_{R_{1(2)}}} - \overline{\tilde{z}^c_{R_{1(1)}}} \geq 4$ in fig. 6.4B), which in turn permits the populations to differentiate and match the trait values favored by local selection in each population. If the difference in optimal phenotype is too great, but the range of values still overlaps, the species will evolve a very low dispersal propensity but differentiate little between populations so that one is a source and one is a sink.

PHENOTYPIC PLASTICITY

Up until now, the models of traits I have utilized assume that an individual will express the same phenotypic trait in all ecological environments in which it develops. However, this is not true for all traits in all species. Some traits are phenotypically plastic, meaning that individuals will express different values for the same trait when they experience different ecological conditions. For our purposes, the most important class of plastic traits are those that fall in the category of *adaptive phenotypic plasticity*; this occurs when the individual expresses a trait value at or very near the phenotypic optimum for the ecological milieu in which it finds itself (Scheiner 1993, Gotthard and Nylin 1995, Ghalambor et al. 2007).

The evolution of plasticity is complicated, and beyond the scope of the presentation I want to make—we have enough moving parts already. However, we can intuit a few general issues in regard to allowing traits to be adaptively plastic (see Scheiner 2016 for a cogent analysis of how habitat choice and phenotypic plasticity may interact). Specifically, if an individual can express the correct trait value for every ecological setting, many of the conflicts among the evolution of dispersal and local adaptation disappear, or are at least greatly diminished. In particular, the consequences of trait averaging because of dispersal would disappear. The species would evolve to match the trait expression of the optimal phenotype everywhere, and its problem of movement would collapse back to simply being an issue of pure spatiotemporal fitness variation affecting the evolution of dispersal. In fact, this species would experience fitness variation very differently from a nonplastic species in the same environment. For example, if $\tilde{z}^c_{R_{1(j)}}$ varied through time within a population or between populations, but the fitness values at each $\tilde{z}^c_{R_{1(j)}}$ were all the same, a species with nonplastic traits would adapt as we have explored in this chapter. However, a species with a perfectly adaptively plastic trait would experience no fitness variation whatsoever because its trait value would always match $\tilde{z}^c_{R_{1(j)}}$, and this species would also maximize $\ln(\bar{W}_1)$ (Scheiner 2016).

Thus, adaptive plasticity of trait expression implies that species in local communities should deviate little from what the local selection regime favors, and dispersal patterns should evolve simply on the basis of spatiotemporal variation in the heights of adaptive peaks they occupy on the fitness topography.

A LANDSCAPE OF LOCAL COMMUNITIES

When we consider movement of individuals among local communities, the issue immediately arises of whether there is such a thing as a "biological community." Is a community an integrated, organizational unit of coexisting species that would assemble at a given site regardless of initial conditions? From this perspective, dispersal simply gets species to all the communities where they can coexist, and local structure is not overwhelmed by the flux of individuals across community borders. Or is a community simply the collection of species that are found at a particular site with no real internal organization or structure? In this view, the distribution of each species is set by its own unique set of ecological needs and driven by their movement patterns across the landscape.

The integrity and cohesiveness of local biological communities as a concept has been a simmering topic of discussion for over a century. In the early part of the twentieth century, ecologists hotly debated whether communities represented integrated units or simply artificial collections of species that happened to be found together (e.g., Clements 1916, Gleason 1926, Phillips 1935, Tansley 1935). For example, Clements analogized the ordered stages of plant succession in a community to the development of an organism and argued that a local community is a functional unit:

In both the individual and the community the clue to development is function, as the record of development is structure. Thus, succession is preeminently a process the progress of which is expressed in certain initial and intermediate structures or stages, but is finally recorded in the structure of the climax formation. The process is complex and often obscure, and its component functions yield only to persistent investigation and experiment. In consequence, the student of succession must recognize clearly that developmental stages, like the climax, are only a record of what has already happened. Each stage is, temporarily at least, a stable structure, and the actual processes can be revealed only by following the development of one stage into the succeeding one (Clements 1916, p. 7).

In fact, some argued that "the biotic community is a complex organism" (Phillips 1935, p. 505). In contrast, Gleason felt that local assemblages of species are merely the result of dispersal and chance:

The sole conclusion we can draw from all the foregoing considerations is that the vegetation of an area is merely the resultant of two factors, the fluctuating and fortuitous immigration of plants and an equally fluctuating and variable environment. As a result, there is no inherent reason why two areas of the earth's surface should bear precisely the same vegetation, nor any reason for adhering to our old ideas of the definiteness and distinctness of plant associations. As a matter of fact, two areas of the earth's surface do bear precisely the same vegetation, except as a matter of chance, and that chance may be broken in another year by a continuance of the same variable migration and fluctuating environment which produced it (Gleason 1926, pp. 23–24).

This discourse continues to the present day (e.g., Callaway 1997, Ricklefs 2008).

Why do these disparate views of communities exist? The "community as an organized unit" outlook arises from the myriad repeatable patterns of species assemblages found across the landscape at many different spatial scales. At the scale of continents, different biomes based on vegetation composition with identifiable ecological characteristics are clearly apparent (Whittaker 1975, Lomolino et al. 2010). On a more local scale, repeatable plant assemblages are found at different places along altitudinal gradients from the bottoms to the tops of various slopes within a mountain range (Whittaker 1956; Whittaker and Niering 1965, 1975; Niering and Lowe 1984), or on different topographic aspects or different soil types (Weiher and Keddy 1995) within a region. Moreover, these different plant assemblages harbor and support various assemblages of microbes, fungi, and animals.

Such regular patterning of species assemblages is not unique to terrestrial habitats on broad scales. The "river continuum" is a biological concept because regular and predictable changes in species composition and functional group organization occur as one moves along a stream network (Vannote et al. 1980). Likewise, all the animal taxa with representatives inhabiting standing waters have characteristic species assemblages at different points along the hydroperiod/predator gradient that exists among ponds and lakes (Brooks and Dodson 1965, Vanni 1988, Well-born et al. 1996, Richardson 2001). Even over very small spatial scales, these patterns are apparent. Around the globe, organisms inhabiting the intertidal areas of marine coastlines sort themselves over a few meters into regular patterns of zonation (Connell 1961; Payne 1966; Menge 1976, 1995; Navarrete et al. 2000). Even across the human body, regular patterning occurs in the microbiome (Human Microbiome Project 2012).

We also know that membership in each of these assemblages is not simply the confluence of the physiological and dispersal limits of the collection of species found there. Species interactions among the possible members permit some and exclude others, and many different types of species interactions are involved in defining community membership: competition (e.g., Hairston 1951, Connell 1961, Tilman 1977, Hairston 1980), herbivory and predation (e.g., Payne 1966, Harper 1969, Sprules 1972, Vanni 1988), facilitation (e.g., Bertness 1991, Callaway 1994), and mutualism (e.g., Bronstein 1988, Thompson 2005). Thus, local assemblages are not "fortuitous" collections of species thrown together by the vagaries of history and stochasticity experienced by individual species. Rather, local species assemblage structures must be understood by the interactions that occur among species, as well as by physiological and dispersal limits, which immediately validates the concept of a *local community* of interacting species.

However, each local community of interacting species is not an integrated, cohesive unit and by no means can be considered a "complex organism" of its own. I think some of the best evidence that communities are not cohesive units is the disaggregation and recombination that has occurred following past climate change events (e.g., Davis 1983, Coope 1994, Graham et al. 1996, Jackson and Overpeck 2000). Consequently, abrupt and wholesale changes in species composition do not occur when moving among local communities that exist at different points along environmental gradients. In fact, species distributions often appear to be random when considered in this context. For example, Whittaker mapped the distributions of plant species along altitudinal gradients in various mountain ranges and could find no sharp transitions in species composition, even though different assemblages are clearly discernable at different points along the gradient (Whittaker 1956, Whittaker and Niering 1965, Whittaker 1975, Whittaker and Niering 1975). This is not because dispersal links communities together, but rather because the traits of species make them differentially successful in various constellations of abiotic and biotic conditions.

Odonate species inhabiting standing bodies of water from vernal ponds to large permanent lakes offer a valuable example to illustrate how patterns of community membership can be quite different among closely related species but still completely validate the existence of unique communities. Like all other animal taxa that live in these water bodies (e.g., Wellborn et al. 1996), characteristic odonate assemblages develop at different points along this gradient from vernal ponds to large, permanent lakes because of local abiotic and biotic conditions (fig. 6.5). For example, *Lestes* species subdivide the entire length of this gradient into four ranges based on the frequency with which the pond completely dries and the predators found there; the four categories are vernal ponds with no predators, ponds with dragonflies that dry infrequently, permanent pond and lakes with dragonflies, and permanent ponds and lakes with fish (fig. 6.5). Unique *Lestes* assemblages are found in each of these four pond types (Stoks and McPeek 2003). Likewise, unique assemblages of *Enallagma* species are found in permanent water bodies with either dragonfly or fish as top predators, but *Enallagma* species are not found in vernal or temporary ponds because their life cycle is not compatible with pond drying (Johnson and Crowley 1980; McPeek 1990b, 1998).

Enallagma and *Lestes* species segregate along this ecological gradient because they have evolved to possess traits that make them successful at evading the predators (either dragonflies or fish) with which they coexist, and the adaptations that make them successful against one predator come at the expense of success against the other (McPeek 1990a, McPeek et al. 1996, Stoks and McPeek 2006). Thus, their distributions are limited to narrow ranges along the gradient. Moreover, *Enallagma* and *Lestes* species largely feed on different prey because of their dissimilar sizes. Thus, *Enallagma* and *Lestes* species form an apparent competition community module with either fish or dragonflies as their linking predators in each permanent pond or lake (fig. 6.5).

In contrast, the same *Ischnura* species are common in all ponds and lakes except vernal ponds (Fig. 6.5) (McPeek 1998). These *Ischnura* species coexist with *Enallagma* and *Lestes* species in each pond and lake by having traits that make them differentially successful at competing for resources and evading predators (fig. 6.5). *Ischnura* and *Enallagma* species feed on the same resources and therefore are resource competitors, but *Ischnura* is much more efficient at converting consumed prey into its own biomass (McPeek 2004). However, *Ischnura* species experience significantly higher mortality rates from both fish and dragonflies than *Enallagma* (McPeek 1998). Consequently, *Ischnura* and *Enallagma* form a diamond community module in each lake where *Enallagma* species are present, and *Ischnura* and *Lestes* species form an apparent competition module (fig. 6.5). The functional group that *Ischnura* species occupy in each local community (namely, being effective competitors for small prey but poor at predator avoidance) is present in a much broader array of communities along this gradient than the functional groups occupied by *Enallagma* (inefficient competitors for small prey but effective at avoiding the local predator) or *Lestes* (efficient at competing for large prey and effective at avoiding the local predator) species.

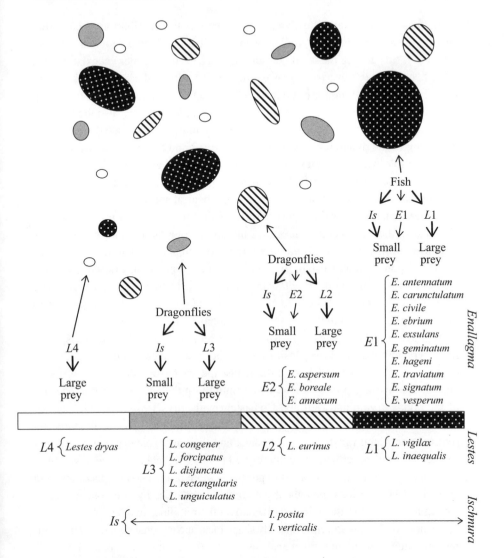

FIGURE 6.5. Schematic representation of the four general categories of local community types for *Enallagma*, *Lestes*, and *Ischnura* assemblages in the metacommunity of ponds and lakes in the New England region of North America. The *lower reaches* of the figure give the checklist of species that can be found in vernal ponds (identified by the *open circle*), temporary ponds that have large, active dragonflies as the top predators (*gray fill*), permanent ponds and lakes with large, active dragonflies as the top predators (*diagonal striped*), and permanent ponds with fish as the top predators (*black stippled*) (McPeek 1990b, 1998; Stoks and McPeek 2006). Above each checklist is a schematic diagram of the community module in which these species are embedded for each lake type. The upper part of the figure is a hypothetical map of a metacommunity of these local community types with each pond or lake type identified by the different fill patterns.

As these odonate taxa illustrate, what influences the distribution of each species along an environmental gradient is the distribution of the ecological opportunity that it fills among the local communities. These three genera coexist in each pond because they have traits that cause them to satisfy the criteria laid out in chapter 2 for the community modules they occupy. The trait differences among species that make them successful in these functional groups are also the result of past adaptation to these ecological roles (Stoks and McPeek 2006), presumably according to the dynamics described in chapters 3 and 5. Moreover, within each lake, multiple neutral species in each genus occupy their respective functional groups (Siepielski et al. 2010). Thus, these odonate assemblages are exemplars of communities of simultaneously coexisting and neutral species.

In addition, these genera appear to differ in dispersal propensities that are consistent with the fitness consequences of moving among these ponds and lakes, as described in this chapter. *Enallagma* species are unable to determine as adults whether a pond contains fish or dragonflies as the top predator. Consequently, the landscape for a dispersing *Enallagma* adult is akin to figure 6.1C, where many potential immigration sites would give this individual very low (essentially zero) fitness (e.g., an individual of a species adapted to living with dragonflies that immigrates to a lake containing fish). Consistent with this, *Enallagma* species have very low dispersal propensities from their natal lakes (McPeek 1989). In contrast, most ponds and lakes on the landscape are hospitable to *Ischnura* species, and temporary ponds can vary substantially in their fitness returns because of the ponds drying and refilling. Consistent with this, *Ischnura* species appear to have a much higher dispersal propensity than *Enallagma*; *Ischnura* adults can routinely be found far from water bodies, and they are typically one of the first species to colonize refilled or newly formed ponds (McPeek, pers. observation).

In this chapter I have attempted to provide a more refined and all-encompassing analysis of how various movement patterns and strategies by component species among local communities might evolve, affect their abilities to adapt to local environmental conditions, and consequently shape local and regional patterns of community structure. This more refined analysis identifies the simple fact that individuals moving among local communities does not necessarily cause significant perturbations to local community structure. Only certain forms of nonadaptive movement will disturb local structure. Also, interacting species in the same metacommunity may have very different movement patterns favored by natural selection. If the spatiotemporal structure of the ecological opportunities available to species in various functional groups are different across the metacommunity, they should evolve dissimilar dispersal propensities in accordance with those differences. However, there can be no doubt that movement of individuals among local communities has strong influences on both local and regional community structure.

CHAPTER SEVEN

Which Ways Forward?

I hope that the next time you walk through the woods, you have a deeper appreciation of the entangled bank into which you have strolled. Why are those species here? Why are other species found somewhere else, but not here? What was here 100,000 years ago, 1 million years ago, or 65 million years ago, and how did those species influence what is here today? I hope that this work has provided you with a framework to organize those questions, see the relationships among those questions, and consider what the possible answers might be. I would like to end by offering some final thoughts on what I think are the important insights that have emerged regarding the major topics discussed in this book.

While it is typically associated only with evolution by natural selection, fitness is also *the* central metric of both population and community ecology. The major organizing concepts in these disciplines explicitly address the various processes that determine the fitnesses of individuals, populations, and species. Species can maintain a population in a local area where the demographic input of individuals (i.e., births and immigration) are greater than the demographic loss of individuals (i.e., deaths and emigration), when the species is rare. These demographic rates are influenced by both abiotic factors and the other species present both locally and regionally. Moreover, the values and properties of those abiotic factors and other species change over time, which causes dynamics in these demographic rates, and thus in fitness at all these levels.

Hutchinson (1958) developed his definition of the niche explicitly to demarcate the environmental conditions that permit a species to maintain a population indefinitely. His description of a niche focused on the abiotic factors that are physiologically limiting to a species, as well as the resources that circumscribe where a species can and cannot maintain a population. He also explicitly recognized that the limits of a species' distribution across environmental gradients of these factors depended on the presence or absence of other species—for him, the important other species were resource competitors. Missing from this conception of the niche are the abundances of species engaged in other types of species interactions, and all the trait values that determine their performances in these interactions with the abiotic environment and other species.

Alternatively, the various conceptions of the adaptive landscape focus on this relationship between the properties of individuals and populations, and their associated fitnesses, but explicitly ignore how abiotic conditions and the abundances and phenotypes of species determine these relationships. Sewall Wright's many variations of the adaptive landscape all focused on the genotypes of individuals (Wright 1988). Simpson (1944) did describe the relationship between fitness and the phenotypes of individuals using this landscape metaphor, and he recognized that populations and species are evolving via natural selection on a "dynamic selection landscape" that changes as environmental conditions change (fig. 13 in Simpson 1944). What is missing from most current conceptions of this fitness topography is the explicit dependence of its shape on various environmental factors and the dynamical nature of the relationship that Simpson recognized.

The theory pioneered by Lande (1982), Iwasa et al. (1991) and Abrams et al. (1993) for understanding the joint dynamics of species' abundances and traits in a community context (e.g., equations (3.9) and (3.10)) provides the conceptual framework for unifying Hutchinson's (1958) conception of the niche with Simpson's (1944) representation of fitness topographies, and unifies many of the concepts I have discussed throughout the book. As I described at the end of chapter 3, the average fitness of individuals in a population (i.e., $dN/Ndt = \ln(\bar{W}_N)$) maps onto a system of axes describing the abiotic factors (one subset of axes) for that site, the various species' abundances (a second subset), and mean traits (a third subset) of all species in the community. Note that one can conceptualize the trait axes to include axes for the types of traits discussed in chapter 3 that influence how individuals interact with one another within local communities, and axes that describe dispersal propensities among other populations within the metacommunity as in chapter 6. The niche and the adaptive landscape are simply representations of this more general system, but each ignores a different subset of these axes.

Hutchinson's niche is simply the N-dimensional hypervolume defined by the abiotic axis system where $\ln(\bar{W}_N) > 0$, but this volume cannot be determined without knowing the abundances and traits of all the other species that are present. Presumably, Hutchinson would have placed the system on the species' abundance and trait axes at their current values.

The fitness landscape is also simply the fitness surface over the trait axes of the species in question when the abiotic factor axes and the trait and abundance axes for the other species are at their current values (fig. 3.13). This fitness landscape changes as the abundances and traits of all the species in the community change (e.g., figs. 3.3 and 5.1, and see the associated animations for these figures), but this changing landscape can be understood by recognizing that it is simply a more limited perspective on this higher dimensional fitness surface (fig. 3.13).

Other ecological and evolutionary concepts I have discussed are easily understood within this conceptual framework. Invasibility is a question of whether

$\ln(\bar{W}_N) > 0$ at the point where the trait of the species in question is at its current mean value, its abundance is near zero, and all the other abiotic factors and species' abundances and traits are at the values where they would equilibrate when this species is absent. We can expand this to discuss an *ecological volume of invasibility* defined by the entire volume having $\ln(\bar{W}_N) > 0$ when the species is extremely rare and at its current mean trait values. This ecological volume of invasibility would contain all combinations of abiotic factors and other species' abundances and traits that would permit this species to invade the community without any evolution. One might think of this volume as the space of ecological opportunities available to a species with a particular phenotype. In addition, perhaps this would be a more robust definition for the current niche of this species.

Once we have also defined the domains of attraction around the various equilibria for the community (see the end of chapter 3), we can then define an *evolutionary volume of invasibility* as the entire volume of this space that is within one of these domains of attraction having a nonzero equilibrium abundance for this particular species. This evolutionary volume of invasibility defines the conditions in which a founding population can invade and adapt when allowing the joint action of abundance and trait dynamics (e.g., a habitat shift). This volume also defines the space of *ecological and evolutionary opportunities* for a species. Dispersal limitation could be evaluated with the concept. That is, could this species establish and maintain a local population were it able to disperse to this location?

The unified ecological and evolutionary fitness surfaces (e.g., fig. 3.13) can also be decomposed into their constituent fitness components to explore how their overall shapes came about, to determine which fitness components are driving selection in the short term, and to quantify the trade-offs that are struck by species to persist in the community. Examination at this level also provides the mechanistic insights to understand the structure of the resulting community, and how it changes as species come and go in the community and as the abiotic environment changes. Even for systems that have arrived at a stable abundance and trait equilibrium, thinking about communities in this joint ecological and evolutionary framework sharpens our focus and understanding of what the included species' critical attributes are, how those attributes generate mechanisms of species interactions, and how they can or cannot live together in the short and long term.

COMMUNITY MEMBERSHIP AND STRUCTURE

A fundamental tenet of population biology is that a species can maintain a population at a given locality if the sum of its per capita birth and immigration rates exceed the sum of its per capita death and emigration rates when it is rare at this locale. Community ecology considers the identical issues to determine whether a species will be present, but considers the mechanisms that determine these demographic rates. What birth rate is supported by the abundances of resources and mutualistic partners at this locale? What death rate is inflicted by the abundances of enemies (i.e., herbivores, predators, disease organisms) at this locale? What other populations are supported in the area that may provide immigrants to or attract emigrants from this locale? In fact, the invasibility criterion for coexistence is simply a statement of these local demographic conditions considering only per capita birth and death rates for each species in a community context. These are the questions that directly address the mechanistic understanding for why invasibility is or is not possible for each species living together in a local community and across the landscape or seascape of communities.

These are not trivial details appended to population biology. The demography of each species is determined by how its traits define its performance in the community's web of interactions. Whether one species can satisfy this criterion depends on the identities and abundances of the other species. Is this species embedded in a community module with a predator or an intraguild predator? What other species utilize the same resources? Do mutualist partners have predators of their own? If a species cannot satisfy its invasibility criterion, what other species are depleting resources/mutualistic partners or inflating enemy abundances to levels that make its local demography unfavorable? Conversely, if this species is able to support a population in this community, what other types of species does its presence and abundance preclude or foster, either directly or indirectly? These are the mechanistic insights that community ecology explores.

All this falls under the rubric of "coexistence," and I have used this word almost interchangeably with invasibility, but coexistence does not exactly capture the spirit of what is being explored. Coexistence implies a pairwise comparison of species: can these two species coexist with one another? Whereas invasibility is a question about each species considered individually: could this species reinvade if it were removed and the rest of the community allowed to come to its long-term dynamical state? Invasibility encompasses all the other species in the community and not just one specific comparison. Moreover, the reason a species may be precluded is typically because of the combined effects of resources/mutualists and enemies.

Also, much of the way we think about coexistence contains baggage held over in community ecology's lexicon from a time when entire communities were presumed to be structured primarily or exclusively by competition for resources. Hutchinson's (1958) conception of the niche is the implied metaphor for coexistence, with all the attendant baggage that this metaphor conjures in one's mind. The idea of fundamental and realized niches only have meaning in the context of resource competition (Hutchinson 1958). How does Hutchinson's niche help us understand coexistence of a predator and prey, mutualist partners, or an intraguild predator and prey and their shared resource (McPeek 2014, 2014b)? This baggage even applies to how we name things; the "apparent competition" community module (fig. 2.1) has that name because the expected experimental result of manipulating one prey's abundance on the other prey (i.e., the two "apparent competitors") is identical to the expected experimental result of manipulating one resource competitor's abundance on the other resource competitor (Holt 1977). Moreover, platitudes about coexistence being promoted by "intraspecific competition being stronger than interspecific competition" also derive from this mindset. How does this axiom help us understand the coexistence of anything other than pure resource competitors? In my opinion, this thinking greatly hinders our understanding of the processes that structure biological communities.

We need to expunge this baggage of considering resource competition as the primary organizing process, and instead embrace a much richer understanding of how the variety of species interactions and the myriad configurations of those interactions into various community modules shape communities. The theoretical body of work that undergirds this richer perspective has been developing for over 50 years (Rosenzweig and MacArthur 1963, Paine 1966, Levin 1970, May 1973, Abrams 1975, DeAngelis et al. 1975, Murdoch and Oaten 1975, Schoener 1976, Holt 1977, Pimm 1982, Tilman 1982, Abrams 1983, Hastings and Powell 1991, Holt et al. 1994, Leibold 1996, Polis and Strong 1996, Holt and Polis 1997, Chesson 2000, Diehl and Feißel 2000, Holland et al. 2002, Bronstein et al. 2003, Murdoch et al. 2003, Wilson et al. 2003, Abrams 2004, Křivan and Schmitz 2004, Křivan and Diehl 2005, Křivan 2007, Chesson and Kuang 2008, McCann 2011). Chapter 2 in this book provides only a glimpse into this much richer, dynamic approach to understanding the ecological mechanisms structuring communities, and only a glimpse at a much richer set of species interactions (e.g., this analysis could not consider detritivores, nutrient cycling, pathogens, etc.). However, I hope that this presentation has developed a strong intuition in the reader that the critical insight is to understand what demographic balance is struck by the multifarious interactions in which each species engages.

Understanding community structure through invasibility as the demographic portrait of each species separately also immediately highlights the fact that not all species in a community are "coexisting." I have always found the fact that we typically teach students about coexisting species in community ecology classes, sink species in population biology classes, and walking-dead species in paleobiology classes to be intellectual dissonance, as though these three types of species only exist in these different realms. Walking-dead species that are being driven locally extinct by local demographic conditions, either because of change in the underlying environmental conditions or the invasion or extermination of other community members, will not simply disappear as soon as their fortunes change. It may take a very long time (e.g., O'Dea and Jackson 2009) for the last individual of a species to die. Sink species also by definition do not satisfy the local invasibility criterion, but are maintained by immigration in a demographic process that is frequently ignored by community ecologists. The presence of sink species in a community also necessitates a broadening of perspective to consider the patchwork of communities at the regional scale—the metacommunity (Holyoak et al. 2005). A sink species may exist in a community at the edge of that species' range (e.g., fig. 4.5B) or a community in an alternative habitat that is interspersed with its filial habitat across the landscape on the interior of the species' range (e.g., fig. 4.5C). Both species types will consume resources, interact with mutualists, and inflate the abundances of enemies, thereby affecting the abundances of all the other species in the community.

Likewise, the existence of neutral species in communities is a fact (e.g., Siepielski et al. 2010, Shinen and Navarrete 2014), but entire communities are not composed of neutral species as depicted in Hubbell's (2001) treatment of the subject. A group of neutral species will constitute one functional group, which is itself a component embedded in the entire interaction web of a community (Leibold and McPeek 2006). Moreover, each functional group in an interaction web may contain multiple neutral species. Thus, the entire enterprise of testing for "niche versus neutral" is just as silly as the debate that took place in community ecology about "competition versus predation" in the 1970s and 1980s. No community will consist of all neutral species, but finding signatures of some coexistence mechanism operating among some members of a community does not negate the possible existence of neutral species being embedded in that interaction network. As with all these arguments about dichotomies, it always turns out to be both in the end.

Moreover, all the arguments against even the possible existence of neutral species are based on false premises. Ecological speciation is not the only mechanism creating new species; many modes of speciation result in largely or completely ecologically undifferentiated daughter species, which means that neutral species may be created and introduced into communities and regional assemblages at substantial rates (chapter 4). In addition, invasibility need not be an issue, because communities have been periodically disaggregated and recombined in the past, which would mix allopatrically produced neutral species together at initially high relative frequencies for all of them (chapter 4). Probably most important for dispelling our niche-based intuition, natural selection only favors the divergence of ecologically similar or identical species in particular circumstance (chapter 5).

Therefore, searching for patterns in community data using models that approximate community structure based on only one type of species or one type of species interaction (e.g., broken sticks, limiting similarity, community matrices, or all neutral) is unsatisfying at best. If we are to continue to explore community structure at a grander level, more sophisticated models that account for all these types of species being present are needed. This is also why I have focused mainly on community modules in both my empirical and theoretical work. I see greater progress being made and greater insights being derived from the mechanistic understanding that comes from these more focused experimental inquiries of components of the larger system.

ADAPTATION

Another fundamental tenet of community ecology should be that species are not static, immutable entities, either on a local or regional scale. Species evolve in response to one another, and this coevolution contributes to the final ecological structure of a community (Travis et al. 2013). The last 30 years has seen a great flourishing of ideas with the intellectual amalgamation of community ecology with evolutionary adaptation and diversification. Two recent studies exemplify this amalgamation. Guppies (*Poecilia reticulata*) occupy communities with different combinations of predators, including those in which they have no predators; guppy populations adapt to these various communities through traits influencing predator escape, feeding, and life history (Reznick 1982, Reznick and Endler 1982). Bassar et al. (2012) experimentally replicated guppy populations in communities with and without their predators, simulating their natural conditions in different communities. In adapting to these various predation regimes, guppies evolved different diet preferences for algae versus invertebrates, and these evolved differences resulted in very different abundances of diatoms and invertebrates in the communities (Bassar et al. 2012). Likewise, Urban and Richardson (2015) showed that spotted salamanders (*Ambystoma maculatum*) in communities where they lived with gape-limited predators (marbled salamanders, *Ambystoma opacum*) evolved responses to modulate foraging rate to the prevalence of the gape-limited predators, and their foraging intensity changed zooplankton abundances. In contrast, spotted salamanders from populations lacking gape-limited predators altered their foraging rates only as conspecific density was altered. Both of these studies considered the same species embedded in different communities and showed how the evolutionary response of this species alters community structure. Moreover, they both illustrate how selection from one species interaction (i.e., predation in these cases) can cause responses in other traits (i.e., foraging in these cases) that will affect multiple types of species interactions simultaneously.

Natural selection occurs when a population of individuals interacts with its environment, both the abiotic features and species present. Every link in the interaction web of a community is a potential agent of phenotypic selection acting on a particular species. For each interaction, the resulting selection gradient associated with each fitness component depends on both the phenotypes and the abundances of the other species involved. Moreover, the shape of the overall fitness surface experienced by a species is determined by the combination of these various underlying selection gradients and the demographic weights of the different selective agents. A species reaches evolutionary equilibrium when both the values of the fitness components and the selection gradients on these components balance. Thus, as we have seen throughout this work, the fitness topography experienced by a species has a dynamic defined by the change in the phenotypes and abundances of the other species in the community. The ecological contexts causing dynamical density- and frequency-dependent fitness surfaces must be incorporated into both empirical characterizations of selection in the wild and theoretical studies of natural selection; it is not just genetics that changes as species evolve and coevolve.

As Lande's (1982) framework emphasizes, overall fitness is composed of various components, and phenotypic selection generated through each species interaction operates on only a subset of these components (see also Arnold and Wade 1984a, 1984b; McPeek 1996a). Because maxima and minima in the overall fitness landscape occur primarily at trait values where a balance is struck among contrasting selection gradients, overall selection is typically optimizing, while the underlying fitness components are each experiencing primarily directional selection (Travis 1989). This may be the reason that directional selection seems so prevalent and stabilizing selection so rare in measures of phenotypic selection in the wild (Endler 1986, Kingsolver et al. 2001, Rieseberg et al. 2002, Siepielski et al. 2009, Kingsolver and Diamond 2011, Kingsolver et al. 2012). Instead of rationalizing why the measure of fitness used in a study is a surrogate for overall fitness, researchers should embrace the fact that they are measuring selection on fitness components, and then expand phenotypic selection studies to explore the balances that are struck among the components. Moreover, measures of selection gradients on various fitness components made simultaneously will provide more robust analyses for how the overall fitness surface is determined (e.g., fig. 3.3).

SPECIATION

Ecologists tend to focus only on the mechanisms of speciation that involve ecological differentiation. I think this too results primarily from the historical baggage associated with niche filling and the assumption of coexistence. As I hope was clear in chapter 4, myriad speciation mechanisms can directly produce new coexisting species (via habitat shifts) and walking-dead species (via hybridization and polyploidy) in a community. In addition, ecologically equivalent species can be produced allopatrically (via the development of Bateson-Dobzhansky-Muller incompatibilities or character displacement speciation) and subsequently be mixed together to result in a set of neutral species. Thus, species with all levels of ecological differentiation can be introduced into communities through speciation.

The relative importance of different speciation mechanisms also varies among the taxa that constitute any particular community. Take as an example the communities that inhabit freshwater lakes. Centrarchid fishes and daphnid cladocerans seem to have speciated primarily through ecological means, given that each species in these taxa occupy very different functional positions in the community (Werner et al. 1977, Mittelbach 1984, Leibold 1991, Leibold and Tessier 1991, Tessier and Leibold 1997). Functional diversity in the community will be influenced primarily by the ecological speciation processes that generate new types of species.

In contrast, the *Enallagma* damselflies in these same lakes are neutral species occupying one functional position (Siepielski et al. 2010), and speciation was accomplished primarily by reproductive character displacement (McPeek et al. 2008, McPeek et al. 2011). Patterns of species diversity in communities will be shaped not only by the number and diversity of functional positions present in the community, but also by how these nonecological mechanisms add multiple species into the various functional positions and by the speciation mechanisms that predominate in the taxa that fill different functional groups.

DISAGGREGATION AND RECOMBINATION

The climate history of the Quaternary period (~2.6 million years ago to today) was marked by periodic spikes in global average temperature driven by Milankovitch cycles that melted the massive northern hemisphere glaciers, with intervening cold periods when the glaciers reformed (Petit et al. 1999, Zachos et al. 2001, Lisiecki and Raymo 2005). Associated with these periodic spikes in global average temperature were major range movements for species in almost all parts of the globe, which disaggregated and recombined local communities (reviewed in chapter 4). Dynesius and Jansson (2000) hypothesized that such forced range movements would have selected for species with greater levels of vagility and broader general ecological tolerances; they conjectured this may have in turn reduced subsequent speciation rates for mechanisms requiring long periods to develop reproductive isolation (e.g., Bateson-Dobzhansky-Muller incompatibilities), and also lessened subsequent extinction rates. The authors postulated this is because these events serve as filters when species that can respond effectively to such rapid climate change are favored.

The paleontological record also includes examples of taxa with coincident spikes in speciation and extinction. During the Quaternary, many species may have gone extinct because they could not effectively respond to the climate in one of the rapid periods of climate change. Although the speciation rate during the Quaternary did not change appreciably, the mechanisms of speciation may have differed considerably during the brief periods of rapid climate change. Speciation may have occurred as the result of some taxa filling vacated ecological opportunities. However, I would suggest that much of the speciation was associated with taxa being thrown into new ecological and social settings that occurred only during these periods of climatic upheaval. For example, the ecological speciation of sticklebacks when invading freshwater lakes is one clear example of a lineage adapting to a new community in a new habitat during such a climatic event (Schluter and McPhail 1993, Hatfield and Schluter 1999, Rundle et al. 2000). *Enallagma* damselflies underwent two radiations in North America over the last 250,000 years to produce 18 extant species from 2 progenitors, and many of the individual speciation events date to global temperature spikes (Turgeon et al. 2005, M. A. McPeek, unpubl. data). The *Enallagma* radiations involved both ecological speciation via multiple habitat shifts from lakes where fish are the top predators to lakes where dragonflies are the top predators, and reproductive character displacement speciation to produce a large collection of neutral species in fish lakes and a smaller collection of neutral species in dragonfly lakes (McPeek and Brown 2000, Turgeon et al. 2005, Stoks and McPeek 2006, McPeek et al. 2008).

Not every taxon and not every part of the world would have necessarily been affected by this climate history in the same ways. For example, the overall rate of diversification among birds does not seem to have heightened during the Quaternary, but many speciation events did occur within the last 250,000 years, and speciation seems to have been more pronounced among high-latitude taxa where such climatic upheaval would have had the most pronounced effects (Johnson and Cicero 2004, Weir and Schluter 2004, Zink et al. 2004, Lovette 2005). In contrast, the high level of plant species richness and endemism in South Africa's Cape Floristic Region may be attributed to the area's relative climatic stability throughout the Quaternary (Meadows and Sugden 1993, Goldblatt 1997, Cowling et al. 1998). However, while a few groups of galling insects do appear to also be more diverse, an overall increase in herbivorous insects did not result from the greater levels of plant species richness (Wright and Samways 1998, Giliomee 2003).

Community ecologists and evolutionary biologists must start incorporating recent Earth history and the attendant ecological and evolutionary consequences into how they view the world around them and how they determine the causes of patterns in the distributions and abundances of species. Earth has not been a static ecological entity even in the very recent past. Moreover, today's distributions and abundances may provide few clues as to the evolutionary history of individual species (e.g., range shifts through the species' history) or the assembly of local communities. Community disaggregation and recombination has clearly been a recurring part of the temperate and boreal biotas, and probably has played substantial roles in most tropical terrestrial environments and in the marine realm. Consequently, we must stop using invasibility as the dominant metaphor for how most real communities may have been assembled over evolutionary time.

TRAITS

As I hope the presentations in chapters 3 and 5 make clear, the traits that determine the performance of species in interactions with others are fundamental to structuring communities; they define the trajectories of species as they coevolve, and affect the likelihood of ecologically similar species differentiating from one another. One area that is critical to understanding community structure and the responses of communities to perturbations (though I have ignored it here) is how individuals may alter their traits in an adaptively plastic fashion to changing ecological conditions within their lifetimes. These responses and their consequences fall under the rubric of *trait-mediated indirect effects* (e.g., Werner and Peacor 2003, Schmitz et al. 2004, Ohgushi et al. 2013). While I have considered trait changes in an evolutionary context, as mentioned briefly in chapter 3, these same models of trait evolution can also be used to model phenotypically plastic responses of individuals to changing conditions (e.g., Abrams 1992; Abrams et al. 1993; Křivan 1998, 2000, 2003, 2013). Therefore, many of the results of evolutionary responses here should translate to taxa with individuals that are capable of expressing adaptively plastic responses to their surroundings. One very fruitful area for future research will be the synergisms and antagonisms between phenotypic plasticity and dispersal strategies (e.g., Scheiner 2016).

Empirical studies of trait-mediated indirect interactions provide excellent guides for how traits can be productively integrated into mechanistic studies of communities (Werner and Peacor 2003, Schmitz et al. 2004, Ohgushi et al. 2013). However, traits are also important in shaping the dynamics of species interactions even if none of the species are phenotypically plastic; species differences in ecological performance are fundamentally determined by their phenotypic trait differences. Therefore, species occupy different functional positions because they have varying trait values, and these values make species successful in different species interactions. *Anolis* lizard species are differentially successful at foraging on prey in different parts of a tree because they differ in body and leg shapes (Losos 1990a, 1990b). The various sunfish species in eastern North American lakes feed on different prey because of their differences in mouth shape and pharyngeal jaw morphology (Werner 1977, Werner et al. 1977; Mittelbach 1981, 1984; Wainwright et al. 1991). Benthic and limnetic stickleback (*Gasterosteus* spp.) species specialize in different functional positions in their lake communities because of differences in gill raker length, gape width, and body size (Schluter and McPhail 1993). *Enallagma* and *Ischnura* species coexist at the intermediate-trophic-level functional positions in a diamond-community module in various types of lakes. This is possible because *Enallagma* have traits that enable them to reduce mortality from their shared predators but have others that result in slow growth and development, whereas due to traits of their own, *Ischnura* grow faster but experience greater mortality rates from predators (McPeek 1998, 2004).

Likewise, traits are what make species capable of satisfying the invasibility criteria in a particular local community, and so define which local communities a species can and cannot potentially occupy. Plant species that grow in serpentine soils have traits making them more drought resistant and better able to deal with high heavy metal concentrations, but these traits make them less effective competitors on nonserpentine soils, where they are excluded (Anacker 2014). Many butterfly species segregate onto different food plants because of their differential abilities to deal with secondary metabolic toxins produced by various plants (Ehrlich and Raven 1964). In contrast, *Timema* stick insects segregate onto various host plants, because different striping patterns provide the best camouflage from predators on particular plants (Nosil et al. 2002, Nosil and Crespi 2006, Nosil 2007). One group of *Enallagma* (damselfly) species can only live in lakes with large dragonfly predators because the former are able to swim rapidly from attacking dragonflies due to large caudal lamellae and abdominal muscles with biochemical properties enabling short bursts of strenuous exertion (McPeek 1995; McPeek et al. 1996; McPeek 1999, 2000). In contrast, *Enallagma* species that live in lakes with fish move very little and are thus cryptic; they do not swim from attacking predators (McPeek 1990a, 2000; Stoks et al. 2003). As these focused examples highlight, trait differences among species are the mechanistic basis of what is now called *habitat filtering* (Weiher and Keddy 2001).

The results of these intensive, mechanistic studies of how species' traits determine performance in various interactions suggest that very broad-scale analyses of traits in communities may provide very little insight into either habitat filtering or the means of coexistence. Analyses of trait distributions in a community context only makes sense if those traits are substantially involved in determining the demographic performances of species in the community context. Otherwise, one is simply analyzing random numbers. Coexisting and neutral species are present because their traits make them successful at occupying different functional positions within the community, and as we have seen throughout this analysis, different functional positions are defined by the species having various traits and trait values. For example, prey species in North American lakes dominated by sunfishes use a multitude of antipredator strategies to thwart sunfish predation. Frogs and toads have noxious, distasteful chemicals in their skin (e.g., Werner and McPeek 1994). Snails have thick shells that are difficult to crush (e.g., Osenberg and Mittelbach 1989, Crowl and Covich 1990). Some caddisflies build cases for camouflage (e.g., Stuart and Currie 2001). Some odonates move very little (e.g., McPeek 1990a, Stoks et al. 2003, McPeek 2004). Some cladocerans shuttle in and out of cold, deep waters of the hypolimnion where fish cannot go (e.g., Leibold 1990, 1991; Leibold and Tessier 1991). Other odonates grow as rapidly as possible to emergence as adults and escape the fish (e.g., McPeek 1998, 2004). Thus, I see little utility in large-scale analyses of a handful of easily measured traits, or analyses that assume that the same trait or small set of traits are important to all the species in a community. Such studies cannot get close to the mechanisms shaping community structure.

The analyses of trait evolution in chapters 3 and 5 only considered one trait, but in reality many traits will influence performances in the various interactions in which a species must engage (see any of the examples discussed above), and different traits will be important for different taxa. My goal in this work was to highlight how diverse sorts of traits can shape adaptation and the development of community structure through evolution, and to explore the dynamics of fitness in species interactions during coevolution. Every real species has many ecologically important traits influencing all the fitness components that shape their overall fitness, and the dynamics I have explored here are compounded across these many traits. Thus, any representation of the fitness surface depicted in figure 3.13 would include multiple trait axes for each species. Obviously, the genetic architecture underlying these traits will also influence what adaptations evolve. However, true understanding will only arrive when we combine studies of the genetic architecture of traits with studies of fitness dynamics generated by species interactions and coevolution.

DIFFERENTIATION AND COMMUNITY STRUCTURE

Another presumption often made by community ecologists is that ecologically similar species should always differentiate from one another to reduce competition for some ecological opportunity, particularly when that opportunity is an exploited resource. This presumption is based on the notion that natural selection should always act to reduce interspecific competition (relative to intraspecific competition, which is presumed to not change in strength) and thus foster the coexistence of these competing species (Brown and Wilson 1956, Lawler and Maynard Smith 1976, Roughgarden 1976, Connell 1980, Slatkin 1980, Taper and Case 1985). The problem with this presumption is that to accomplish such differentiation, the species must somehow end up in the domains of attraction for different adaptive peaks. Previous theoretical analyses of coevolution driven by species interactions have noted that selection does not always drive species apart and may actually cause species to converge in phenotype instead of diverge (e.g., Roughgarden 1976; Slatkin 1980; Abrams 1986, 1987, 1990; Brown and Vincent 1992; Abrams and Chen 2002; Abrams 2003); however, the fact that convergence may happen is typically mentioned briefly and then forgotten in favor of focusing on the conditions that foster divergence. This preference is also a residue of our historical baggage.

The results presented in chapter 5 highlight a number of important issues about the likelihood of differentiation of ecologically similar or identical species. One clear prediction to emerge from this analysis is that differentiation is more likely when the strengths of selection gradients generated by species interactions in the community module are greater than those primarily generated intraspecifically. Here too, understanding why differentiation occurs requires the realization that the overall fitness of each species is composed of various components, and the differentiating species are trading off fitness components in different ways. We would again greatly benefit by taking a more granular approach to the study of fitness and natural selection that focuses on multiple components simultaneously.

The dynamics of differentiation are also driven by the dynamics of the fitness surfaces of the coevolving species. No peak shifts in the Wrightian sense (Wright 1932) are required, which means that species that ultimately diverge may begin under the influence of the same adaptive peak and still easily differentiate if conditions are correct. The process of differentiation may also proceed in some rather counterintuitive ways. For example, in the first case I presented of two apparent competitors differentiating (fig. 5.1), the process required that one resource initially evolve to increase its susceptibility to the consumer. I think this is surprising to most people—why would a resource evolve to have a higher death rate from a consumer?—until one understands the differentiation in terms of relative selection gradients on various fitness components.

The likelihood of differentiation also varies depending on the type of traits involved in the species interactions, which leads to another prediction. Since differentiation occurs over the greatest area of parameter space with dependent traits and, in particular, bidirectional-dependent traits, these should be the ones most frequently observed as underlying the dissimilarities between closely related but ecologically differentiated species. In contrast, differentiation occurs in only a limited area of parameter space with unidirectional-independent traits, and so these may be expected to be overrepresented as the ecologically important traits in undifferentiated species. For example, beak morphology in Darwin's finches (Grant and Grant 1982) and gape width/gill raker morphology in sticklebacks (Schluter and McPhail 1993) are clearly both bidirectional-dependent traits, but the many neutral *Enallagma* species that inhabit fish lakes across eastern North America reduce predation by being relatively inactive (McPeek 1990a, Stoks et al. 2003), which is a unidirectional-independent trait. Perhaps these are not coincidences, but rather representative of broader patterns of how various types of traits influence the likelihood of species differentiation in communities.

If differentiation occurs, the structure of the resulting community module also depends on the types of traits involved in species interactions. Trophic specialization of predators toward different prey is only possible if a bidirectional-dependent trait defines their interactions. Thus, ideas such as limiting similarity only apply when bidirectional traits shape species interactions. With unidirectional traits, the community structure of interaction strengths should be nested and hierarchical with species having similar rankings.

The analyses of chapter 5, along with previous studies (e.g., Abrams 1986, 1987, 1990; Brown and Vincent 1992; Abrams and Chen 2002; Abrams 2003), show that differentiation of ecologically similar species is not inevitable and that even ecologically different species may converge in their phenotypes. It is not surprising that reproductive character displacement speciation would result in a set of neutral species; in other words, species would not differentiate. However, natural selection may also favor convergence of initially dissimilar species to make neutral species (e.g., species in different genera, as in Shinen and Navarrete 2014).

Interactions among species result in a much richer set of possible outcomes and a much broader array of possible effects on overall community structure than we have typically considered. Moreover, our intuition about these outcomes and effects are often wrong. Considering the explicit and mechanistic dynamics that result from species interactions in a community context will allow us to expand our understanding into the community structures we document in nature.

LANDSCAPE STRUCTURE

The metacommunity perspective (Holyoak et al. 2005) has crystalized the reality that dispersal among local communities can significantly perturb community structure away from what local processes would favor in the absence of that dispersal. This includes perturbing the abundances of coexisting and neutral species as well as introducing sink species for explicit consideration. However, community ecologists must become more sophisticated about how and why species move among communities and the evolutionary forces that operate on the traits that influence their movement.

As the analyses in chapter 6 showed, the main perturbations that result from dispersal are to homogenize both abundances and traits across the metacommunity. Depending on the distributions of abundances and traits for the various species, this means that some species will be inflated in abundance and others depressed. However, if dispersal strategies can adapt in ways that result in an ideal free distribution (e.g., balanced dispersal or habitat selection), perturbations to abundances away from what local demography favors should be minimal. Thus, the degree to which individual species will be altered will depend the dispersal strategies they employ. In addition, if species can express adaptively plastic responses in ecologically important traits to local conditions, deviations to trait distributions favored by local selection regimes should be minimal as well.

Different species will not necessarily express the same types of movement patterns among local communities and thus may operate on very different spatial scales within the same metacommunity. This will depend on the fitness returns experienced across the collection of local communities. For example, since *Ischnura* damselfly species can satisfy the invasibility criteria in a much broader array of communities than each *Enallagma* or *Lestes* species (fig. 6.5), these two groups of species (i.e., *Ischnura* versus *Enallagma* and *Lestes*) should have very different dispersal strategies among local communities and operate on very disparate spatial scales. Each lake would support *Enallagma* and *Lestes* populations with only trickles of dispersal in and out of them (McPeek 1989), whereas *Ischnura* should move at much higher rates among lakes (M. A. McPeek, pers. observation). In such cases, species like *Ischnura* are essentially link species among communities found in what are different habitats for other species, because of the distribution of ecological opportunities across the landscape. Such differences in species' dispersal strategies will also make the definition of the boundary of a particular metacommunity just as difficult as defining the boundary of any single local community.

PHYLOGENY

Some may find it odd that I am at the end of a book on evolutionary community ecology and have not mentioned the role of phylogeny in all of this, particularly given my use of phylogenies in understanding community assembly and structure (e.g., McPeek and Brown 2000, Webb et al. 2002, Stoks and McPeek 2006). This is not an oversight. A phylogeny is the historical record of how a clade of species diversified. Therefore, it is a window on how the species we study today got here and the framework for the narrative of how a clade has diversified and adapted. A phylogeny is, in effect, a partial glimpse at Gould's (1989) tape of life. It is an emergent property of the clade and the various evolutionary histories of all the species descended from the common ancestor of interest in all the far flung corners of the world where these species are found.

Phylogenies are not, however, traits of species that influence their ecologies. The ecological performance of a species today is determined by the specific distribution of traits that it presently possesses, and not by what traits its lineage had in the past or other species in the clade have now. Certainly, the trait distribution that a species possesses today will be influenced by the trait distributions of its ancestors (that's why trait reconstruction techniques are so useful). However, both the species' performance today and where it will evolve to in the next time interval only depend on what trait values it has now. The place of a species in the phylogeny of its clade does not determine whether it can satisfy the invasibility criterion of a community or how it will evolve in response to the interaction network in which it is embedded. In addition, the number of close relatives in the community does not influence whether the species can satisfy its invasibility criterion; what matters are the trait values of these other species. Thus, phylogenies provide no insight into the mechanistic processes that structure communities or shape coevolution.

Phylogenetic affinities should also not be used as a surrogate for trait distributions among species. In some extremely broad general perspectives, more closely related species may be phenotypically more similar, but this is by no means a hard-and-fast rule that can be applied across huge swaths of biotic diversity. Just remember how ecologically different *Daphnia* species are and how ecologically similar *Enallagma* species are; then it is easy to see the fallacy of making such an assumption. This is why I find inferring mechanisms structuring communities based on phylogenetic diversity indices so dangerous. These indices are now easily calculated, given the prevalence of phylogenetic information, but they provide a false sense of insight based on the mountain of assumptions that must be made to make inferences.

This is not to say that community ecologists should not utilize phylogenetic information. Far from it! I believe that community ecologists should use phylogenetic, phylogeographic, and paleontological information in more expansive and mechanistic studies of community assembly. Phylogenetic analyses are critical tools for reconstructing where a species came from, how it adapted, and how it got to the communities in which it is found today (e.g., Schluter and McPhail 1993, Losos et al. 1998, Stoks and McPeek 2006). Comparing the phylogenies for multiple taxa in a community can also provide insights into community assembly, such as the sources of species and the timing of when species may have entered the community. By reconstructing trait evolution across multiple interacting clades, hypotheses about coevolutionary responses among species can also be tested.

I hope that this synthesis has made clear that these are all simplifications that focus on different facets of a biological community. The simplifications we make by ignoring either the ecological or evolutionary dynamics of a community often force us to conclusions that are unjustified (e.g., "invasibility" as the metaphor for community assembly, static fitness surfaces) or ignore what may be possible (e.g., neutral species, peak shifts driven by natural selection). I hope this synthesis also highlights that we have had the theoretical machinery to overcome these simplifications for a long time, and we can make great progress by not making these simplifying assumptions anymore. Consequently, whether we call something a niche, invasibility, an adaptive peak, an ecological opportunity, or whatever, we are all talking about the same thing.

Literature Cited

Abrahamson, W. G., and A. E. Weis. 1997. Evolutionary ecology across three trophic levels: Goldenrods, gallmakers, and natural enemies. Princeton University Press, Princeton, NJ.

Abrams, P. 1975. Limiting similarity and the form of the competition coefficient. Theoretical Population Biology 8:356–375.

Abrams, P. 1983. The theory of limiting similarity. Annual Review of Ecology and Systematics 14:359–376.

Abrams, P. A. 1986. Character displacement and niche shift analyzed using consumer-resource models of competition. Theoretical Population Biology 29:107–160.

Abrams, P. A. 1987. Alternative models of character displacement and niche shift. 2. Displacement when there is competition for a single resource. American Naturalist 130:271–282.

Abrams, P. A. 1990. Adaptive responses of generalist herbivores to competition: Convergence and divergence. Evolutionary Ecology 4:103–114.

Abrams, P. A. 1992. Adaptive foraging by predators as a cause of predator-prey cycles. Evolutionary Ecology 6:56–72.

Abrams, P. A. 1999. Is predator-mediated coexistence possible in unstable systems? Ecology 80:608–621.

Abrams, P. A. 2000. Character shifts of prey species that share predators. American Naturalist 156:S45–S61.

Abrams, P. A. 2003. Can adaptive evolution or behaviour lead to diversification of traits determining a trade-off between foraging gain and predation risk? Evolutionary Ecology Research 5:653–670.

Abrams, P. A. 2004. Trait-initiated indirect effects due to changes in consumption rates in simple food webs. Ecology 85:1029–1038.

Abrams, P. A. 2006. The effects of switching behavior on the evolutionary diversification of generalist consumers. American Naturalist 168:645–659.

Abrams, P. A. 2007. Habitat choice in predator-prey systems: Spatial instability due to interacting adaptive movements. American Naturalist 169:581–594.

Abrams, P. A., and X. Chen. 2002. The evolution of traits affecting resource acquisition and predator vulnerability: Character displacement under real and apparent competition. American Naturalist 160:692–704.

Abrams, P. A., R. Cressman, and V. Křivan. 2007. The role of behavioral dynamics in determining the patch distributions of interacting species. American Naturalist 169:505–518.

Abrams, P. A., and P. A. Fung. 2010a. The impact of adaptive defense on top-down and bottom-up effects in systems with intraguild predation. Evolutionary Ecology Research 12:307–325.

Abrams, P. A., and S. R. Fung. 2010b. Prey persistence and abundance in systems with intraguild predation and type-2 functional responses. Theoretical Population Biology **264**:1033–1042.

Abrams, P. A., R. D. Holt, and J. D. Roth. 1998. Apparent competition or apparent mutualism? Shared predation when populations cycle. Ecology **79**:201–212.

Abrams, P. A., and H. Matsuda. 1996. Positive indirect effects between prey species that share predators. Ecology **77**:610–616.

Abrams, P. A., and H. Matsuda. 1997a. Fitness minimization and dynamic instability as a consequence of predator-prey coevolution. Evolutionary Ecology **11**:1–20.

Abrams, P. A., and H. Matsuda. 1997b. Prey adaptation as a cause of predator-prey cycles. Evolution **51**:1742–1750.

Abrams, P. A., H. Matsuda, and Y. Harada. 1993. Evolutionary unstable fitness maxima and stable fitness minima of continuous traits. Evolutionary Ecology **7**:465–487.

Abrams, P. A., and C. Rueffler. 2009. Coexistence and limiting similarity of consumer species competing for a linear array of resources. Ecology **90**:812–822.

Adler, P. B., J. HilleRisLambers, and J. M. Levine. 2007. A niche for neutrality. Ecology Letters **10**:95–104.

Allee, W. C., and E. S. Bowen. 1932. Studies in animal aggregations: Mass protection against colloidal silver among goldfishes. Journal of Experimental Biology **61**:185–207.

Allesina, S., D. Alonso, and M. Pascual. 2008. A general model for food web structure. Science **320**:658–661.

Alroy, J. 2008. Dynamics of origination and extinction in the marine fossil record. Proceedings of the National Academy of Sciences, USA **105** Supplement 1:11536–11542.

Amarasekare, P. 1998. Interactions between local dynamics and dispersal: Insights from single species models. Theoretical Population Biology **53**:44–59.

Amarasekare, P. 2004. The role of density-dependent dispersal in source–sink dynamics. Journal of Theoretical Biology **226**:159–168.

Amarasekare, P. 2007. Spatial dynamics of communities with intraguild predation: The role of dispersal strategies. American Naturalist **170**:819–831.

Amarasekare, P. 2008. Coexistence of intraguild predators and prey in resource-rich environments. Ecology **89**:2786–2797.

Anacker, B. L. 2014. The nature of serpentine endemism. American Journal of Botany **101**:219–224.

Andersson, M. 1994. Sexual selection. Princeton University Press, Princeton, NJ.

Angert, A. L. 2006. Demography of central and marginal populations of monkeyflowers (*Mimulus cardinalis* and *M. lewisii*). Ecology **87**:2014–2025.

Arditi, R., and L. R. Ginzburg. 2012. How species interact: Altering the standard view on trophic ecology. Oxford University Press, New York, NY.

Arditi, R., C. Lobry, and T. Sari. 2015. Is dispersal always beneficial to carrying capacity? New insights from the multi-patch logistic equation. Theoretical Population Biology **106**:45–59.

Armsworth, P. R., and J. E. Roughgarden. 2008. The structure of clines with fitness-dependent dispersal. American Naturalist **172**:648–657.

Arnold, S. J. 1983. Sexual selection: The interface of theory and empiricism. Pages 67–108 *in* P. Bateson, ed. Mate choice. Cambridge University Press, Cambridge, UK.

Arnold, S. J., and M. J. Wade. 1984a. On the measurement of natural and sexual selection: Applications. Evolution **38**:720–734.

Arnold, S. J., and M. J. Wade. 1984b. On the measurement of natural and sexual selection: Theory. Evolution **38**:709–719.

Arnqvist, G. 1998. Comparative evidence for the evolution of genitalia by sexual selection. Nature **393**:784–786.

Arnqvist, G., and L. Rowe. 2005. Sexual conflict. Princeton University Press, Princeton, NJ.

Asmussen, M. A. 1983. Evolution of dispersal in density regulated populations: A haploid model. Theoretical Population Biology **23**:281–299.

Avise, J. C. 2000. Phylogeography: The history and formation of species. Harvard University Press, Cambridge, MA.

Badgley, C., T. M. Smiley, and J. A. Finarelli. 2014. Great Basin mammal diversity in relation to landscape history. Journal of Mammalogy **95**:1090–1106.

Balkau, B. J., and M. W. Feldman. 1973. Selection for migration modification. Genetics **74**:171–174.

Barfield, M., R. D. Holt, and R. Gomulkiewicz. 2011. Evolution in stage-structured populations. American Naturalist **177**:397–409.

Barraclough, T. G., P. H. Harvey, and S. Nee. 1995. Sexual selection and taxonomic diversity in passerine birds. Proceedings of the Royal Society of London B, Biological Sciences **259**:211–215.

Bartoli, G., M. Sarnthein, M. Weinelt, H. Erlenkeuser, D. Garbe-Schönberg, and D. W. Lea. 2005. Final closure of Panama and the onset of northern hemisphere glaciation. Earth and Planetary Science Letters **237**:33–44.

Barton, N. H. 1988. Speciation. Pages 185–218 *in* A. A. Myers and P. S. Giller, eds. Analytical biogeography: An integrated approach to the study of animal and plant distributions. Chapman and Hall, New York, NY.

Barton, N. H., and G. M. Hewitt. 1985. Analysis of hybrid zones. Annual Review of Ecology and Systematics **16**:113–148.

Barton, N. H., and S. Rouhani. 1987. The frequency of shifts between alternative equilibria. Journal of Theoretical Biology **125**:397–418.

Bascompte, J., P. Jordano, C. J. Melián, and J. M. Olesen. 2003. The nested assembly of plant- animal mutualistic networks. Proceedings of the National Academy of Sciences USA **100**:9383–9387.

Bascompte, J., P. Jordano, and J. M. Olesen. 2006. Asymmetric coevolutionary networks facilitate biodiversity maintenance. Science **312**:431–433.

Bassar, R. D., R. Ferriere, A. López-Sepulcre, M. C. Marshall, J. Travis, C. M. Pringle, and D. N. Reznick. 2012. Direct and indirect ecosystem effects of evolutionary adaptation in the Trinidadian guppy (*Poecilia reticulata*). American Naturalist **180**:167–185.

Bateson, P. 1983. Mate choice. Cambridge University Press, Cambridge, UK.

Bateson, W. 1909. Heredity and variation in modern lights. Pages 85–101 *in* A. C. Seward, ed. Darwin and science. Cambridge University Press, Cambridge, UK.

Bauer, S., J. Samietz, and U. Berger. 2005. Sexual harassment in heterogeneous landscapes can mediate population regulation in a grasshopper. Behavioral Ecology **16**:239–246.

Beddington, J. R. 1975. Mutual interference between parasites or predators and its effect on searching efficiency. Journal of Animal Ecology **44**:331–340.

Berlocher, S. H., and J. L. Feder. 2002. Sympatric speciation in phytophagous insects: Moving beyond controversy? Annual Review of Entomology **47**:773–815.

Bertness, M. D. 1991. Interspecific interactions among high marsh perennials in a New England salt marsh. Ecology **72**:125–137.

Bickford, D., D. J. Lohman, N. S. Sodhi, P. K. Ng, R. Meier, K. Winker, K. K. Ingram, and I. Das. 2007. Cryptic species as a window on diversity and conservation. Trends in Ecology & Evolution **22**:148–155.

Blüthgen, N., J. Fründ, D. P. Vázquez, and F. Menzel. 2008. What do interaction network metrics tell us about specialization and biological traits? Ecology **89**:3387–3399.

Boake, C.R.B., M. P. DeAngelis, and D. K. Andreadis. 1997. Is sexual selection and species recognition a continuum? Mating behavior of the stalk-eyed fly *Drosophila heteroneura*. Proceedings of the National Academy of Sciences, USA **94**:12442–12445.

Bolnick, D. I. 2006. Multi-species outcomes in a common model of sympatric speciation. Journal of Theoretical Biology **241**:734–744.

Bolnick, D. I., and B. M. Fitzpatrick. 2007. Sympatric speciation: Models and empirical evidence. Annual Review of Ecology, Evolution, and Systematics **38**:459–487.

Boucher, D. H. 1985. Lotka-Volterra models of mutualism and positive density-dependence. Ecological Modelling **27**:251–270.

Boughman, J. W. 2001. Divergent sexual selection enhances reproductive isolation in sticklebacks. Nature **411**:944–948.

Boughman, J. W. 2002. How sensory drive can promote speciation. Trends in Ecology & Evolution **17**:571–577.

Bowers, J. E., B. A. Chapman, J. K. Ruong, and A. H. Paterson. 2003. Unraveling angiosperm genome evolution by phylogenetic analysis of chromosomal duplication events. Nature **422**:433–438.

Bowers, M. A., and J. H. Brown. 1982. Body size and coexistence in desert rodents: Chance or community structure? Ecology **63**:391–400.

Bronstein, J. L. 1988. Mutualism, antagonism, and the fig-pollinator interaction. Ecology **69**:1298–1302.

Bronstein, J. L. 1994. Our current understanding of mutualism. Quarterly Review of Biology **69**:31–51.

Bronstein, J. L., W. G. Wilson, and W. F. Morris. 2003. Ecological dynamics of mutualist/antagonist communities. American Naturalist **162**:S24–S39.

Brooks, J. L., and S. I. Dodson. 1965. Predation, body size, and composition of plankton. Science **150**:28–35.

Brown, J. D., and R. J. O'Neill. 2010. Chromosomes, conflict, and epigenetics: Chromosomal speciation revisited. Annual Review of Genomics and Human Genetics **11**:291–316.

Brown, J. H. 1995. Macroecology. University of Chicago Press, Chicago, IL.

Brown, J. H., and G. A. Lieberman. 1973. Resource utilization and coexistence of seed-eating desert rodents in sand dune habitats. Ecology **54**:788–797.

Brown, J. S. 1988. Patch use as an indicator of habitat preference, predation risk, and competition. Behavioral Ecology and Sociobiology **22**:37–47.

Brown, J. S. 1989. Desert rodent community structure: A test of four mechanisms of coexistence. Ecological Monographs **59**:1–20.

Brown, J. S., and T. L. Vincent. 1992. Organization of predator-prey communities as an evolutionary game. Evolution **46**:1269–1283.

Brown, W. L., Jr., and E. O. Wilson. 1956. Character displacement. Systematic Zoology **5**:49–64.

Buerkle, C. A., R. J. Morris, M. A. Asmussen, and L. H. Rieseberg. 2000. The likelihood of homoploid hybrid speciation. Heredity **84**:441–451.

Bush, G. L. 1969. Sympatric host race formation and speciation in frugivorous flies of the genus. *Rhagoletis*. Evolution **23**:237–251.

Callahan, M. S., and M. A. McPeek. 2016. Multi-locus phylogeny and divergence time estimates of *Enallagma* damselflies (Odonata: Coenagrionidae). Molecular Phylogenetics and Evolution **94**:182–195.

Callaway, R. M. 1994. Facilitation and interfering effects of *Arthrocnemum subterminale* on winter annuals. Ecology **75**:681–686.

Callaway, R. M. 1995. Positive interactions among plants. The Botanical Review **61**:306–349.

Callaway, R. M. 1997. Positive interactions in plant communities and the individualistic-continuum concept. Oecologia **112**:143–149.

Cantrell, R. S., C. Cosner, D. L. DeAngelis, and V. Padron. 2007. The ideal free distribution as an evolutionarily stable strategy. Journal of Biological Dynamics **1**:249–271.

Cantrell, R. S., C. Cosner, and Y. Lou. 2010. Evolution of dispersal and the ideal free distribution. Mathematical Biosciences and Engineering **7**:17–36.

Cantrell, R. S., C. Cosner, Y. Lou, and C. Xie. 2013. Random dispersal versus fitness-dependent dispersal. Journal of Differential Equations **254**:2905–2941.

Carson, H. L., and K. Y. Kaneshiro. 1976. *Drosophila* in Hawaii: Systematics and ecological genetics. Annual Review of Ecology and Systematics **7**:311–346.

Casas, J. J., and P. H. Langton. 2008. Chironomid species richness of a permanent and a temporary Mediterranean stream: A long-term comparative study. Journal of the North American Benthological Society **27**:746–759.

Case, T. J. 1999. An illustrated guide to theoretical ecology. Oxford University Press, New York, NY.

Castellano, S., and P. Cermelli. 2006. Reconciling sexual selection to species recognition: A process-based model of mating decision. Journal of Theoretical Biology **242**:529–538.

Caswell, H. 1989. Analysis of life table response experiments. I. Decomposition of effects on population growth rate. Ecological Modeling **46**:221–237.

Caswell, H. 1996. Analysis of life table response experiments. II. Alternative parameterizations for size- and stage-structured models Ecological Modeling **88**:73–82.

Caswell, H. 2001. Matrix population models: Construction, analysis, and interpretation, 2nd ed. Sinauer Associates, Sunderland, MA.

Caswell, H. 2010. Life table response experiment analysis of the stochastic growth rate. Journal of Ecology **98**:324–333.

Caswell, H., and M. G. Neubert. 1998. Chaos and closure terms in plankton food chain models. Journal of Plankton Research **20**:1837–1845.

Charlesworth, B. 1994. Evolution in age-structured populations. Cambridge University Press, New York.

Charlesworth, B., R. Lande, and M. Slatkin. 1982. A neo-Darwinian commentary on macroevolution. Evolution **36**:474–498.

Chase, J. M., and M. A. Leibold. 2003. Ecological niches: Linking classical and contemporary approaches. University of Chicago Press, Chicago, IL.

Cheng, H., A. Sinha, F. W. Cruz, X. Wang, R. L. Edwards, F. M. d'Horta, C. C. Ribas, et al. 2013. Climate change patterns in Amazonia and biodiversity. Nature Communications **4**:1411–1416.

Chesson, P. 2000. Mechanisms of maintenance of species diversity. Annual Review of Ecology and Systematics 31:343–366.

Chesson, P., and N. Huntly. 1997. The roles of harsh and fluctuating conditions in the dynamics of ecological communities. American Naturalist 150:519–553.

Chesson, P., and J. J. Kuang. 2008. The interaction between predation and competition. Nature 456:235–238.

Chevin, L.-M., and R. Lande. 2009. When do adaptive plasticity and genetic evolution prevent extinction of a density-regulated population? Evolution 64:1143–1150.

Christiansen, F. B. 1975. Hard and soft selection in a subdivided population. American Naturalist 109:11–16.

Clements, F. E. 1916. Plant succession: Analysis of the development of vegetation. No. 242, Carnegie Institution of Washington Publication, Washington, DC.

Cohen, D., and S. A. Levin. 1991. Dispersal in patchy environments: The effects of temporal and spatial structure. Theoretical Population Biology 39:63–99.

Comins, H. N., W. D. Hamilton, and R. M. May. 1980. Evolutionarily stable dispersal strategies. Journal of Theoretical Biology 82:205–230.

Connell, J. H. 1961. The influence of interspecific competition and other factors on the distribution of the barnacle *Chthamalus stellatus*. Ecology 42:710–723.

Connell, J. H. 1978. Diversity in tropical rainforests and coral reefs. Science 199:1302–1310.

Connell, J. H. 1980. Diversity and the coevolution of competitors, or the ghost of competition past. Oikos 35:131–138.

Consortium, C.S.A. 2005. Initial sequence of the chimpanzee genome and comparison with the human genome. Nature 437:69–87.

Coope, G. R. 1979. Late Cenozoic fossil Coleoptera: Evolution, biogeography, and ecology. Annual Review of Ecology and Systematics 10:247–267.

Coope, G. R. 1994. The response of insect faunas to glacial-interglacial climatic fluctuations. Philosophical Transactions of the Royal Society of London B, Biological Sciences 344:19–26.

Cornell, H. V., and R. H. Karlson. 1996. Species richness of reef-building corals determined by local and regional processes. Journal of Animal Ecology 65:233–241.

Cornell, H. V., and J. H. Lawton. 1992. Species interactions, local and regional processes, and limits to the richness of ecological communities: A theoretical perspective. Journal of Animal Ecology 61:1–12.

Correa-Metrio, A., M. B. Bush, D. A. Hodell, M. Brenner, J. Escobar, and T. Guilderson. 2012. The influence of abrupt climate change on the ice-age vegetation of the Central American lowlands. Journal of Biogeography 39:497–509.

Correa-Metrio, A., J. A. Meave, S. Lozano-Garcia, and M. B. Bush. 2014. Environmental determinism and neutrality in vegetation at millennial time scales. Journal of Vegetation Science 25:627–635.

Cowling, R. M., P. W. Rundel, P. G. Desmet, and K. J. Esler. 1998. Extraordinary high regional- scale plant diversity in southern African arid lands: Subcontinental and global comparisons. Diversity and Distributions 4:27–36.

Coyne, J. A., and H. A. Orr. 2004. Speciation. Sinauer Associates, Sunderland, MA.

Cressman, R., and V. Křivan. 2006. Migration dynamics for the ideal free distribution. American Naturalist 168:384–397.

Cressman, R., and V. Křivan. 2010. The ideal free distribution as an evolutionarily stable state in density-dependent population games. Oikos 119:1231–1242.

Cressman, R., V. Křivan, and J. Garay. 2004. Ideal free distributions, evolutionary games, and population dynamics in multi-species environments. American Naturalist 164: 473–489.

Crombie, A. C. 1947. Interspecific competition. Journal of Animal Ecology 16:44–73.

Cronin, T. M., D. M. DeMartino, G. S. Dwyer, and J. Rodriguez-Lazaro. 1999. Deep-sea ostracod species diversity: Response to late Quaternary climate change. Marine Micropaleontology 37:231–249.

Cronin, T. M., and M. E. Raymo. 1997. Orbital forcing of deep-sea benthic diversity. Nature 385:624–627.

Crowder, L. B., and W. E. Cooper. 1982. Habitat structural complexity and the interaction of bluegills and their prey. Ecology 63:1802–1813.

Crowl, T. A., and A. P. Covich. 1990. Predator-induced life-history shifts in a freshwater snail: A chemical mediated and phenotypically plastic response. Science 247: 949–951.

Darwin, C. 1859. On the origin of species by means of natural selection, or the preservation of favoured races in the struggle for life. John Murray, London, UK.

Darwin, C. 1868. The variation of animals and plants under domestication, 2 vols. John Murray, London, UK.

Darwin, C. 1871. The descent of man, and selection in relation to sex, 2 vols. John Murray, London, UK.

Darwin, C., and A. R. Wallace. 1858. On the tendency of species to form varieties; And on the perpetuation of varieties and species by natural means of selection. Journal of the Proceedings of the Linnean Society of London. Zoology 3:45–62.

Davis, M. A. 2009. Invasion biology. Oxford University Press, New York, NY.

Davis, M. B. 1983. Quaternary history of deciduous forests of eastern North America and Europe. Annals of the Missouri Botanical Garden 70:550–563.

Dayton, P. K. 1971. Competition, disturbance, and community organization: The provision and subsequent utilization of space in a rocky intertidal community. Ecological Monographs 41:351–389.

de Mazancourt, C., and U. Dieckmann. 2004. Trade-off geometries and frequency-dependent selection. American Naturalist 164:765–778.

de Queiroz, K. 1998. The general lineage concept of species, species criteria, and the process of speciation. Pages 57–75 in D. J. Howard and S. H. Berlocher, eds. Endless forms: Species and speciation. Oxford University Press, New York, NY.

de Queiroz, K. 2007. Species concepts and species delimitation. Systematic Biology 56: 879–886.

Dean, A. M. 1983. A simple model of mutualism. American Naturalist 121:409–417.

DeAngelis, D. L., R. A. Goldstein, and R. V. O'Neill. 1975. A model for trophic interaction. Ecology 56:881–892.

Dehal, P., and J. L. Boore. 2005. Two rounds of whole genome duplication in the ancestral vertebrate. PLoSBiology 3:e14. doi:10.1371/journal.pbio.0030314.

Delcourt, H. R., and P. A. Delcourt. 1991. Quaternary ecology: A paleoecological perspective. Chapman and Hall, New York, NY.

Dempster, E. R. 1955. Maintenance of genetic heterogeneity. Cold Spring Harbor Symposium of Quantitative Biology 20:25–32.

Dieckmann, U., and M. Doebeli. 1999. On the origin of species by sympatric speciation. Nature 400:354–357.

Dieckmann, U., and R. Law. 1996. The dynamical theory of coevolution: A derivation from stochastic ecological processes. Journal of Mathematical Biology **34**:579–612.

Dieckmann, U., P. Marrow, and R. Law. 1995. Evolutionary cycling in predator-prey interactions: population dynamics and the Red Queen. Journal of Theoretical Biology **176**:91–102.

Diehl, S., and M. Feißel. 2000. Effects of enrichment on three-level food chains with omnivory. American Naturalist **155**:200–218.

Dobzhansky, T. 1937a. Genetics and the origin of species. Columbia University Press, New York, NY.

Dobzhansky, T. 1937b. What is a species? Scientia **61**:280–286.

Dobzhansky, T. 1940. Speciation as a stage in evolutionary divergence. American Naturalist **74**:312–321.

Dodson, S. I. 1970. Complementary feeding niches maintained by size-selective predation. Limnology and Oceanography **15**:131–137.

Doebeli, M. 2011. Adaptive diversification. Princeton University Press, Princeton, NJ.

Doebeli, M., and U. Dieckmann. 2000. Evolutionary branching and sympatric speciation caused by different types of ecological interactions. American Naturalist **156**:S77–S101.

Doebeli, M., and U. Dieckmann. 2003. Speciation along environmental gradients. Nature **421**:259–264.

Doebeli, M., and G. D. Ruxton. 1997. Evolution of dispersal rates in metapopulation models: Branching and cyclic dynamics in phenotype space. Evolution **51**:1730–1741.

Doncaster, C. P., J. Clobert, B. Doligez, L. Gustafsson, and E. Danchin. 1997. Balanced dispersal between spatially varying local populations: An alternative to the source-sink model. American Naturalist **150**:425–445.

Donoghue, P.C.J., J. N. Keating, and A. Smith. 2014. Early vertebrate evolution. Palaeontology **57**:879–893.

Duarte, J., C. Januário, and N. Martins. 2008. Chaos in ecology: The topological entropy of a tritrophic food chain model. Discrete Dynamics in Nature and Society **2008**:1–12.

Dufour, L. 1844. Anatomie générale des Dipteres. Annales des Sciences Naturelles Zoologie et Biologie Animale **1**:244–264.

Dunne, J. A., R. J. Williams, and N. D. Martinez. 2002. Food-web structure and network theory: The role of connectance and size. Proceedings of the National Academy of Sciences, USA **99**:12917–12922.

Dutrillaux, B. 1979. Chromosomal evolution in primates: Tentative phylogeny from *Microcebus murinus* (Prosimian) to man. Human Genetics **48**:251–314.

Dynesius, M., and R. Jansson. 2000. Evolutionary consequences of changes in species' geographic distributions driven by Milankovitch climate oscillations. Proceedings of the National Academy of Sciences, USA **97**:9115–9120.

Eberhard, W. G. 1985. Sexual selection and animal genitalia. Harvard University Press, Cambridge, MA.

Ehrlich, P. R., and P. H. Raven. 1964. Butterflies and plants: A study in coevolution. Evolution **18**:586–608.

Eldredge, N., and S. J. Gould. 1972. Punctuated equilibrium: An alternative to phyletic gradualism. Pages 82–115 *in* T.J.M. Schopf, ed. Models in paleontology. W. H. Freeman, San Francisco, CA.

Elton, C. S. 1927. Animal ecology. MacMillan, New York, NY.

Elton, C. S. 1958. The ecology of invasions by animals and plants. University of Chicago Press, Chicago, IL.

Endler, J. A. 1977. Geographic variation, speciation, and clines. Princeton University Press, Princeton, NJ.

Endler, J. A. 1986. Natural selection in the wild. Princeton University Press, Princeton, NJ.

Endler, J. A. 1992. Signals, signal conditions, and the direction of evolution. American Naturalist 139:S125–S153.

Endler, J. A., and A. L. Basolo. 1998. Sensory ecology, receiver biases and sexual selection. Trends in Ecology & Evolution 13:415–420.

Erwin, D. H. 2000. Macroevolution is more than repeated rounds of microevolution. Evolution & Development 2:78–84.

Eshel, I. 1981a. On the survival probability of a slightly advantageous mutant gene with a general distribution of progeny size—a branching process model. Journal of Mathematical Biology 12:355–362.

Eshel, I. 1981b. On the survival probability of a slightly advantageous mutant gene with a general distribution of progeny size—a branching process model. Journal of Mathematical Biology 12:355–362.

Faith, J. T., and A. K. Behrensmeyer. 2013. Climate change and faunal turnover: Testing the mechanics of the turnover-pulse hypothesis with South African fossil data. Paleobiology 39:609–627.

Falconer, D. S., and T.F.C. Mackay. 1996. Introduction to quantitative genetics, 4th ed. Longmans Green, Harrow, Essex, UK.

Fauth, J. E., J. Bernardo, M. Camara, W. J. Resetarits Jr., J. Van Buskirk, and S. A. McCollum. 1996. Simplifying the jargon of community ecology: A conceptual approach. American Naturalist 147:282–286.

Feder, J. L., S. H. Berlocher, J. B. Roethele, H. Dambroski, J. J. Smith, W. L. Perry, V. V. Gavrilovic, et al. 2003. Allopatric genetic origins for sympatric host plant shifts and race formation in *Rhagoletis*. Proceedings of the National Academy of Sciences, USA 100:10314–10319.

Feder, J. L., S. B. Opp, B. Wlazlo, K. Reynolds, and W. Go. 1994. Host fidelity is an effective premating barrier between sympatric races in the apple maggot fly. Proceedings of the National Academy of Sciences, USA 91:7990–7994.

Fisher, R. A. 1930. The genetical theory of natural selection. Clarendon, Oxford, UK.

Fishman, M. A., and L. Hadany. 2010. Plant–pollinator population dynamics. Theoretical Population Biology 78:270–277.

Foote, M. 2000. Origination and extinction components of taxonomic diversity: Paleozoic and post-Paleozoic dynamics. Paleobiology 26:578–605.

Foote, M. 2005. Pulsed origination and extinction in the marine realm. Paleobiology 31:6–20.

Foote, M., J. S. Crampton, A. G. Beu, B. A. Marshall, R. A. Cooper, P. A. Maxwell, and I. Matcham. 2007. Rise and fall of species occupancy in Cenozoic fossil mollusks. Science 318:1131–1134.

Fortuna, M. A., D. B. Stouffer, J. M. Olesen, P. Jordano, D. Mouillot, B. R. Krasnov, R. Poulin, and J. Bascompte. 2010. Nestedness versus modularity in ecological networks: Two sides of the same coin? Journal of Animal Ecology 79:811–817.

Fox, L. R. 1975. Cannibalism in natural populations. Annual Reviews of Ecology and Systematics 6:87–106.

Fretwell, S. D., and H. L. Lucas Jr. 1969. On territorial behavior and other factors influencing habitat distribution in birds. Acta Biotheoretica **19**:45–52.

Fuller, R. C., D. Houle, and J. Travis. 2005. Sensory bias as an explanation for the evolution of mate preferences. American Naturalist **166**:437–446.

Fussmann, G. F., S. P. Ellner, K. W. Shertzer, and N. G. Hairston Jr. 2000. Crossing the Hopf bifurcation in a live predator-prey system. Science **290**:1358–1360.

Fussmann, G. F., M. Loreau, and P. A. Abrams. 2007. Eco-evolutionary dynamics of communities and ecosystems. Functional Ecology **21**:465–477.

Futuyma, D. J., and G. C. Mayer. 1980. Non-allopatric speciation in animals. Systematic Zoology **29**:254–271.

Gadgil, M. 1971. Dispersal: Population consequences and evolution. Ecology **52**:253–261.

García-Ramos, G., and M. Kirkpatrick. 1997. Genetic models of adaptation and gene flow in peripheral populations. Evolution **51**:21–28.

Gatto, M. 1991. Some remarks on models of plankton densities in lakes. American Naturalist **137**:264–267.

Gause, G. F. 1934. The struggle for existence. Williams and Wilkins, Baltimore, MD.

Gause, G. F., O. K. Nastukova, and W. W. Alpatov. 1934. The influence of biologically conditioned media on the growth of a mixed population of *Paramecium caudatum* and *P. aurelia*. Journal of Animal Ecology **3**:222–230.

Gavrilets, S. 2000a. Rapid evolution of reproductive barriers by sexual conflict. Nature **403**:886– 889.

Gavrilets, S. 2000b. Waiting time to parapatric speciation. Proceedings of the Royal Society of London B, Biological Sciences **267**:2483–2492.

Gavrilets, S. 2004. Fitness landscapes and the origin of species. Princeton University Press, Princeton, NJ.

Gavrilets, S. 2006. The Maynard Smith model of sympatric speciation. Journal of Theoretical Biology **239**:172–182.

Gavrilets, S., and D. Waxman. 2002. Sympatric speciation by sexual conflict. Proceedings of the National Academy of Sciences, USA **99**:10533–10538.

Gerhardt, H. C. 2005. Acoustic spectral preferences in two cryptic species of grey treefrogs: Implications for mate choice and sensory mechanisms. Animal Behaviour **70**:39–48.

Geritz, S.A.H., E. Kisdi, G. Meszéna, and J.A.J. Metz. 1998. Evolutionary singular strategies and the adaptive growth and branching of the evolutionary tree. Evolutionary Ecology **12**:35–57.

Geritz, S.A.H., J.A.J. Metz, E. Kisdi, and G. Meszéna. 1997. Dynamics of adaptation and evolutionary branching. Physical Review Letters **78**:2024–2027.

Ghalambor, C. K., J. K. McKay, S. P. Carroll, and D. N. Reznick. 2007. Adaptive versus non- adaptive phenotypic plasticity and the potential for contemporary adaptation in new environments. Functional Ecology **21**:394–407.

Giliomee, J. H. 2003. Insect diversity in the Cape Floristic Region. African Journal of Ecology **41**:237–244.

Gilliam, J. F., and D. F. Fraser. 1987. Habitat selection under predation hazard: A test of a model with foraging minnows. Ecology **68**:1856–1862.

Gilpin, M. E. 1975. Group selection in predator-prey communities. Princeton University Press, Princeton, NJ.

Gíslason, D., M. M. Ferguson, S. Skúlason, and S. S. Snorrason. 1999. Rapid and coupled phenotypic and genetic divergence in Icelandic Arctic char (*Salvelinus alpinus*). Canadian Journal of Fisheries and Aquatic Sciences **56**:2229–2234.

Gleason, H. A. 1926. The individualistic concept of the plant association. Bulletin of the Torrey Botanical Club **53**:7–26.

Goh, B. S. 1979. Stability in models of mutualism. American Naturalist **113**:261–275.

Goldberg, D. E., R. Turkington, L. O. Whittaker, and A. R. Dyer. 2001. Density dependence in an annual plant community: Variation among life history stages. Ecological Monographs **71**:423–446.

Goldblatt, P. 1997. Floristic diversity in the Cape Flora of South Africa. Biodiversity and Conservation **6**:359–377.

Gomulkiewicz, R., and R. D. Holt. 1995. When does evolution by natural selection prevent extinction? Evolution **49**:201–207.

Gomulkiewicz, R., R. D. Holt, and M. Barfield. 1999. The effects of density dependence and immigration on local adaptation and niche evolution in a black-hole sink environment. Theoretical Population Biology **55**:283–296.

Goodall, D. W. 1963. The continuum and the individualistic association. Vegetatio **11**:297–316.

Gotthard, K., and S. Nylin. 1995. Adaptive plasticity and plasticity as an adaptation: A selective review of plasticity in animal morphology and life history. Oikos **74**:3–17.

Gould, S. J. 1989. Wonderful life: The Burgess shale and the nature of history. Norton, New York, NY.

Graham, R. W., E. L. Lundelius Jr., M. A. Graham, E. K. Schroeder, R. S. Toomey III, E. Anderson, A. D. Barnosky, et al. 1996. Spatial response of mammals to late Quaternary environmental fluctuations. Science **272**:1601–1606.

Grant, B. R., and P. R. Grant. 1982. Niche shifts and competition in Darwin's finches: *Geospiza conirostris* and congeners. Evolution **36**:637–657.

Grant, P. R., and B. R. Grant. 2002. Unpredictable evolution in a 30-year study of Darwin's finches. Science **296**:707–711.

Grant, P. R., and B. R. Grant. 2006. Evolution of character displacement in Darwin's finches. Science **313**:224–226.

Grant, V. 1971. Plant speciation. Columbia University Press, New York, NY.

Gross, B. L., and L. H. Rieseberg. 2005. The ecological genetics of homoploid hybrid speciation. Journal of Heredity **96**:241–252.

Grover, J. P. 1994. Assembly rules for communities of nutrient-limited plants and specialist herbivores. American Naturalist **143**:258–282.

Hadid, Y., S. Tzur, T. Pavlicek, R. Sumbera, J. Skliba, M. Lovy, O. Fragman-Sapir, et al. 2013. Possible incipient sympatric ecological speciation in blind mole rats (*Spalax*). Proceedings of the National Academy of Sciences, USA **110**:2587–2592.

Hairston, N. G. 1951. Interspecific competition and its probable influence upon the vertical distribution of Appalachian salamanders in the genus *Pelthodon*. Ecology **32**:266–274.

Hairston, N. G. 1980. The experimental test of an analysis of field distributions: Competition in terrestrial salamanders. Ecology **61**:817–826.

Haller, B. C., and A. P. Hendry. 2014. Solving the paradox of stasis: Squashed stabilizing selection and the limits of detection. Evolution **68**:483–500.

Hamilton, W. D., and R. M. May. 1977. Dispersal in stable habitats. Nature **269**:578–581.

Hamilton, W. D., and M. Zuk. 1982. Heritable true fitness and bright birds: A role for parasites? Science **218**:384–387.

Hanski, I. 1998. Metapopulation dynamics. Nature **396**:41–49.

Harmon, L. J., J. B. Losos, T. Jonathan Davies, R. G. Gillespie, J. L. Gittleman, W. Bryan Jennings, K. H. Kozak, et al. 2010. Early bursts of body size and shape evolution are rare in comparative data. Evolution **64**:2385–2396.

Harmon, L. J., B. Matthews, S. Des Roches, J. M. Chase, J. B. Shurin, and D. Schluter. 2009. Evolutionary diversification in stickleback affects ecosystem functioning. Nature **458**:1167–1170.

Harper, J. L. 1969. The role of predation in vegetational diversity. Brookhaven Symposium in Biology **22**:48–62.

Harrison, R. G. 1992. Hybrid zones and the evolutionary process. Oxford University Press, New York, NY.

Harrison, R. G., and D. M. Rand. 1989. Mosaic hybrid zones and the nature of species boundaries. Pages 111–133 *in* D. Otte and J. A. Endler, eds. Speciation and its consequences. Sinauer Associates, Sunderland, MA.

Hassell, M. 1978. The dynamics of arthropod predator-prey systems. Princeton University Press, Princeton, NJ.

Hastings, A. 1983. Can spatial variation alone lead to selection for dispersal? Theoretical Population Biology **24**:244–251.

Hastings, A., and T. Powell. 1991. Chaos in a three-species food chain. Ecology **72**: 896–903.

Hatfield, T., and D. Schluter. 1999. Ecological speciation in sticklebacks: Environment-dependent hybrid fitness. Evolution **53**:866–873.

Hausdorf, B. 2011. Progress toward a general species concept. Evolution **65**:923–931.

Hays, J. D., J. Imbrie, and N. J. Shackleton. 1976. Variations in the Earth's orbit: Pacemaker of the ice ages. Science **194**:1121–1132.

Heaton, T. H. 1990. Quaternary mammals of the Great Basin: Extinct giants, Pleistocene relicts, and recent immigrants. Pages 422–466 *in* R. M. Ross and W. D. Allmon, eds. Causes of evolution: A paleontological perspective. University of Chicago Press, Chicago, IL.

Hebert, P. D., E. H. Penton, J. M. Burns, D. H. Janzen, and W. Hallwachs. 2004. Ten species in one: DNA barcoding reveals cryptic species in the neotropical skipper butterfly *Astraptes fulgerator*. Proceedings of the National Academy of Sciences, USA **101**:14812–14817.

Hendry, A. P., and M. T. Kinnison. 1999. Perspective: The pace of modern life: Measuring rates of contemporary microevolution. Evolution **53**:1637–1653.

Hendry, A. P., P. Nosil, and L. H. Rieseberg. 2007. The speed of ecological speciation. Functional Ecology **21**:455–464.

Hewitt, G. 2000. The genetic legacy of the Quaternary ice ages. Nature **405**:907–913.

Hewitt, G. M. 1999. Post-glacial re-colonization of European biota. Biological Journal of the Linnean Society **68**:87–112.

Higashi, M., G. Takimoto, and N. Yamamura. 1999. Sympatric speciation by sexual selection. Nature **402**:523–526.

Higgie, M., and M. W. Blows. 2008. The evolution of reproductive character displacement conflicts with how sexual selection operates within a species. Evolution **62**: 1192–1203.

Higgie, M., S. Chenoweth, and M. W. Blows. 2000. Natural selection and the reinforcement of mate recognition. Science **290**:519–521.

Hirsch, M. W., S. Smale, and R. L. Devaney. 2012. Differential equations, dynamical systems, and an introduction to chaos, 3rd ed. Academic Press, New York, NY.

Hochberg, M. E., and R. D. Holt. 1995. Refuge evolution and the population dynamics of coupled host-parasitoid associations. Evolutionary Ecology **9**:633–661.

Hoekstra, H. E., J. M. Hoekstra, D. Berrigan, S. N. Vignieri, A. Hoang, C. E. Hill, P. Beerli, and J. G. Kingsolver. 2001. Strength and tempo of directional selection in the wild. Proceedings of the National Academy of Sciences USA **98**:9157–9160.

Holland, J. N., and D. L. DeAngelis. 2010. A consumer-resource approach to the density-dependent population dynamics of mutualism. Ecology **91**:1286–1295.

Holland, J. N., D. L. DeAngelis, and J. L. Bronstein. 2002. Population dynamics and mutualism: Functional responses of benefits and costs. American Naturalist **159**:231–244.

Holland, J. N., J. H. Ness, A. Boyle, and J. L. Bronstein. 2005. Mutualisms as consumer-resource interactions. Pages 17–33 *in* P. Barbosa, ed. Ecology of predator-prey interactions. Oxford University Press, New York, NY.

Holling, C. S. 1959. The components of predation as revealed by a study of small mammal predation of the European pine sawfly. Canadian Entomologist **91**:209–223.

Holt, R. D. 1977. Predation, apparent competition, and structure of prey communities. Theoretical Population Biology **12**:197–229.

Holt, R. D. 1984. Spatial heterogeneity, indirect interactions, and the coexistence of prey species. American Naturalist **124**:377–406.

Holt, R. D. 1985. Population dynamics in two-patch environments: Some anomalous consequences of an optimal habitat distribution. Theoretical Population Biology **28**: 181–208.

Holt, R. D. 1996. Adaptive evolution in source-sink environments: Direct and indirect effects of density-dependence on niche evolution. Oikos **75**:182–192.

Holt, R. D. 1997a. Community modules. Pages 333–349 *in* A. C. Gange and V. K. Brown, eds. Multitrophic interactions in terrestrial ecosystems. Blackwell Science, London, UK.

Holt, R. D. 1997b. On the evolutionary stability of sink populations. Evolutionary Ecology **11**:723–731.

Holt, R. D., and M. Barfield. 2001. On the relationship between the ideal free distribution and the evolution of dispersal. Pages 83–95 *in* A.D.J. Clobert, E. Canchin, and J. Nichols, eds. Dispersal. Oxford University Press, New York, NY.

Holt, R. D., and R. Gomulkiewicz. 1997. How does immigration influence local adaptation? A reexamination of a familiar paradigm. American Naturalist **149**:563–572.

Holt, R. D., J. Grover, and D. Tilman. 1994. Simple rules for interspecific dominance in systems with exploitative and apparent competition. American Naturalist **144**:741–771.

Holt, R. D., and M. A. McPeek. 1996. Chaotic population dynamics favors the evolution of dispersal. American Naturalist **148**:709–718.

Holt, R. D., and G. A. Polis. 1997. A theoretical framework for intraguild predation. American Naturalist **149**:745–764.

Holyoak, M., M. A. Leibold, and R. D. Holt, eds. 2005. Metacommunities: Spatial dynamics and ecological communities. University of Chicago Press, Chicago, IL.

Horth, L. 2007. Sensory genes and mate choice: Evidence that duplications, mutations, and adaptive evolution alter variation in mating cue genes and their receptors. Genomics **90**:159–175.

Hosken, D. J., and P. Stockley. 2004. Sexual selection and genital evolution. Trends in Ecology & Evolution **19**:87–93.

Hoskin, C. J., and M. Higgie. 2010. Speciation via species interactions: The divergence of mating traits within species. Ecology Letters **13**:409–420.

Hoskin, C. J., M. Higgie, K. R. McDonald, and C. Moritz. 2005. Reinforcement drives rapid allopatric speciation. Nature **437**:1353–1356.

Hou, Z., B. Sket, C. Fiser, and S. Li. 2011. Eocene habitat shift from saline to freshwater promoted Tethyan amphipod diversification. Proceedings of the National Academy of Sciences, USA **108**:14533–14538.

Howard, D. J. 1993. Reinforcement: Origin, dynamics, and fate of an evolutionary hypothesis. Pages 46–69 *in* R. G. Harrison, ed. Hybrid zones and the evolutionary process. Oxford University Press, Oxford, UK.

Hubbell, S. P. 2001. The unified neutral theory of biodiversity and biogeography. Princeton University Press, Princeton, NJ.

Huber, B. A. 2003. Rapid evolution and species-specificity of arthropod genitalia: Fact or artifact? Organisms, Diversity and Evolution **3**:63–71.

Human Microbiome Project. 2012. Structure, function and diversity of the healthy human microbiome. Nature **486**:207–214.

Huntley, B., and T. Webb III. 1989. Migration: Species' response to climatic variations caused by changes in the earth's orbit. Journal of Biogeography **16**:5–19.

Hutchinson, G. E. 1958. Concluding remarks. Cold Spring Harbor Symposium of Quantitative Biology **22**:415–427.

Hutchinson, G. E. 1959. Homage to Santa Rosalia or why are there so many animals? American Naturalist **93**:145–159.

Hutchinson, G. E. 1965. The ecological theater and the evolutionary play. Yale University Press, New Haven, CT.

Ijdo, J. W., A. Baldini, D. C. Ward, S. T. Reeders, and R. A. Wells. 1991. Origin of human chromosome 2: An ancestral telomere-telomere fusion. Proceedings of the National Academy of Sciences, USA **88**:9051–9055.

Ings, T. C., J. M. Montoya, J. Bascompte, N. Blüthgen, L. Brown, C. F. Dormann, F. Edwards, et al. 2009. Ecological networks—Beyond food webs. Journal of Animal Ecology **78**:253–269.

Iwasa, Y., F. Michor, and M. A. Nowak. 2004. Evolutionary dynamics of invasion and escape. Journal of Theoretical Biology **226**:205–214.

Iwasa, Y., A. Pomiankowski, and S. Nee. 1991. The evolution of costly mate preferences. II. The "handicap" principle. Evolution **45**:1431–1442.

Jablonski, D. 2008. Colloquium paper: Extinction and the spatial dynamics of biodiversity. Proceedings of the National Academy of Sciences, USA **105** Supplement 1:11528–11535.

Jackson, J.B.C., and K. G. Johnson. 2000. Life in the last few million years. Paleobiology **26**:221–235.

Jackson, S. T., and J. L. Blois. 2015. Community ecology in a changing environment: Perspectives from the Quaternary. Proceedings of the National Academy of Sciences, USA **112**:4915–4921.

Jackson, S. T., and J. T. Overpeck. 2000. Responses of plant populations and communities to environmental changes of the late Quaternary. Paleobiology **26**:194–220.

Jiao, Y., N. J. Wickett, S. Ayyampalayam, A. S. Chanderbali, L. Landherr, P. E. Ralph, L. P. Tomsho, et al. 2011. Ancestral polyploidy in seed plants and angiosperms. Nature **473**:97–100.

Johansson, B. G., and T. M. Jones. 2007. The role of chemical communication in mate choice. Biological Reviews **82**:265–289.

Johnson, C. A., and P. Amarasekare. 2013. Competition for benefits can promote the persistence of mutualistic interactions. Journal of Theoretical Biology **328**:54–64.

Johnson, D. M., and P. H. Crowley. 1980. Habitat and seasonal segregation among coexisting odonate larvae. Odonatologica **9**:297–308.

Johnson, N. K., and C. Cicero. 2004. New mitochondrial DNA data affirm the importance of Pleistocene speciation in North American birds. Evolution **58**:1122–1130.

Jones, E. I., J. L. Bronstein, and R. Ferriere. 2012. The fundamental role of competition in the ecology and evolution of mutualisms. Annals of the New York Academy of Science **1256**:66–88.

Jordano, P., J. Bascompte, and J. M. Olesen. 2003. Invariant properties in coevolutionary networks of plant-animal interactions. Ecology Letters **6**:69–81.

Kawecki, T. J. 1995. Demography of source-sink populations and the evolution of ecological niches. Evolutionary Ecology **9**:38–44.

Kawecki, T. J., and P. A. Abrams. 1999. Character displacement mediated by the accumulation of mutations affecting resource consumption abilities. Evolutionary Ecology Research **1**:173–188.

Kelly, J. K., and M.A.F. Noor. 1996. Speciation by reinforcement: A model derived from studies of *Drosophila*. Genetics **143**:1485–1497.

King, M. 1993. Species evolution: The role of chromosome change. Cambridge University Press, Cambridge, UK.

Kingsolver, J. G., and S. E. Diamond. 2011. Phenotypic selection in natural populations: What limits directional selection? American Naturalist **177**:346–357.

Kingsolver, J. G., S. E. Diamond, A. M. Siepielski, and S. M. Carlson. 2012. Synthetic analyses of phenotypic selection in natural populations: Lessons, limitations and future directions. Evolutionary Ecology **26**:1101–1118.

Kingsolver, J. G., H. E. Hoekstra, J. M. Hoekstra, D. Berrigan, S. N. Vignieri, C. E. Hill, A. Hoang, P. Gibert, and P. Beerli. 2001. The strength of selection in natural populations. American Naturalist **157**:245–261.

Kirkpatrick, M. 1982. Sexual selection and the evolution of female choice. Evolution **36**:1–12.

Kirkpatrick, M. 1982. Quantum evolution and punctuated equilibria in continuous genetic characters. American Naturalist **119**:833–848.

Kirkpatrick, M., and N. H. Barton. 1997. Evolution of a species' range. American Naturalist **150**:1–23.

Kirkpatrick, M., and V. Ravigné. 2002. Speciation by natural and sexual selection: Models and experiments. American Naturalist **159**:S22–S35.

Kokko, H., R. Brooks, M. D. Jennions, and J. Morley. 2003. The evolution of mate choice and mating biases. Proceedings of the Royal Society of London Series B, Biological Sciences **270**:653–664.

Kokko, H., and A. López-Sepulcre. 2007. The ecogenetic link between demography and evolution: Can we bridge the gap between theory and data? Ecology Letters **10**:773–782.

Kondrashov, A. S., and F. A. Kondrashov. 1999. Interactions among quantitative traits in the course of sympatric speciation. Nature **400**:351–354.

Kondrashov, A. S., and M. V. Mina. 1986. Sympatric speciation: When is it possible? Biological Journal of the Linnean Society **27**:201–223.

Kozak, G. M., G. Roland, C. Rankhorn, A. Falater, E. L. Berdan, and R. C. Fuller. 2015. Behavioral isolation due to cascade reinforcement in *Lucania* killifish. American Naturalist **185**:491–506.

Kramer, A. M., B. Dennis, A. M. Liebhold, and J. M. Drake. 2009. The evidence for Allee effects. Population Ecology **51**:341–354.

Krause, A. E., K. A. Frank, D. M. Mason, R. E. Ulanowicz, and W. F. Taylor. 2003. Compartments revealed in food-web structure. Nature **426**:282–285.

Křivan, V. 1998. Effects of optimal antipredator behavior of prey on predator-prey dynamics: The role of refuges. Theoretical Population Biology **53**:131–142.

Křivan, V. 2000. Optimal intraguild foraging and population stability. Theoretical Population Biology **58**:79–94.

Křivan, V. 2003. Competitive co-existence caused by adaptive predators. Evolutionary Ecology Research **5**:1163–1182.

Křivan, V. 2007. The Lotka-Volterra predator-prey model with foraging-predation risk trade-offs. American Naturalist **170**:771–782.

Křivan, V. 2013. Behavioral refuges and predator-prey coexistence. Journal of Theoretical Biology **339**:112–121.

Křivan, V. 2014. Competition in di- and tri-trophic food web modules. Journal of Theoretical Biology **343**:127–137.

Křivan, V., R. Cressman, and C. Schneider. 2008. The ideal free distribution: A review and synthesis of the game-theoretic perspective. Theoretical Population Biology **73**:403–425.

Křivan, V., and S. Diehl. 2005. Adaptive omnivory and species coexistence in tri-trophic food webs. Theoretical Population Biology **67**:85–99.

Křivan, V., and J. Eisner. 2003. Optimal foraging and predator–prey dynamics III. Theoretical Population Biology **63**:269–279.

Křivan, V., and J. Eisner. 2006. The effect of the Holling type II functional response on apparent competition. Theoretical Population Biology **70**:421–430.

Křivan, V., and O. J. Schmitz. 2004. Trait and density mediated indirect interactions in simple food webs. Oikos **107**:239–250.

Krukenberg, A. R. 1951. Intraspecific variability in the response of certain native plant species to serpentine soil. American Journal of Botany **38**:408–419.

Kuno, E. 1981. Dispersal and the persistence of populations in unstable habitats: A theoretical note. Oecologia **49**:123–126.

Lande, R. 1979. Quantitative genetic analysis of multivariate evolution, applied to brain:body size allometry. Evolution **33**:402–416.

Lande, R. 1980. Genetic variation and phenotypic evolution during allopatric speciation. American Naturalist **116**:463–479.

Lande, R. 1981. Models of speciation by sexual selection on polygenic traits. Proceedings of the National Academy of Sciences, USA **78**:3721–3725.

Lande, R. 1982. A quantitative genetic theory of life history evolution. Ecology **63**:607–615.

Lande, R. 2007. Expected relative fitness and the adaptive topography of fluctuating selection. Evolution **61**:1835–1846.

Lande, R. 2008. Adaptive topography of fluctuating selection in a Mendelian population. Journal of Evolutionary Biology **21**:1096–1105.

Lande, R., and S. J. Arnold. 1983. The measurement of selection on correlated characters. Evolution **37**:1210–1226.

Lande, R., S. Engen, and B.-E. Sæther. 2003. Stochastic population dynamics in ecology and conservation. Oxford University Press, New York, NY.

Lande, R., S. Engen, and B.-E. Sæther. 2009. An evolutionary maximum principle for density- dependent population dynamics in a fluctuating environment. Philosophical Transactions of the Royal Society of London B, Biological Sciences 364:1511–1518.

Langerhans, R. B., M. E. Gifford, and E. O. Joseph. 2007. Ecological speciation in *Gambusia* fishes. Evolution 61:2056–2074.

Lawler, L. R., and J. Maynard Smith. 1976. The coevolution and stability of competing species. American Naturalist 110:79–99.

Ledón-Rettig, C. C., D. W. Pfennig, A. J. Chunco, and I. Dworkin. 2014. Cryptic genetic variation in natural populations: A predictive framework. Integrative and Comparative Biology 54:783–793.

Lee, C. E. 1999. Rapid and repeated invasions of fresh water by the copepod *Eurytemora affinis*. Evolution 53:1423–1434.

Lee, C. E. 2000. Global phylogeography of a cryptic copepod species complex and reproductive isolation between genetically proximate "populations." Evolution 54:2014–2027.

Leibold, M. A. 1990. Resources and predators can affect the vertical distributions of zooplankton. Limnology and Oceanography 35:938–944.

Leibold, M. A. 1991. Trophic interactions and habitat segregation between competing *Daphnia* species. Oecologia 86:510–520.

Leibold, M. A. 1996. A graphical model of keystone predators in food webs: Trophic regulation of abundance, incidence and diversity patterns in communities. American Naturalist 147:784–812.

Leibold, M. A., and M. A. McPeek. 2006. Coexistence of the niche and neutral perspectives in community ecology. Ecology 87:1399–1410.

Leibold, M. A., and A. J. Tessier. 1991. Contrasting patterns of body size for *Daphnia* species that segregate by habitat. Oecologia 86:342–348.

Lemmon, E. M. 2009. Diversification of conspecific signals in sympatry: Geographic overlap drives multidimensional reproductive character displacement in frogs. Evolution 63:1155–1170.

Levene, H. 1953. Genetic equilibrium when more than one ecological niche is available. American Naturalist 87:331–333.

Levin, D. A. 1983. Polyploidy and novelty in flowering plants. American Naturalist 122:1–25.

Levin, S. A. 1970. Community equilibria and stability, and the extension of the competitive exclusion principle. American Naturalist 104:413–423.

Levin, S. A., D. Cohen, and A. Hastings. 1984. Dispersal strategies in patchy environments. Theoretical Population Biology 26:165–191.

Levine, J. M., and J. HilleRisLambers. 2009. The importance of niches for the maintenance of species diversity. Nature 461:254–257.

Lewontin, R. C., and D. Cohen. 1969. On population growth in a randomly varying environment. Proceedings of the National Academy of Science, USA 62:1056–1060.

Lima, S. L., and L. M. Dill. 1990. Behavioral decisions made under the risk of predation: A review and prospectus. Canadian Journal of Zoology 68:619–640.

Liow, L. H., and N. C. Stenseth. 2007. The rise and fall of species: Implications for macroevolutionary and macroecological studies. Proceedings of the Royal Society of London B, Biological Sciences 274:2745–2752.

Lisiecki, L. E., and M. E. Raymo. 2005. A Pliocene-Pleistocene stack of 57 globally distributed benthic $\delta^{18}O$ records. Paleoceanography 20, PA1003. doi:10.1029/2004PA001071.

Littlejohn, M. J., and J. J. Loftus-Hills. 1968. An experimental evaluation of premating isolation in the *Hyla ewingi* complex (Anura: Hylidae). Evolution **22**:659–663.

Livingstone, K., and L. Rieseberg. 2004. Chromosomal evolution and speciation: A recombination-based approach. New Phytologist **161**:107–112.

Lochmiller, R. L. 1996. Immunocompetence and animal population regulation. Oikos **76**: 594–602.

Lofstedt, C. 1993. Moth pheromone genetics and evolution. Philosophical Transactions of the Royal Society of London B, Biological Sciences **340**:167–177.

Lomolino, M. V., B. R. Riddle, R. J. Whittaker, and J. H. Brown. 2010. Biogeography. Sinauer Associates, Sunderland, MA.

Losos, J. B. 1990a. Ecomorphology, performance capability, and scaling of West Indian *Anolis* lizards; An evolutionary analysis. Ecological Monographs **60**:369–388.

Losos, J. B. 1990b. The evolution of form and function: Morphology and locomotor performance in West Indian *Anolis* lizards. Evolution **44**:1189–1203.

Losos, J. B., and R. E. Glor. 2003. Phylogenetic comparative methods and the geography of speciation. Trends in Ecology & Evolution **18**:220–227.

Losos, J. B., T. R. Jackman, A. Larson, K. de Queiroz, and L. Rodrígues-Schettino. 1998. Contingency and determinism in replicated adaptive radiations of island lizards. Science **279**:2115–2118.

Lotka, A. J. 1932a. Contribution to the mathematical theory of capture. I. Conditions for capture. Proceedings of the National Academy of Sciences, USA **18**:172–178.

Lotka, A. J. 1932b. The growth of mixed populations: Two species competing for a common food supply. Journal of the Washington Academy of Sciences **22**:461–469.

Loverdo, C., and J. O. Lloyd-Smith. 2013. Evolutionary invasion and escape in the presence of deleterious mutations. PLoSOne **8**:e68179. doi:10.1371/journal.pone.0068179.

Lovette, I. J. 2005. Glacial cycles and the tempo of avian speciation. Trends in Ecology & Evolution **20**:57–59.

Lowe, W. H., and M. A. McPeek. 2014. Is dispersal neutral? Trends in Ecology and Evolution **29**:444–450.

Lu, G., and L. Bernatchez. 1999. Correlated trophic specialization and genetic divergence in sympatric lake whitefish ecotypes (*Coregonus clupeaformis*): Support for the ecological speciation hypothesis. Evolution **53**:1491–1505.

Lynch, M., and B. Walsh. 1998. Genetics and analysis of quantitative traits. Sinauer Associates, Sunderland, MA.

M'Gonigle, L. K., R. Mazzucco, S. P. Otto, and U. Dieckmann. 2012. Sexual selection enables long-term coexistence despite ecological equivalence. Nature **484**:506–509.

Mable, B. K. 2004. 'Why polyploidy is rarer in animals than in plants': Myths and mechanisms. Biological Journal of the Linnean Society **82**:453–466.

MacArthur, R. H. 1965. Patterns of species diversity. Cambridge Philosophical Society Biological Review **40**:510–533.

MacArthur, R. H. 1970. Species packing and competitive equilibrium for many species. Theoretical Population Biology **1**:1–11.

MacArthur, R. H. 1972. Geographical ecology. Princeton University Press, Princeton, NJ.

MacArthur, R. H., and R. Levins. 1967. The limiting similarity, convergence, and divergence of coexisting species. American Naturalist **101**:377–385.

Maguire, K. C., D. Nieto-Lugilde, M. C. Fitzpatrick, J. W. Williams, and J. L. Blois. 2015. Modeling species and community responses to past, present, and future episodes of climatic and ecological change. Annual Review of Ecology, Evolution, and Systematics **46**:343–368.

Mallet, J. 2005. Hybridization as an invasion of the genome. Trends in Ecology & Evolution **20**:229–237.

Mallet, J. 2007. Hybrid speciation. Nature **446**:279–283.

Mallet, J. 2008. Hybridization, ecological races and the nature of species: Empirical evidence for the ease of speciation. Philosophical Transactions of the Royal Society of London B, Biological Sciences **363**:2971–2986.

Manceau, M., V. S. Domingues, C. R. Linnen, E. B. Rosenblum, and H. E. Hoekstra. 2010. Convergence in pigmentation at multiple levels: Mutations, genes and function. Philosophical Transactions of the Royal Society of London B, Biological Sciences **365**:2439–2450.

Marra, P. P., K. Lampe, and B. Tedford. 1995. An analysis of daily corticosterone profiles in two species of *Zonotrichia* under captive and natural conditions. Wilson Bulletin **107**:296–304.

Martin, O. Y., and D. J. Hosken. 2003. The evolution of reproductive isolation through sexual conflict. Nature **423**:979–992.

Martinez, N. D. 1991. Artifacts or attributes? Effects of resolution on the Little Rock Lake food web. Ecological Monographs **61**:367–392.

Massol, F., and F. Debarre. 2015. Evolution of dispersal in spatially and temporally variable environments: The importance of life cycles. Evolution **69**:1925–1937.

May, R. M. 1973. Stability and complexity in model ecosystems. Princeton University Press, Princeton, NJ.

May, R. M. 1981. Models for two interacting populations. Pages 78–104 *in* R. M. May, ed. Theoretical ecology. Sinauer Associates, Sunderland, MA.

Mayden, R. L. 1997. A hierarchy of species concepts: The denouement in the saga of the species problem. Pages 381–424 *in* M. F. Claridge, H. A. Dawah, and M. R. Wilson, eds. Species: The units of biodiversity. Chapman and Hall, New York, NY.

Maynard Smith, J. 1966. Sympatric speciation. American Naturalist **100**:637–650.

Maynard Smith, J. 1983. Evolution and the theory of games. Cambridge University Press, Cambridge, UK.

Mayr, E. 1940. Speciation phenomena in birds. American Naturalist **74**:249–278.

Mayr, E. 1942. Systematics and the origin of species. Columbia University Press, New York, NY.

Mayr, E. 1963. Animal species and evolution. Belknap of Harvard University Press, Cambridge, MA.

McCann, K., and P. Yodzis. 1994. Biological conditions for chaos in a three-species food chain. Ecology **75**:561–564.

McCann, K. S. 2011. Food webs. Princeton University Press, Princeton, NJ.

McKinnon, J. S., S. Mori, B. K. Blackman, L. David, D. M. Kingsley, L. Jamieson, J. Chou, and D. Schluter. 2004. Evidence for ecology's role in speciation. Nature **429**:294–298.

McNamara, J. M., and S. R. Dall. 2011. The evolution of unconditional strategies via the 'multiplier effect.' Ecology Letters **14**:237–243.

McPeek, M. A. 1989. Differential dispersal tendencies among *Enallagma* damselflies (Odonata) inhabiting different habitats. Oikos **56**:187–195.

McPeek, M. A. 1990a. Behavioral differences between *Enallagma* species (Odonata) influencing differential vulnerability to predators. Ecology **71**:1714–1726.

McPeek, M. A. 1990b. Determination of species composition in the *Enallagma* damselfly assemblages of permanent lakes. Ecology **71**:83–98.

McPeek, M. A. 1995. Morphological evolution mediated by behavior in the damselflies of two communities. Evolution **49**:749–769.

McPeek, M. A. 1996a. Linking local species interactions to rates of speciation in communities. Ecology **77**:1355–1366.

McPeek, M. A. 1996b. Trade-offs, food web structure, and the coexistence of habitat specialists and generalists. American Naturalist **148**:S124–S138.

McPeek, M. A. 1998. The consequences of changing the top predator in a food web: A comparative experimental approach. Ecological Monographs **68**:1–23.

McPeek, M. A. 1999. Biochemical evolution associated with antipredator adaptation in damselflies. Evolution **53**:1835–1845.

McPeek, M. A. 2000. Predisposed to adapt? Clade-level differences in characters affecting swimming performance in damselflies. Evolution **54**:2072–2080.

McPeek, M. A. 2004. The growth/predation risk trade-off: So what is the mechanism? American Naturalist **163**:E88–E111.

McPeek, M. A. 2007. The macroevolutionary consequences of ecological differences among species. Palaeontology **50**:111–129.

McPeek, M. A. 2008. The ecological dynamics of clade diversification and community assembly. American Naturalist **172**:E270–284.

McPeek, M. A. 2012. Intraspecific density dependence and a guild of consumers coexisting on one resource. Ecology **93**:2728–2735.

McPeek, M. A. 2014a. Keystone and intraguild predation, intraspecific density dependence, and a guild of coexisting consumers. American Naturalist **183**:E1–E16.

McPeek, M. A. 2014b. Limiting factors, competitive exclusion, and a more expansive view of species coexistence. American Naturalist **183**:iii–iv.

McPeek, M. A. 2017. The ecological dynamics of natural selection: Traits and the coevolution of community structure. American Naturalist, in press.

McPeek, M. A., and J. M. Brown. 2000. Building a regional species pool: Diversification of the *Enallagma* damselflies in eastern North America. Ecology **81**:904–920.

McPeek, M. A., B. L. Cook, and W. C. McComb. 1983. Habitat selection by small mammals in an urban woodlot. Transactions of the Kentucky Academy of Science **44**:68–73.

McPeek, M. A., and S. Gavrilets. 2006. The evolution of female mating preferences: Differentiation from species with promiscuous males can promote speciation. Evolution **60**:1967–1980.

McPeek, M. A., and R. Gomulkiewicz. 2005. Assembling and depleting species richness in metacommunities: Insights from ecology, population genetics and macroevolution. Pages 355–373 *in* M. A. Leibold, M. Holyoak, and R. D. Holt, eds. Metacommunities: Spatial dynamics and ecological communities. University of Chicago Press, Chicago, IL.

McPeek, M. A., M. Grace, and J.M.L. Richardson. 2001a. Physiological and behavioral responses to predators shape the growth/predation risk trade-off in damselflies. Ecology **82**:1535–1545.

McPeek, M. A., and R. D. Holt. 1992. The evolution of dispersal in spatially and temporally varying environments. American Naturalist **140**:1010–1027.

McPeek, M. A., and B. L. Peckarsky. 1998. Life histories and the strengths of species interactions: Combining mortality, growth, and fecundity effects. Ecology **79**:867–879.

McPeek, M. A., N. L. Rodenhouse, R. T. Holmes, and T. W. Sherry. 2001b. A general model of site-dependent population regulation: Population-level regulation without individual-level interactions. Oikos **94**:417–424.

McPeek, M. A., A. K. Schrot, and J. M. Brown. 1996. Adaptation to predators in a new community: Swimming performance and predator avoidance in damselflies. Ecology **77**:617–629.

McPeek, M. A., L. Shen, and H. Farid. 2009. The correlated evolution of three-dimensional reproductive structures between male and female damselflies. Evolution **63**:73–83.

McPeek, M. A., L. Shen, J. Z. Torrey, and H. Farid. 2008. The tempo and mode of three-dimensional morphological evolution in male reproductive structures. American Naturalist **171**:E158–E178.

McPeek, M. A., L. B. Symes, D. M. Zong, and C. L. McPeek. 2011. Species recognition and patterns of population variation in the reproductive structures of a damselfly genus. Evolution **65**:419–428.

Meadows, M. E., and J. M. Sugden. 1993. The late Quaternary palaeoecology of a floristic kingdom: The southwestern Cape South Africa. Palaeogeography, Palaeoclimatology, Palaeoecology **101**:271–281.

Mendelson, T. C., and K. L. Shaw. 2012. The (mis)concept of species recognition. Trends in Ecology & Evolution **27**:421–427.

Menge, B. A. 1976. Organization of the New England rocky intertidal community: Role of predation, competition, and environmental heterogeneity. Ecological Monographs **46**:355–393.

Menge, B. A. 1995. Indirect effects in marine rocky intertidal webs: Patterns and importance. Ecological Monographs **65**:21–74.

Merilä, J., B. C. Sheldon, and L. E. Kruuk. 2001. Explaining stasis: Microevolutionary studies of natural populations. Genetica **112–113**:199–222.

Meszéna, G., I. Czibula, and S. Geritz. 1997. Adaptive dynamics in a 2-patch environment: A toy model for allopatric and parapatric speciation. Journal of Biological Systems **5**:265–284.

Metz, J.A.J., T. J. de Jong, and P.G.L. Klinkhamer. 1983. What are the advantages of dispersing; A paper by Kuno explained and extended. Oecologia **57**:166–169.

Metz, J.A.J., and M. Gyllenberg. 2001. How should we define fitness in structured metapopulation models? Including an application to the calculation of evolutionary stable dispersal strategies. Proceedings of the Royal Society B, Biological Sciences **268**:499–508.

Michaelis, L., and M. L. Menten. 1913. Die Kinetik der Invertinwirkung. Biochemische Zeitschrift **49**:333–369.

Milligan, B. G. 1985. Evolutionary divergence and character displacement in two phenotypically-variable, competing species. Evolution **39**:1207–1222.

Milligan, B. G. 1986. Punctuated evolution induced by ecological change. American Naturalist **127**:522–532.

Mittelbach, G. G. 1981. Foraging efficiency and body size: A study of optimal diet and habitat use by bluegills. Ecology **62**:1370–1386.

Mittelbach, G. G. 1984. Predation and resource partitioning in two sunfishes (Centrarchidae). Ecology **65**:499–513.

Moody, A. L., A. I. Houston, and J. M. McNamara. 1996. Ideal free distributions under predation risk. Behavioral Ecology and Sociobiology **38**:131–143.

Morris, W. F., J. L. Bronstein, and W. G. Wilson. 2003. Three-way coexistence in obligate mutualist-exploiter interactions: The potential role of competition. American Naturalist **161**:860–875.

Motro, U. 1982. Optimal rates of dispersal I. Haploid populations. Theoretical Population Biology **21**:394–411.

Mouquet, N., and M. Loreau. 2003. Community patterns in source-sink metacommunities. American Naturalist **162**:544–557.

Mousseau, T. A., and D. A. Roff. 1987. Natural selection and the heritability of fitness components. Heredity **59**:1841–1197.

Muller, H. J. 1925. Why polyploidy is rarer in animals than in plants. American Naturalist **59**:346–353.

Muller, H. J. 1942. Isolating mechanisms, evolution and temperature. Biological Symposia **6**:71–125.

Murdoch, W. W., C. J. Briggs, and R. M. Nisbet. 2003. Consumer-resource dynamics. Princeton University Press, Princeton, NJ.

Murdoch, W. W., C. J. Briggs, R. M. Nisbet, W.S.C. Gurney, and A. Stewart-Oaten. 1992. Aggregation and stability in metapopulation models. American Naturalist **140**:41–58.

Murdoch, W. W., and A. Oaten. 1975. Predation and population stability. Advances in Ecological Research **9**:2–132.

Murdoch, W. W., and A. Stewart-Oaten. 1989. Aggregation by parasitoids and predators: Effects on equilibrium and stability. American Naturalist **134**:288–310.

Nagylaki, T. 1992. Introduction to theoretical population genetics. Springer, Berlin, Germany.

Navarrete, S. A., B. A. Menge, and B. A. Daley. 2000. Species interactions in intertidal food webs: Prey or predator regulation of intermediate predators? Ecology **81**:2264–2277.

Navarro, A., and N. H. Barton. 2003. Chromosomal speciation and molecular divergence-accelerated evolution in rearranged chromosomes. Science **300**:321–324.

Neubert, M. G., T. Klanjscek, and H. Caswell. 2004. Reactivity and transient dynamics of predator–prey and food web models. Ecological Modelling **179**:29–38.

Neubert, M. G., P. Klepac, and P. van den Driessche. 2002. Stabilizing dispersal delays in predator-prey metapopulation models. Theoretical Population Biology **61**:339–347.

Niering, W. A., and C. H. Lowe. 1984. Vegetation of the Santa Catalina Mountains: Community types and dynamics. Vegetatio **58**:3–28.

Noor, M.A.F. 1999. Reinforcement and other consequences of sympatry. Heredity **83**: 503–508.

Nosil, P. 2007. Divergent host plant adaptation and reproductive isolation between ecotypes of *Timema cristinae* walking sticks. American Naturalist **169**:151–162.

Nosil, P. 2012. Ecological speciation. Oxford University Press, New York, NY.

Nosil, P., and B. J. Crespi. 2006. Experimental evidence that predation promotes divergence in adaptive radiation. Proceedings of the National Academy of Sciences, USA **103**:9090–9095.

Nosil, P., B. J. Crespi, and C. P. Sandoval. 2002. Host-plant adaptation drives the parallel evolution of reproductive isolation. Nature **417**:440–443.

Nosil, P., S. P. Egan, and D. J. Funk. 2008. Heterogeneous genomic differentiation between walking-stick ecotypes: "Isolation by adaptation" and multiple roles for divergent selection. Evolution **62**:316–336.

Nosil, P., T. H. Vines, and D. J. Funk. 2005. Perspective: Reproductive isolation caused by natural selection against immigrants from divergent habitats. Evolution 59:705.

O'Dea, A., and J.B.C. Jackson. 2009. Environmental change drove macroevolution in cupuladriid bryozoans. Proceedings of the Royal Society of London B, Biological Sciences 276:3629–3634.

O'Dea, A., J.B.C. Jackson, H. Fortunato, J. T. Smith, L. D'Croz, K. G. Johnson, and J. A. Todd. 2007. Environmental change preceded Caribbean extinction by 2 million years. Proceedings of the National Academy of Sciences, USA 104:5501–5506.

O'Donald, P. 1980. Genetic models of sexual selection. Cambridge University Press, Cambridge, UK.

Ohgushi, T., O. Schmitz, and R. D. Holt. 2013. Trait-mediated indirect interactions: Ecological and evolutionary perspectives. Cambridge University Press, New York, NY.

Orr, H. A. 1990. "Why polyploidy is rarer in animals than in plants" revisited. American Naturalist 136:759–770.

Orr, H. A., and D. C. Presgraves. 2000. Speciation by postzygotic isolation: Forces, genes and molecules. Bioessays 22:1085–1094.

Orr, H. A., and M. Turelli. 2001. The evolution of post-zygotic isolation: Accumulating Dobzhansky-Muller incompatibilities. Evolution 55:1085–1094.

Orr, H. A., and R. L. Unckless. 2008. Population extinction and the genetics of adaptation. American Naturalist 172:160–169.

Ortiz-Barrientos, D., A. Grealy, and P. Nosil. 2009. The genetics and ecology of reinforcement: Implications for the evolution of prezygotic isolation in sympatry and beyond. Annals of the New York Academy of Sciences 1168:156–182.

Osenberg, C. W., and G. G. Mittelbach. 1989. Effects of body size on the predator-prey interaction between pumpkinseed sunfish and gastropods. Ecological Monographs 59:405–432.

Østbye, K., P. A. Amundsen, L. Bernatchez, A. Klemetsen, R. Knudsen, R. Kristoffersen, T. F. Naesje, and K. Hindar. 2006. Parallel evolution of ecomorphological traits in the European whitefish Coregonus lavaretus (L.) species complex during postglacial times. Molecular Ecology 15:3983–4001.

Ostfeld, R. S., F. Keesing, and V. T. Eviner. 2008. Infectious disease ecology: Effects of ecosystems on disease and of disease on ecosystems. Princeton University Press, Princeton, NJ.

Otte, D. 1989. Speciation in Hawaiian crickets. Pages 482–526 in D. Otte and J. A. Endler, eds. Speciation and its consequences. Sinauer Associates, Sunderland, MA.

Otto, S. P. 2007. The evolutionary consequences of polyploidy. Cell 131:452–462.

Otto, S. P., and J. Whitton. 2000. Polyploid incidence and evolution. Annual Review of Genetics 34:401–437.

Overpeck, J. T., R. S. Webb, and T. Webb III. 1992. Mapping eastern North American vegetation change over the past 18 ka: No-analogs and the future. Geology 20:1071–1074.

Owens, I.P.F., P. M. Bennett, and P. H. Harvey. 1999. Species richness among birds: Body size, life history, sexual selection or ecology? Proceedings of the Royal Society of London B, Biological Sciences 255:37–45.

Ozgul, A., S. Tuljapurkar, T. G. Benton, J. M. Pemberton, T. H. Clutton-Brock, and T. Coulson. 2009. The dynamics of phenotypic change and the shrinking sheep of St. Kilda. Science 325:464–467.

Pacala, S. W., and J. Roughgarden. 1982. The evolution of resource partitioning in a multi-dimensional resource space. Theoretical Population Biology **22**:127–145.

Padron, V., and M. C. Trevisan. 2006. Environmentally induced dispersal under heterogeneous logistic growth. Mathematical Biosciences **199**:160–174.

Paine, R. T. 1966. Food web complexity and species diversity. American Naturalist **100**:65–75.

Paine, R. T. 1974. Intertidal community structure: Experimental studies on the relationship between a dominant competitor and its principal prey. Oecologia **15**:93–120.

Paine, R. T. 1980. Food webs: Linkage, interaction strength and community infrastructure. Journal of Animal Ecology **49**:666–685.

Paine, R. T. 1992. Food-web analysis through field measurement of per capita interaction strength. Nature **355**:73–75.

Paine, R. T., and S. A. Levin. 1981. Intertidal landscapes: Disturbance and the dynamics of pattern. Ecological Monographs **51**:145–178.

Palike, H., R. D. Norris, J. O. Herrle, P. A. Wilson, H. K. Coxall, C. H. Lear, N. J. Shackleton, et al. 2006. The heartbeat of the Oligocene climate system. Science **314**:1894–1898.

Panhuis, T. M., R. Butlin, M. Zuk, and T. Tregenza. 2001. Sexual selection and speciation. Trends in Ecology & Evolution **16**:364–371.

Parisod, C., R. Holderegger, and C. Brochmann. 2010. Evolutionary consequences of auto-polyploidy. New Phytologist **186**:5–17.

Park, T. 1948. Experimental studies of interspecific competition. I. Competition between populations of the flour beetles *Tribolium confusum* Duval and *Tribolium castaneum* Herbst. Ecological Monographs **18**:267–307.

Park, T., E. V. Gregg, and C. Z. Lutherman. 1941. Studies in population physiology: X, interspecific competition in populations of granary beetles. Physiological Zoology **14**:395–430.

Parker, G. A., and W. J. Sutherland. 1986. Ideal free distributions when individuals differ in competitive ability—Phenotype-limited ideal free models. Animal Behaviour **34**:1222–1242.

Parvinen, K. 2005. Evolutionary suicide. Acta Biotheoretica **53:241–264**.

Paterson, H.E.H. 1985. The recognition concept of species. Pages 21–29 *in* E. S. Vrba, ed. Species and speciation. Transvaal Museum, Pretoria, South Africa.

Paulson, D. R. 1974. Reproductive isolation in damselflies. Systematic Zoology **23**:40–49.

Payne, R. T. 1966. Food web complexity and species diversity. American Naturalist **100**:65–75.

Pease, C. M., R. Lande, and J. J. Bull. 1989. A model of population growth, dispersal and evolution in a changing environment. Ecology **70**:1657–1664.

Petit, J. R., J. Jouzel, D. Raynaud, N. I. Barkov, J.-M. Barnola, I. Basile, M. Bender, et al. 1999. Climate and atmospheric history of the past 420,000 years from the Vostok ice core, Antarctica. Nature **399**:429–436.

Pfennig, D. W., M. A. Wund, E. C. Snell-Rood, T. Cruickshank, C. D. Schlichting, and A. P. Moczek. 2010. Phenotypic plasticity's impacts on diversification and speciation. Trends in Ecology & Evolution **25**:459–467.

Pfennig, K. S. 1998. The evolution of mate choice and the potential for conflict between species and mate-quality recognition. Proceedings of the Royal Society of London B, Biological Sciences **265**:1743–1748.

Pfennig, K. S., and A. M. Rice. 2014. Reinforcement generates reproductive isolation between neighbouring conspecific populations of spadefoot toads. Proceedings of the Royal Society of London B, Biological Sciences **281**:20140949. doi:10.1098/rspb.2014.0949.

Pfennig, K. S., and M. J. Ryan. 2006. Reproductive character displacement generates reproductive isolation among conspecific populations: An artificial neural network study. Proceedings of the Royal Society of London B, Biological Sciences **273**:1361–1368.

Pfenninger, M., and K. Schwenk. 2007. Cryptic animal species are homogeneously distributed among taxa and biogeographical regions. BMC Evolutionary Biology **7**:121. doi:10.1186/1471–2148–7–121.

Phelps, S. M., S. A. Rand, and M. J. Ryan. 2006. A cognitive framework for mate choice and species recognition. American Naturalist **167**:28–42.

Phillips, J. 1935. Succession, development, the climax, and the complex organism: An analysis of concepts: Part III. The complex organism: Conclusions. Journal of Ecology **23**:488–508.

Pickett, S.T.A., and P. S. White. 1985. The ecology of natural disturbance and patch dynamics. Academic Press, New York, NY.

Pimm, S. L. 1979. Sympatric speciation: A simulation model. Biological Journal of the Linnean Society **11**:131–139.

Pimm, S. L. 1982. Food webs. Chapman and Hall, London, UK.

Pimm, S. L., and J. H. Lawton. 1980. Are food webs divided into compartments? Journal of Animal Ecology **49**:879–898.

Polis, G. A. 1981. The evolution and dynamics of intraspecific predation. Annual Review of Ecology and Systematics **12**:225–251.

Polis, G. A., and D. R. Strong. 1996. Food web complexity and community dynamics. American Naturalist **147**:813–846.

Ponel, P., J. Orgeas, M. J. Samways, V. Andrieu-Ponel, J.-L. de Beaulieu, M. Reille, P. Roche, and T. Tatoni. 2003. 110,000 years of Quaternary beetle diversity change. Biodiversity and Conservation **12**:2077–2089.

Porretta, D., and S. Urbanelli. 2012. Evolution of premating reproductive isolation among conspecific populations of the sea rock-pool beetle *Ochthebius urbanelliae* driven by reinforcing natural selection. Evolution **66**:1284–1295.

Price, T. 1998. Sexual selection and natural selection in bird speciation. Proceedings of the Royal Society of London B, Biological Sciences **353**:251–260.

Price, T. 2008. Speciation in birds. Roberts, Greenwood Village, CO.

Price, T., I. J. Lovette, E. Bermingham, H. L. Gibbs, and A. D. Richman. 2000. The imprint of history on communities of North American and Asian warblers. American Naturalist **156**:354–367.

Price, T. D., and M. Kirkpatrick. 2009. Evolutionarily stable range limits set by interspecific competition. Proceedings of the Royal Society B **276**:1429–1434.

Prothero, D. R. 2014. Species longevity in North American fossil mammals. Integrative Zoology **9**:383–393.

Pulliam, H. R. 1988. Sources, sinks, and population regulation. American Naturalist **132**:652–661.

Pulliam, H. R., and B. J. Danielson. 1991. Sources, sinks, and habitat selection—a landscape perspective on population dynamics. American Naturalist **137**:S50–S66.

Punyasena, S. W., F. E. Mayle, and J. C. McElwain. 2008. Quantitative estimates of glacial and Holocene temperature and precipitation change in lowland Amazonian Bolivia. Geology **36**:667–670.

Ramsey, J., and T. S. Ramsey. 2014. Ecological studies of polyploidy in the 100 years following its discovery. Philosophical Transactions of the Royal Society of London B, Biological Sciences **369**:20130352. doi:10.1098/rstb.2013.0352.

Ramsey, J., and D. W. Schemske. 1998. Pathways, mechanisms, and rates of polyploid formation in flowering plants. Annual Review of Ecology and Systematics **29**:467–501.

Ramsey, J., and D. W. Schemske. 2002. Neopolyploidy in flowering plants. Annual Review of Ecology and Systematics **33**:589–639.

Raup, D. M. 1992. Extinction: Bad genes or bad luck? W. W. Norton, New York, NY.

Rausher, M. D. 1992. The measurement of selection on quantitative traits: Biases due to the environmental covariances between traits and fitness. Evolution **46**:616–625.

Revilla, T. 2002. Effects of intraguild predation on resource competition. Journal of Theoretical Biology **214**:49–62.

Reynolds, S. A., and C. E. Brassil. 2013. When can a single-species, density-dependent model capture the dynamics of a consumer-resource system? Journal of Theoretical Biology **339**:70–83.

Reznick, D. 1982. The impact of predation on life history evolution in Trinidadian guppies: Genetic basis of observed life history patterns. Evolution **36**:1236–1250.

Reznick, D., and J. A. Endler. 1982. The impact of predation on life history evolution in Trinidadian guppies (*Poecilia reticulata*). Evolution **36**:160–177.

Rice, A. M., and D. W. Pfennig. 2010. Does character displacement initiate speciation? Evidence of reduced gene flow between populations experiencing divergent selection. Journal of Evolutionary Biology **23**:854–865.

Richards-Zawacki, C. L., and M. E. Cummings. 2011. Intraspecific reproductive character displacement in a polymorphic poison dart frog, *Dendrobates pumilio*. Evolution **65**:259–267.

Richardson, D. M., and P. M. Pysek. 2006. Plant invasions: Merging the concepts of species invasiveness and community invasibility. Progress in Physical Geography **30**:409–431.

Richardson, J.M.L. 2001. The relative roles of adaptation and phylogeny in determination of larval traits in diversifying anuran lineages. American Naturalist **157**:282–299.

Ricklefs, R. E. 1987. Community diversity: Relative roles of local and regional processes. Science **235**:167–171.

Ricklefs, R. E. 1989. Speciation and diversity: The integration of local and regional processes. Pages 599–622 *in* D. Otte and J. A. Endler, eds. Speciation and its consequences. Sinauer and Associates, Sunderland, MA.

Ricklefs, R. E. 2008. Disintegration of the ecological community. American Naturalist **172**:741–750.

Rieseberg, L. H. 1991. Homoploid reticulate evolution in *Helianthus* (Asteraceae): Evidence from ribosomal genes. American Journal of Botany **78**:1218–1237.

Rieseberg, L. H. 1997. Hybrid origin of plant species. Annual Review of Ecology and Systematics **28**:359–389.

Rieseberg, L. H. 2001. Chromosomal rearrangements and speciation. Trends in Ecology & Evolution **16**:351–358.

Rieseberg, L. H., and N. C. Ellstrand. 1993. What can molecular and morphological markers tell us about plant hybridization? Critical Reviews in Plant Sciences **13**:213–241.

Rieseberg, L. H., O. Raymond, D. M. Rosenthal, Z. Lai, K. Livingston, T. Nakazato, J. L. Durphy, et al. 2003. Major ecological transitions in wild sunflowers facilitated by hybridization. Science **301**:1211–1216.

Rieseberg, L. H., A. Widmer, A. M. Arntz, and J. M. Burke. 2002. Directional selection is the primary cause of phenotypic diversification. Proceedings of the National Academy of Sciences, USA 99:12242–12245.

Rieseberg, L. H., and J. H. Willis. 2007. Plant speciation. Science 317:910–914.

Ritchie, M. G. 2007. Sexual selection and speciation. Annual Review of Ecology, Evolution, and Systematics 38:79–102.

Robertson, H. M., and H.E.H. Paterson. 1982. Mate recognition and mechanical isolation in *Enallagma* damselflies (Odonata: Coenagrionidae). Evolution 36:243–250.

Roff, D. A. 1975. Population stability and the evolution of dispersal in a heterogeneous environment. Oecologia 19:217–237.

Rohani, P., R. M. May, and M. P. Hassell. 1996. Metapopulations and equilibrium stability: The effects of spatial structure. Journal of Theoretical Biology 181:97–109.

Ronce, O. 2007. How does it feel to be like a rolling stone? Ten questions about dispersal evolution. Annual Review of Ecology, Evolution, and Systematics 38:231–253.

Ronce, O., and M. Kirkpatrick. 2001. When sources become sinks: Migrational meltdown in heterogeneous habitats. Evolution 55:1520–1531.

Rosenzweig, M. L. 1969. Why the prey curve has a hump. American Naturalist 103:81–87.

Rosenzweig, M. L. 1978. Competitive speciation. Biological Journal of the Linnean Society 10:275–289.

Rosenzweig, M. L., and R. H. MacArthur. 1963. Graphical representation and stability conditions of predator-prey interactions. American Naturalist 97:209–223.

Roughgarden, J. 1974. Species packing and the competition function with illustrations from coral reef fish. Theoretical Population Biology 5:163–186.

Roughgarden, J. 1976. Resource partitioning among competing species—A coevolutionary approach. Theoretical Population Biology 9:388–424.

Rouhani, S., and N. H. Barton. 1987a. The probability of peak shifts in a founder population. Journal of Theoretical Biology 126:51–62.

Rouhani, S., and N. H. Barton. 1987b. Speciation and the "shifting balance" in a continuous population. Theoretical Population Biology 31:465–492.

Roy, M., R. D. Holt, and M. Barfield. 2005. Temporal autocorrelation can enhance the persistence and abundance of metapopulations comprised of coupled sinks. American Naturalist 166:246–261.

Rudolf, V.H.W. 2007. The interaction of cannibalism and omnivory: Consequences for community dynamics. Ecology 88:2697–2705.

Rundle, H. D., L. Nagel, J. W. Boughman, and D. Schluter. 2000. Natural selection and parallel speciation in sympatric sticklebacks. Science 287:306–308.

Rundle, H. D., and P. Nosil. 2005. Ecological speciation. Ecology Letters 8:336–352.

Ruxton, G. D. 1996a. Density-dependent migration and stability in a system of linked populations. Bulletin of Mathematical Biology 58:643–660.

Ruxton, G. D. 1996b. Dispersal and chaos in spatially structured models: An individual-level approach. Journal of Animal Ecology 65:161–169.

Ruxton, G. D., and S. Humphries. 1999. Multiple ideal free distributions of unequal competitors. Evolutionary Ecology Research 1:635–640.

Ruxton, G. D., and P. Rohani. 1999. Fitness-dependent dispersal in metapopulations and its consequences for persistence and synchrony. Journal of Animal Ecology 68:530–539.

Ryan, M. J., and A. Keddy-Hector. 1992. Directional patterns of female mate choice and the role of sensory biases. American Naturalist 139:S4–S35.

Ryan, M. J., and A. S. Rand. 1993. Species recognition and sexual selection as a unitary problem in animal communication. Evolution **47**:647–657.

Salzburger, W. 2009. The interaction of sexually and naturally selected traits in the adaptive radiations of cichlid fishes. Molecular Ecology **18**:169–185.

Scheiner, S. M. 1993. Genetics and evolution of phenotypic plasticity. Annual Review of Ecology and Systematics **24**:35–68.

Scheiner, S. M. 2016. Habitat choice and temporal variation alter the balance between adaptation by genetic differentiation, a jack-of-all-trades strategy, and phenotypic plasticity. American Naturalist **187**:633–646.

Scheiner, S. M., R. Gomulkiewicz, and R. D. Holt. 2015. Genetics of phenotypic plasticity. XIV. Coevolution. American Naturalist **185**:594–609.

Schemske, D. W., G. G. Mittelbach, H. V. Cornell, J. M. Sobel, and K. Roy. 2009. Is there a latitudinal gradient in the importance of biotic interactions? Annual Review of Ecology, Evolution, and Systematics **40**:245–269.

Schluter, D. 2000. The ecology of adaptive radiation. Oxford University Press, New York, NY.

Schluter, D., and P. R. Grant. 1984. Determinants of morphological patterns in communities of Darwin's finches. American Naturalist **123**:175–196.

Schluter, D., and J. D. McPhail. 1993. Ecological character displacement and speciation in sticklebacks. American Naturalist **140**:85–108.

Schmitz, O. J., V. Křivan, and O. Ovadia. 2004. Trophic cascades: The primacy of trait-mediated indirect interactions. Ecology Letters **7**:153–163.

Schoener, T. W. 1976. Alternatives to Lotka-Volterra competition: Models of intermediate complexity. Theoretical Population Biology **10**:309–333.

Schoener, T. W. 2011. The newest synthesis: Understanding the interplay of evolutionary and ecological dynamics. Science **331**:426–429.

Schreiber, S. J. 2012. The evolution of patch selection in stochastic environments. American Naturalist **180**:17–34.

Schreiber, S. J., R. Burger, and D. I. Bolnick. 2011. The community effects of phenotypic and genetic variation within a predator population. Ecology **92**:1582–1593.

Schumer, M., G. G. Rosenthal, and P. Andolfatto. 2014. How common is homoploid hybrid speciation? Evolution **68**:1553–1560.

Schwert, D. P., and A. C. Ashworth. 1988. Late Quaternary history of the northern beetle fauna of North America: A synthesis of fossil and distributional evidence. Memoirs of the Entomological Society of Canada **120**:93–107.

Scordato, E. S., L. B. Symes, T. C. Mendelson, and R. J. Safran. 2014. The role of ecology in speciation by sexual selection: A systematic empirical review. Journal of Heredity **105** Supplement 1:782–794.

Seehausen, O., Y. Terai, I. S. Magalhaes, K. L. Carleton, H. D. Mrosso, R. Miyagi, I. van der Sluijs, et al. 2008. Speciation through sensory drive in cichlid fish. Nature **455**: 620–626.

Seehausen, O., and J.J.M. van Alphen. 1998. The effect of male coloration on female mate choice in closely related Lake Victoria cichlids (*Haplochromis nyererei* complex). Behavioral Ecology and Sociobiology **42**:1–8.

Shapiro, A. M., and A. H. Porter. 1989. The lock-and-key hypothesis: Evolutionary and biosystematic interpretation of insect genitalia. Annual Review of Ecology and Systematics **34**:231–245.

Shaw, K. L. 2000. Interspecific genetics of mate recognition: Inheritance of female acoustic preferences in Hawaiian crickets. Evolution **54**:1303–1312.

Shinen, J. L., and S. A. Navarrete. 2014. Lottery coexistence on rocky shores: Weak niche differentiation or equal competitors engaged in neutral dynamics? American Naturalist **183**:342–362.

Shipley, W., and P. A. Keddy. 1987. The individualistic and community-unit concepts as falsifiable hypotheses. Pages 47–55 *in* I. C. Prentice and E. Maarel, eds. Theory and models in vegetation science. Springer, Dordrecht, Netherlands.

Shmida, A., and S. Ellner. 1984. Coexistence of plant species with similar niches. Vegetatio **58**:29–55.

Siepielski, A. M., J. D. DiBattista, and S. M. Carlson. 2009. It's about time: The temporal dynamics of phenotypic selection in the wild. Ecology Letters **12**:1261–1276.

Siepielski, A. M., J. D. DiBattista, J. A. Evans, and S. M. Carlson. 2011a. Differences in the temporal dynamics of phenotypic selection among fitness components in the wild. Proceedings of the Royal Society B, Biological Sciences **278**:1572–1580.

Siepielski, A. M., K. M. Gotanda, M. B. Morrissey, S. E. Diamond, J. D. DiBattista, and S. M. Carlson. 2013. The spatial patterns of directional phenotypic selection. Ecology Letters **16**:1382–1392.

Siepielski, A. M., K.-L. Hung, E.E.B. Bein, and M. A. McPeek. 2010. Experimental evidence for neutral community dynamics governing an insect assemblage. Ecology **91**:847–857.

Siepielski, A. M., and M. A. McPeek. 2010. On the evidence for species coexistence: A critique of the coexistence program. Ecology **91**:3153–3164.

Siepielski, A. M., and M. A. McPeek. 2013. Niche versus neutrality in structuring the beta diversity of damselfly assemblages. Freshwater Biology **58**:758–768.

Siepielski, A. M., A. N. Mertens, B. L. Wilkinson, and M. A. McPeek. 2011b. Signature of ecological partitioning in the maintenance of damselfly diversity. Journal of Animal Ecology **80**:1163–1173.

Sih, A. 1980. Optimal behavior: Can foragers balance two conflicting demands? Science **210**:1041–1043.

Simberloff, D. 2012. Nature, natives, nativism, and management: Worldviews underlying controversies in invasion biology. Environmental Ethics **34**:5–25.

Simmons, L. W. 2014. Sexual selection and genital evolution. Austral Entomology **53**:1–17.

Simpson, G. G. 1944. Tempo and mode in evolution. Columbia University Press, New York, NY.

Simpson, G. G. 1951. The species concept. Evolution **5**:285–298.

Simpson, G. G. 1961. Principles of animal taxonomy. Columbia University Press, New York, NY. Sites, J. W., and J. C. Marshall. 2004. Operational Criteria for Delimiting Species. Annual Review of Ecology, Evolution, and Systematics **35**:199–227.

Slatkin, M. 1973. Gene flow and selection in a cline. Genetics **75**:733–756.

Slatkin, M. 1980. Ecological character displacement. Ecology **61**:163–177.

Soltis, D. E., P. S. Soltis, D. W. Schemske, J. F. Handcock, J. N. Thompson, B. C. Husband, and W. S. Judd. 2007. Autopolyploidy in angiosperms: Have we grossly underestimated the number of species? Taxon **56**:13–30.

Soltis, P. S., and D. E. Soltis. 2016. Ancient WGD events as drivers of key innovations in angiosperms. Current Opinion in Plant Biology **30**:159–165.

Sousa, W. P. 1979. Experimental investigations of disturbance and ecological succession in a rocky intertidal algal community. Ecological Monographs **49**:227–254.

Sprules, W. G. 1972. Effects of size-selective predation and food competition on high altitude zooplankton communities. Ecology **53**:375–386.

Stanyon, R., M. Rocchi, O. Capozzi, R. Roberto, D. Misceo, M. Ventura, M. F. Cardone, et al. 2008. Primate chromosome evolution: Ancestral karyotypes, marker order and neocentromeres. Chromosome Research 16:17–39.

Stebbins, G. L. 1940. Significance of polyploidy in plant evolution American Naturalist 74:54–66.

Stinchecombe, J. R., M. T. Rutter, D. S. Burdick, M. D. Rausher, and R. Mauricio. 2002. Testing for environmentally induced bias in phenotypic estimates of natural selection: Theory and practice. American Naturalist 160:511–523.

Stoks, R., and M. A. McPeek. 2003. Predators and life histories shape *Lestes* damselfly assemblages along a freshwater habitat gradient. Ecology 84:1576–1587.

Stoks, R., and M. A. McPeek. 2006. A tale of two diversifications: Reciprocal habitat shifts to fill ecological space along the pond permanence gradient. American Naturalist 168:S50–S72.

Stoks, R., M. A. McPeek, and J. L. Mitchell. 2003. Evolution of prey behavior in response to changes in predation regime: Damselflies in fish and dragonfly lakes. Evolution 57:574–585.

Strauss, S. Y., J. A. Lau, and S. P. Carroll. 2006. Evolutionary responses of natives to introduced species: What do introductions tell us about natural communities? Ecology Letters 9:357–374.

Strobbe, F., M. A. McPeek, M. De Block, L. De Meester, and R. Stoks. 2009. Survival selection on escape performance and its underlying phenotypic traits: A case of many-to-one mapping. Journal of Evolutionary Biology 22:1172–1182.

Strobbe, F., M. A. McPeek, M. De Block, and R. Stoks. 2010. Survival selection imposed by predation on a physiological trait underlying escape speed. Functional Ecology 24:1306–1312.

Stuart, A. E., and D. C. Currie. 2001. Using caddisfly (Trichoptera) case-building behaviour in higher level phylogeny reconstruction. Canadian Journal of Zoology 79:1842–1854.

Sved, J. A. 1981a. A two-sex polygenic model for the evolution of premating isolation. II. Computer simulation of experimental selection procedure. Genetics 97:217–235.

Sved, J. A. 1981b. A two-sex polygenic model for the evolution of premating isolation. II. Deterministic theory for natural populations. Genetics 97:197–215.

Tanabe, K., and T. Namba. 2005. Omnivory creates chaos in simple food web models. Ecology 86:3411–3414.

Tanner, J. T. 1966. Effects of population density on growth rates of animal populations. Ecology 47:733–745.

Tansley, A. G. 1935. The use and abuse of vegetational concepts and terms. Ecology 16:284–307.

Taper, M. L., and T. J. Case. 1985. Quantitative genetic models for the coevolution of character displacement. Ecology 66:355–371.

Taper, M. L., and T. J. Case. 1992. Models of character displacement and the theoretical robustness of taxon cycles. Evolution 46:317–333.

Teague, R. 1977. A model of migration modification. Theoretical Population Biology 12:86–94.

Templeton, A. R. 1989. The meaning of species and speciation: A genetic perspective. Pages 3–27 *in* D. Otte and J. Endler, eds. Speciation and its consequences. Sinauer Associates, Sunderland, MA.

Tessier, A. J., and M. A. Leibold. 1997. Habitat use and ecological specialization within lake *Daphnia* populations. Oecologia **109**:561–570.

Thompson, J. N. 1994. The coevolutionary process. University of Chicago Press, Chicago, IL.

Thompson, J. N. 2005. The geographic mosaic of coevolution. University of Chicago Press, Chicago, IL.

Thompson, J. N., S. L. Nuismer, and K. Merg. 2004. Plant polyploidy and the evolutionary ecology of plant/animal interactions. Biological Journal of the Linnean Society **82**:511–519.

Tilman, D. 1977. Resource competition between plankton algae: An experimental and theoretical approach. Ecology **58**:338–348.

Tilman, D. 1982. Resource competition and community structure. Princeton University Press, Princeton, NJ.

Tilman, D. 1994. Competition and biodiversity in spatially structured habitats. Ecology **75**:2–16.

Tilman, D., R. M. May, C. L. Lehman, and M. A. Nowak. 1994. Habitat destruction and the extinction debt. Nature **371**:65–66.

Tilman, D., and S. Pacala. 1993. The maintenance of species richness in plant communities. Pages 13–25 *in* R. E. Ricklefs and D. Schluter, eds. Species diversity in ecological communities: Historical and geographical perspectives. University of Chicago Press, Chicago, IL.

Trauth, M. H., M. A. Maslin, A. L. Deino, and M. R. Strecker. 2005. Late Cenozoic moisture history of East Africa. Science **309**:2051–2053.

Travis, C. C., and W. M. Post. 1979. Dynamics and comparative statics of mutualistic communities. Journal of Theoretical Biology **78**:553–571.

Travis, J. 1989. The role of optimizing selection in natural populations. Annual Review of Ecology and Systematics **20**:279–296.

Travis, J., J. Leips, and F. H. Rodd. 2013. Evolution in population parameters: Density-dependent selection or density-dependent fitness? American Naturalist **181**:S9–S20.

Travis, J., D. Reznick, R. D. Bassar, A. López-Sepulcre, R. Ferriere, and T. Coulson. 2014. Do eco-evo feedbacks help us understand nature? Answers from studies of the Trinidadian guppy. Advances in Ecological Research **50**:1–40.

Tufto, J. 2000. Quantitative genetic models for the balance between migration and stabilizing selection. Genetical Research **76**:285–293.

Tufto, J. 2001. Effects of releasing maladapted individuals: A demographic-evolutionary model. American Naturalist **158**:331–340.

Tufto, J. 2010. Gene flow from domesticated species to wild relatives: Migration load in a model of multivariate selection. Evolution **64**:180–192.

Turelli, M. 1977. Random environments and stochastic calculus. Theoretical Population Biology **12**:140–178.

Turgeon, J., R. Stoks, R. A. Thum, J. M. Brown, and M. A. McPeek. 2005. Simultaneous Quaternary radiations of three damselfly clades across the Holarctic. American Naturalist **165**:E78–E107.

Turner, G. F., and M. T. Burrows. 1995. A model of sympatric speciation by sexual selection. Proceedings of the Royal Society of London B, Biological Sciences **260**:287–292.

Uecker, H., S. P. Otto, and J. Hermisson. 2014. Evolutionary rescue in structured populations. American Naturalist **183**:E17–E35.

Urban, M. C., L. De Meester, M. Vellend, R. Stoks, and J. Vanoverbeke. 2012. A crucial step toward realism: Responses to climate change from an evolving metacommunity perspective. Evolutionary Applications **5**:154–167.

Urban, M. C., and J. L. Richardson. 2015. The evolution of foraging rate across local and geographic gradients in predation risk and competition. American Naturalist **186**: E16–E32.

Valéry, L., H. Fritz, and J.-C. Lefeuvre. 2013. Another call for the end of invasion biology. Oikos **122**:1143–1146.

van Alphen, J. J. M., O. Seehausen, and F. Galis. 2004. Speciation and radiation in African haplochromine cichlids. Cambridge University Press, Cambridge.

Van Devender, T. R. 1986. Climate cadences and the composition of Chihuahuan desert communities: The late Pleistocene packrat midden record. Pages 285–313 *in* J. Diamond and T. J. Case, eds. Community ecology. Harper and Row, New York, NY.

van Doorn, G. S., P. Edelaar, and F. J. Weissing. 2009. On the origin of species by natural and sexual selection. Science **326**:1704–1707.

Vance, R. R. 1980. The effect of dispersal on population size in a temporally varying environment. Theoretical Population Biology **18**:343–362.

Vance, R. R. 1984. The effect of dispersal on population stability in one-species, discrete-space population growth models. American Naturalist **123**:230–254.

Vandermeer, J. H., and D. H. Boucher. 1978. Varieties of mutualistic interaction in population models. Journal of Theoretical Biology **74**:549–558.

Vanni, M. J. 1988. Freshwater zooplankton community structure: Introduction of large invertebrate predators and large herbivores to a small-species community. Canadian Journal of Fisheries and Aquatic Sciences **45**:1758–1770.

Vannote, R. L., G. W. Minshall, K. W. Cummins, J. R. Sedell, and C. E. Cushing. 1980. The river continuum concept. Canadian Journal of Fisheries and Aquatic Sciences **37**:130–137.

Vázquez, D. P., and D. Simberloff. 2003. Changes in interaction biodiversity induced by an introduced ungulate. Ecology Letters **6**:1077–1083.

Via, S. 2001. Sympatric speciation in animals: The ugly duckling grows up. Trends in Ecology & Evolution **16**:381–390.

Visser, A. W., P. Mariani, and S. Pigolotti. 2012. Adaptive behaviour, tri-trophic food-web stability and damping of chaos. Journal of the Royal Society Interface **9**:1373–1380.

Volterra, V. 1926. Variations and fluctuations of the number of individuals in animal species living together. Journal du Conseil Permanent International pour l'Exploration de la Mer 3:3–51.

Wade, M. J., and S. Kalisz. 1990. The causes of natural selection. Evolution **44**:1947–1955.

Wainwright, P. C., G. V. Lauder, C. W. Osenberg, and G. G. Mittelbach. 1991. The functional basis of intraspecific trophic diversification in sunfishes. Pages 515–529 *in* E. Dudley, ed. The unity of evolutionary biology. Dioscorides Press, Portland, OR.

Wang, X., A. S. Auler, R. L. Edwards, H. Cheng, P. A. Cristalli, P. L. Smart, D. A. Richards, and C.-C. Shen. 2004. Wet periods in northeastern Brazil over the past 210 kyr linked to distant climate anomalies. Nature **432**:740–743.

Wang, Y. S., D. L. DeAngelis, and J. N. Holland. 2011. Uni-directional consumer-resource theory characterizing transitions of interaction outcomes. Ecological Complexity **8**: 249–257.

Wang, Y. S., D. L. DeAngelis, and J. N. Holland. 2012. Uni-directional interaction and plant-pollinator-robber coexistence. Bulletin of Mathematical Biology **74**:2142–2164.

Webb, C. 2003. A complete classification of Darwinian extinction in ecological interactions. American Naturalist **161**:181–205.

Webb, C. O., D. D. Ackerly, M. A. McPeek, and M. J. Donoghue. 2002. Phylogenies and Community Ecology. Annual Review of Ecology and Systematics **33**:475–505.

Weiher, E., and P. A. Keddy. 1995. The assembly of experimental wetland plant communities. Oikos **73**:323–335.

Weiher, E., and P. A. Keddy. 2001. Ecological assembly rules: Perspectives, advances, retreats. Cambridge University Press, New York, NY.

Weir, J. T., and D. Schluter. 2004. Ice sheets promote speciation in boreal birds. Proceedings of the Royal Society of London B, Biological Sciences **271**:1881–1887.

Wellborn, G. A., D. K. Skelly, and E. E. Werner. 1996. Mechanisms creating community structure across a freshwater habitat gradient. Annual Review of Ecology and Systematics **27**:337–363.

Wells, M. M., and C. S. Henry. 1992. The role of courtship songs in reproductive isolation among populations of green lacewings of the genus *Chrysoperla* (Neuroptera: Chrysopidae). Evolution **46**:31–42.

Werner, E. E. 1977. Species packing and niche complementarity in three sunfishes. American Naturalist **111**:553–578.

Werner, E. E., and J. F. Gilliam. 1984. The ontogenetic niche and species interactions in size- structured populations. Annual Reviews of Ecology and Systematics **15**: 393–425.

Werner, E. E., D. J. Hall, D. R. Laughlin, D. J. Wagner, L. A. Wilsmann, and F. C. Funk. 1977. Habitat partitioning in a freshwater fish community. Journal of the Fisheries Research Board of Canada **34**:360–370.

Werner, E. E., and M. A. McPeek. 1994. Direct and indirect effects of predators on two anuran species along an environmental gradient. Ecology **75**:1368–1382.

Werner, E. E., and S. D. Peacor. 2003. A review of trait-mediated indirect interactions in ecological communities. Ecology **84**:1083–1100.

West-Eberhard, M. J. 1983. Sexual selection, social competition, and speciation. Quarterly Review of Biology **58**:155–183.

West-Eberhard, M. J. 1989. Phenotypic plasticity and the origins of diversity. Annual Reviews of Ecology and Systematics **20**:249–278.

White, M.J.D. 1969. Chromosomal rearrangements and speciation in animals. Annual Review of Genetics **3**:75–98.

White, M.J.D. 1978. Modes of speciation. W. H. Freeman, San Francisco, CA.

Whitlock, M. C. 1995. Variance-induced peak shifts. Evolution **49**:252–259.

Whitlock, M. C. 1997. Founder effects and peak shifts without genetic drift: Adaptive peak shifts occur easily when environments fluctuate slightly. Evolution **51**: 1044–1048.

Whitlock, M. C., P. C. Phillips, F. B.-G. Moore, and S. J. Tonsor. 1995. Multiple fitness peaks and epistasis. Annual Review of Ecology and Systematics **26**:601–629.

Whittaker, R. H. 1953. A consideration of climax theory: The climax as a population and pattern. Ecological Monographs **23**:41–78.

Whittaker, R. H. 1956. Vegetation of the Great Smoky Mountains. Ecological Monographs **26**:1–80.

Whittaker, R. H. 1975. Communities and ecosystems, 2nd ed. Macmillan, New York, NY.

Whittaker, R. H., and W. A. Niering. 1965. Vegetation of the Santa Catalina Mountains, Arizona: A gradient analysis of the south slope. Ecology **46**:429–452.

Whittaker, R. H., and W. A. Niering. 1975. Vegetation of the Santa Catalina Mountains, Arizona. V. Biomass, production, and diversity along the elevational gradient. Ecology **56**:771–790.

Wilczynski, W., A. C. Keddy-Hector, and M. J. Ryan. 1992. Call patterns and basilar papilla tuning in cricket frogs. I. Differences among populations and between sexes. Brain, Behavior and Evolution **39**:229–237.

Wilkins, J. S. 2009a. Defining species: A sourcebook from antiquity to today. Peter Lang, New York, NY.

Wilkins, J. S. 2009b. Species: A history of the idea. University of California Press, Berkeley, CA. Wilson, A. J., D. Gislason, S. Skulason, S. S. Snorrason, C. E. Adams, G. Alexander, R. G.

Danzmann, and M. M. Ferguson. 2004. Population genetic structure of Arctic charr, *Salvelinus alpinus* from northwest Europe on large and small spatial scales. Molecular Ecology **13**:1129–1142.

Wilson, W. G., W. F. Morris, and J. L. Bronstein. 2003. Coexistence of mutualists and exploiters on spatial landscapes. Ecological Monographs **73**:397–413.

Winemiller, K. O. 1990. Spatial and temporal variation in tropical fish trophic networks. Ecological Monographs **60**:331–367.

Wolin, C. L., and L. R. Lawlor. 1984. Models of facultative mutualism: Density effects. American Naturalist **124**:843–862.

Wolkovich, E. M., S. Allesina, K. L. Cottingham, J. C. Moore, S. A. Sandin, and C. de Mazancourt. 2014. Linking the green and brown worlds: The prevalence and effect of multichannel feeding in food webs. Ecology **95**:3376–3386.

Wood, T. E., N. Takebayashi, M. S. Barker, I. Mayrose, P. B. Greenspoon, and L. H. Rieseberg. 2009. The frequency of polyploid speciation in vascular plants. Proceedings of the National Academy of Sciences, USA **106**:13875–13879.

Wright, D. H. 1989. A simple, stable model of mutualism incorporating handling time. American Naturalist **134**:664–667.

Wright, M. G., and M. J. Samways. 1998. Insect species richness tracking plant species richness in a diverse flora: Gall-insects in the Cape Floristic Region, South Africa. Oecologia **115**:427–433.

Wright, S. 1931. Statistical theory of evolution. Journal of the American Statistical Association **26**:201–208.

Wright, S. 1932. The roles of mutation, inbreeding, crossbreeding and selection in evolution. Proceedings of the 6th International Congress of Genetics **1**:356–366.

Wright, S. 1988. Surfaces of selective value revisited. American Naturalist **131**:115–123.

Wyatt, T. D. 2003. Pheromones and animal behaviour. Cambridge University Press, Cambridge, UK.

Yasuhara, M., G. Hunt, T. M. Cronin, and H. Okahashi. 2009. Temporal latitudinal-gradient dynamics and tropical instability of deep-sea species diversity. Proceedings of the National Academy of Sciences, USA **106**:21717–21720.

Yodzis, P. 1981. The stability of real ecosystems. Nature **289**:674–676.

Yoshida, T., L. E. Jones, S. P. Ellner, G. F. Fussman, and N. G. Hairston Jr. 2003. Rapid evolution drives ecological dynamics in a predator-prey system. Nature **424**:303–306.

Yunis, J. J., and O. Prakash. 1982. The origin of man: A chromosomal pictorial legacy. Science **219**:1525–1530.

Zachos, J., M. Pagani, L. Sloan, E. Thomas, and K. Billups. 2001. Trends, rhythms, and aberrations in global climate 65 Ma to present. Science **292**:686–693.

Zink, R. M., J. Klicka, and B. R. Barber. 2004. The tempo of avian diversification during the Quaternary. Philosophical Transactions of the Royal Society of London B, Biological Sciences **359**:215–219; discussion, pp. 219–220.

Index

MONOGRAPHS IN POPULATION BIOLOGY

EDITED BY SIMON A. LEVIN AND HENRY S. HORN